Advanced Research in Asian Economic Studies – Vol. 5

Future Perspectives on the Economic Development of Asia

Advanced Research in Asian Economic Studies **(ISSN: 1793-0944)**

Series Editor: Manoranjan DUTTA
(Rutgers, The State University of New Jersey, USA)

Published

Vol. 1 Asian Economic Cooperation in the New Millennium: China's Economic Presence
Edited by Calla Wiemer (East Asian Institute, National University of Singapore) & Heping Cao (Peking University, China)

Vol. 2 China's Industrial Revolution and Economic Presence
by Manoranjan Dutta (Rutgers, The State University of New Jersey, USA)

Vol. 3 International Economic Integration and Asia
Edited by Michael G Plummer & Erik Jones (Johns Hopkins University SAIS-Bologna, Italy)

Vol. 4 Economic Dynamism of Asia in the New Millenium: From the Asian Crisis to a New Stage of Growth
Edited by Yoshinori Shimizu (Hitotsubashi University, Japan)

Vol. 5 Future Perspectives on the Economic Development of Asia
by John Malcolm Dowling (Singapore Management University, Singapore)

Advanced Research in Asian Economic Studies – Vol. 5

Future Perspectives on the Economic Development of Asia

John Malcolm Dowling

Singapore Management University, Singapore

World Scientific

NEW JERSEY • LONDON • SINGAPORE • BEIJING • SHANGHAI • HONG KONG • TAIPEI • CHENNAI

Published by

World Scientific Publishing Co. Pte. Ltd.

5 Toh Tuck Link, Singapore 596224

USA office: 27 Warren Street, Suite 401-402, Hackensack, NJ 07601

UK office: 57 Shelton Street, Covent Garden, London WC2H 9HE

Library of Congress Cataloging-in-Publication Data

Dowling, J. Malcolm (John Malcolm)

Future perspectives on the economic development of Asia / by John Malcolm Dowling.

p. cm. -- (Advanced research in Asian economic studies ; 5)

Includes bibliographical references and index.

ISBN-13: 978-981-270-609-6

ISBN-10: 981-270-609-7

1. Technological innovations--Economic aspects--Asia. 2. Asia--Foreign economic relations. 3. Poverty--Asia. 4. Political corruption--Economic aspects--Asia. 5. Sustainable development--Asia. I. Title.

HC415.T4D69 2007

338.95--dc22

2007018080

British Library Cataloguing-in-Publication Data

A catalogue record for this book is available from the British Library.

Typeset by Stallion Press

Email: enquiries@stallionpress.com

Printed in Singapore by World Scientific Printers (S) Pte Ltd

For my wife and my Guru

Preface

When Jan Dutta asked me to write a book about Asia's future for the series that he edits entitled Advanced Research in Asian Economic Studies, my first reaction was "Wow!!" It is such a broad and challenging topic. How do I approach it? And then later, I thought: what are the most important subjects for Asian economists and policy makers to consider in the coming decades?

As I reflected on these questions, I settled on seven major topical issues. Each of them is critical and could easily be the subject of a volume on its own. Together, they represent a broad spectrum of economic, political and social issues that confront Asia in 2007.

Economic growth and technological change is the subject of Chapter 1. Both of these aspects — growth and technology — are critical if Asia is to continue to raise living standards for its citizens. The region has benefited handsomely from international trade and the transfer of technology from industrial countries. This process has to continue and deepen to include more local innovation and research. At the same time, economic efficiency also has to improve if the region is to meet future challenges. The chapter explores potential for further development in key sectors including electronics and information technology, as well as the importance of production platforms and the contribution of total factor productivity to growth. Future projections are developed based on performance in recent decades.

Chapter 2 addresses the subject of regional cooperation. As efforts to reach a new global agreement on trade within the framework of the World Trade Organization have faltered, regional organizations have become more active in Asia, and a series of bilateral regional preferential trade agreements are being negotiated and finalized. These and other regional cooperation initiatives in areas such as infrastructure, transportation and telecommunications are important components of development plans in Asia and are discussed and analyzed in this chapter. Prospects for further regional cooperation are assessed and evaluated as well.

Chapter 3 deals with issues facing the global economy with particular reference to the interface between industrial economies in North America and Europe and the developing countries of Asia. The chapter focuses on several important economic issues including international trade and global imbalances, immigration, health, genetic engineering and environment. It also considers pertinent political problem areas such as issues related to the PRC and Taipei, China, as well as India and Pakistan. The chapter describes developments in the past decade and then moves on to explore possible scenarios for the future.

Chapter 4 explores poverty and income distribution. The main focus of the chapter is on policies to reduce poverty with special reference to South Asia, where the bulk of the poor reside. Initially, the poor are identified and their characteristics are discussed. Policy initiatives and success rates for reducing poverty are reviewed, including macroeconomic adjustments to achieve lower inflation and greater stability as well as micro-policies for achieving more flexible labor markets and human resource development. Poverty reduction strategies and millennium development objectives are discussed, along with prospects for reducing poverty and lifting living standards over the next two decades.

Chapter 5 deals with corruption and the ethical foundations of government. At the outset, the nature of corruption and the role of government in deterring and reducing corruption are discussed. General determinants of corruption are reviewed, followed by a discussion of trends in corruption in Asia. The cost of corruption on growth, investment and to the poor is surveyed, followed by a discussion of methods for controlling corruption, including higher pay for government officials and legal and bureaucratic reform. Individual countries' experiences are reviewed and future prospects are analyzed by regions and economies. General policies for the future are discussed along with governance issues.

Chapter 6 considers the environmental impact of economic activity in Asia. Various forms of pollution including air, water, soil and global warming are discussed in the Asian context. Reasons for the rapid increase in environmental degradation in Asia are analyzed, and costs of pollution and economic mechanisms for reducing it are reviewed. The interaction between pollution, economic growth and structural change is explored, and scientific evidence is presented for possible new technologies that reduce pollution, together with suggestions on how these might be implemented. Key suggestions for reducing pollution and slowing global warming are reviewed and a set of policy initiatives suggested.

Chapter 7 focuses on the two giant Asian economies of China and India. Agriculture, industry, foreign trade and exchange rates (including global developments and WTO) and special economic zone developments and policies as well as environmental and financial sector issues are reviewed for both countries. Labor market and population developments and policies are also discussed. All of these topics are presented from historical perspectives as well as future perspectives. The chapter concludes with a comparison between the two countries from perspectives of aggregate and sectoral economic growth potential, environmental impact, international trade, poverty and income distribution, financial sector performance and savings and investment.

Future prospects for the Asian region cannot be neatly compartmentalized into exclusive and self-contained topics. There are many overlappings and spillovers that create interactions among the different topics in the book. Developing policies and programs to reduce environmental degradation have potential impacts on the level and structure of economic activity. Corruption slows the rate of economic progress and distorts the allocation of resources. Innovation creates potential for new products, greater economic efficiency and the potential for more international trade. Regional cooperation influences the pace and structure of international trade. These relationships are not spelled out in any depth since they are difficult to easily establish and quantify precisely without a fully specified regional econometric or computable general equilibrium model. Furthermore, making future projections are even more tenuous. Rather, the approach taken here is to note within each chapter the potential interactions that may exist with different parts of the economic and political system and to note which chapter the reader can consult for further discussion.

Contents

Chapter 1

Growth, Structural Change and Technology

The Laws of Nature are such that things at their peak must decline and those at their lowest point must rise up, just as the sun and moon follow one another ceaselessly through the skies. Man also follows these laws. ... It is an inexorable Law of Nature that bad must follow good, that decline must follow a rise. To feel that we can rest on our achievements is a dangerous fallacy.

— I Ching (1150 BC)

Introduction

According to the *I Ching*, known as one of the greatest classics of the Chinese and dating back to the origins of the Chinese civilization, only change is constant[1]. The *I Ching* attempts to explain current situations in terms of its natural laws, and so provide some guidance for the future. The natural pattern of law stated in the *I Ching* is for all things to experience a cycle of growth and decline, and the rise and fall of nations have been well-documented and analyzed by historians, politicians and economists.

In this book we concentrate on the recent rise in the fortunes of Asian economies and analyze future prospects for a continuation of this growth cycle. Will decline follow this cycle of growth? Why has

[1] *I Ching* is also known as Book of Change in the West. It has its origins from Tao and some say commentaries from Chinese philosopher Confucius to explain the 64 inter-related hexagrams regarding state of the things in the universe. *I Ching* has been said to be used by modern Japanese corporate leaders such as founder of Panasonic, Matsushita Konosuke, for corporate successes (Cleary, 1995).

Asia been successful? Can Asia continue to grow and prosper in the future? Can Asia make the successful transition from low and middle levels of income to the status of the wealthy? Or will some countries be left behind, caught in the awkward middle-income trap between manufacturing and the knowledge economy (Gill and Kharas, 2006)?

To begin to provide some analysis and answers to these questions, we look at the pattern of economic growth and development in Asia. We first look at the major regions of East Asia, Southeast Asia, and South Asia. At the end of the section on income growth we will have a further look at the Central Asian Republics, a region that had a distinctly different historical experience from the rest of the Asian region.

For East Asia, Southeast Asia, and South Asia the period prior to 1997 has been extensively studied by the World Bank, the Asian Development Bank (ADB), and many academic researchers (for example, see World Bank, 1993; Dowling and Valenzuela, 2004, Chapter 3; and Quibria, 2002). In this chapter we focus on subsequent developments in the past decade as well as the underlying factors that will contribute to growth trajectories in the next two decades.

Income Growth

Within the last decade the Asian region has recovered well from the financial crisis of 1997. Structural reforms have been undertaken in the three crisis countries and also in other economies that were affected by the crisis. Stronger growth in South Asia has been a new feature of the last decade. Rapid growth in China and sustained performance in the rest of East Asia and Southeast Asia has resulted in higher incomes, expanded international trade, and reductions in the level and incidence of poverty. Decade-by-decade growth figures show an appreciable slowdown in growth in East Asia outside of China, particularly in Taiwan and Korea. Hong Kong has been buoyed up in the past few years by continued rapid growth in China (Figure 1.1).

In Southeast Asia (Figure 1.2) Indonesia picked up a bit compared with the previous decade although still lower than growth achieved before the Asian financial crisis in the 1970s and 1980s.

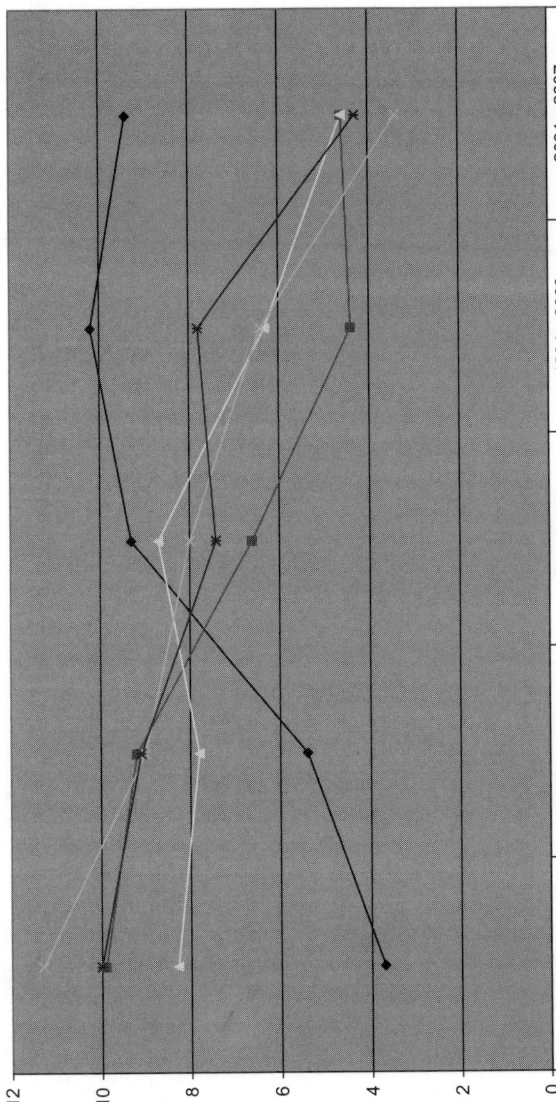

	1961 - 1970	1971 - 1980	1981 - 1990	1991 - 2000	2001 - 2007
China	3.7	5.4	9.3	10.2	9.4
Hong Kong	9.9	9.2	6.6	4.4	4.6
Korea	8.3	7.8	8.7	6.3	4.6
Taiwan	11.3	9	8	6.4	3.4
Singapore	10	9.1	7.4	7.8	4.3

Figure 1.1 Decade Growth — NIE

Source: Data from Naya (2002) and ADB (2006).

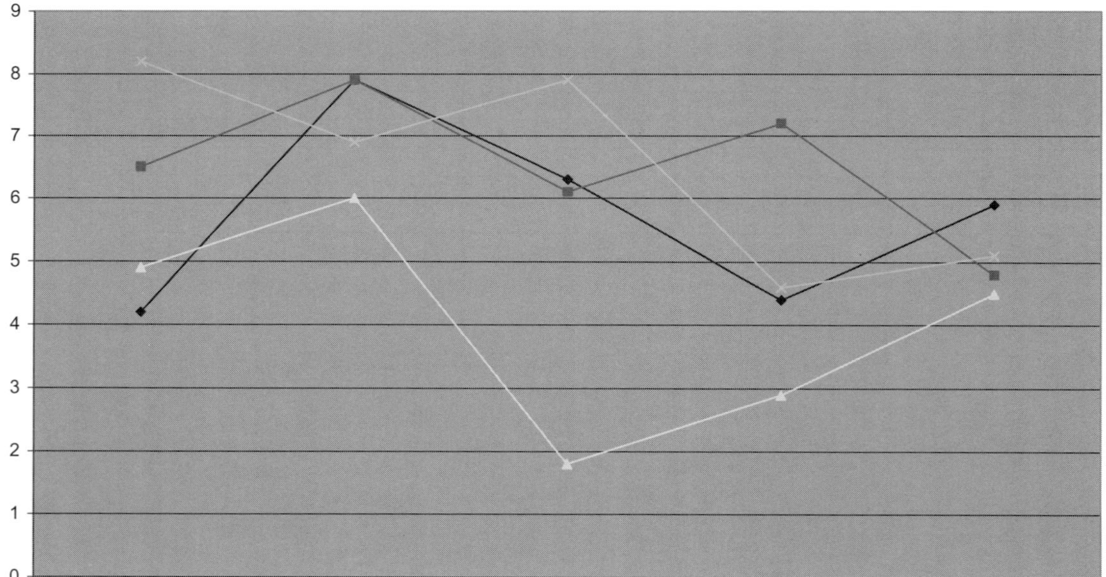

	1961 - 1970	1971 - 1980	1981 - 1990	1991 - 2000	2001 - 2007
Indonesia	4.2	7.9	6.3	4.4	5.9
Malaysia	6.5	7.9	6.1	7.2	4.8
Philippines	4.9	6	1.8	2.9	4.5
Thailand	8.2	6.9	7.9	4.6	5.1

Figure 1.2 Decade Growth — SE Asia

Source: Data from Naya (2002) and ADB (2006).

In Malaysia and Singapore, growth slowed somewhat from the 1990s while it picked up somewhat in the Philippines and Thailand. On balance, overall growth in Southeast Asia was still lower in the new millennium than in the 1970s and 1980s with the exception of the Philippines.

In South Asia (Figure 1.3), growth picked up across the board compared with the historical performance in Bangladesh and India, while improving on the 1990s in Pakistan and Sri Lanka.

Comparing the decade leading up to the financial crisis and the decade since the crisis, the main point to be made is that growth has not slowed appreciably except in a few countries in Southeast Asia and even that slowdown has not been so dramatic. This performance is even more surprising when we note that 2001 was characterized by negative growth in Indonesia, Malaysia, Philippines and Thailand. These negative growth numbers were partly in response to the general global export slowdown which followed the recession in the United States.

Furthermore, performance in India has been much stronger in recent years as liberalization measures have proved beneficial for growth in the services industry and have also revitalized industrial sector growth. Export growth for the Asian region has also been rapid, far exceeding the rate of growth in income. In 2004 and 2005 exports grew even faster than they did in the decade leading up to the crisis for the region as a whole even when India and China were excluded.

Looking at longer term trends, there has been a slowdown in growth compared with that in the 1970s and 1980s in East Asia and less dramatically in Southeast Asia. No such trend is evident in South Asia, where growth has generally accelerated since 1980. This upward trend is particularly striking for Bangladesh, a country which was considered a basket case in the 1970s. This trend in South Asia demonstrates the power of good policies to bring about a renaissance in growth and development.

Further details of the Asian growth experience by region are displayed in Tables 1.1 and 1.2. Over the period 1970 to 2003 the countries of East Asia grew more rapidly than Southeast Asia or South Asia. Even giving the rest of the region a decade head start in the reform process, China outstripped the other regions.

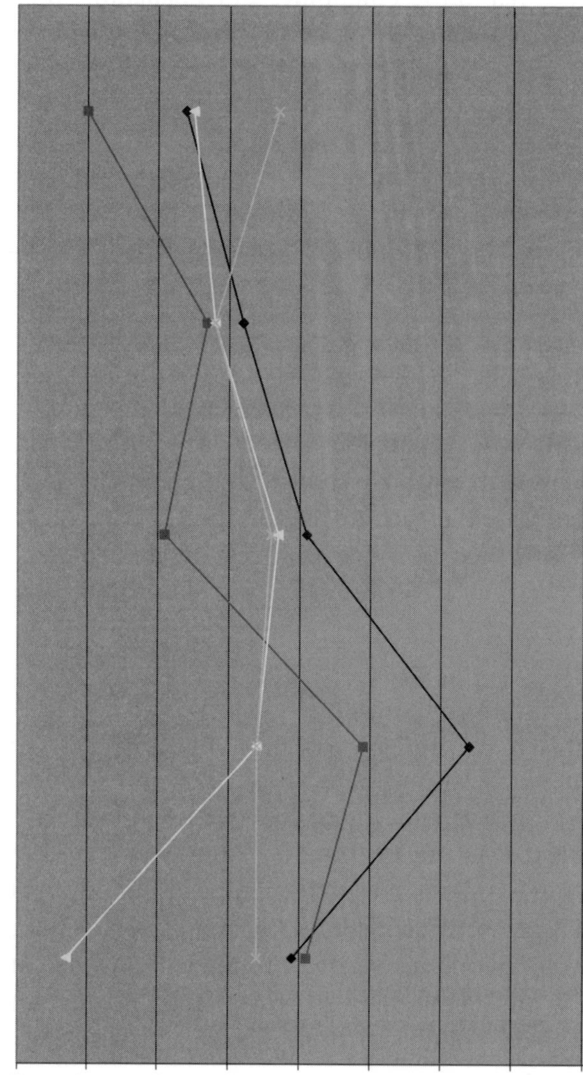

	1961 - 1970	1971 - 1980	1981 - 1990	1991 - 2000	2001 - 2007
Bangladesh	4.1	1.6	3.9	4.8	5.6
India	3.9	3.1	5.9	5.3	7
Pakistan	7.3	4.6	4.3	5.2	5.5
Sri Lanka	4.6	4.6	4.4	5.2	4.3

Figure 1.3 Decade Growth — South Asia

Source: Data from Naya (2002) and ADB (2006).

Table 1.1 Annual Growth Rate of GDP Per Capita, 1970–2003

East Asia excluding China	5.8
China	6.7
Southeast Asia	3.7
South Asia	2.5

Source: United Nations database.

Table 1.2 Real GDP Growth in Developing Asian Economies (%)

Country or Region	1988–1997	1998–2007	2004–2005
Developing Asia	7.9	7.3	8.7
China	9.9	8.9	10.0
India	5.9	6.5	7.8
Excluding India and China	6.3	4.1	5.8

Source: IMF (2006), Table 6; 2006 and 2007 are forecasts.

Overall growth in per capita income since 1970 show that the typical regional growth hierarchy of East Asia first, followed by Southeast Asia and South Asia manifests in these longer series (see Table 1.1). For shorter periods separated by the Asian financial crisis, there has been a modest slowdown in growth in developing Asia and even more so when India and China are excluded (Table 1.2). Singling out the recent past (2004–2005) there has been a further acceleration in growth in all regions.

In the Central Asian Republics (CARs), economic growth was adversely affected by the transition after the breakup of the Soviet Union in the early 1990s[2]. After several years of negative growth there was a turnaround in the mid-1990s and the economies of this region began to grow again (Figure 1.4). It is only recently that their economies have recovered to the levels of output attained before the transition.

There are five basic reasons for the CARs revival — (i) higher international commodity prices and their impact on investment, fiscal

[2]The section on the economic performance of the CARs draws heavily on Dowling and Wignaraja (2006).

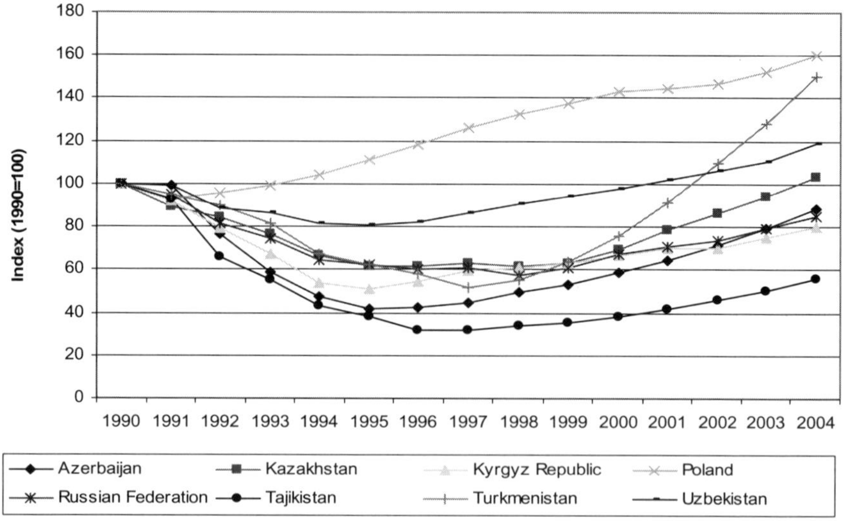

Figure 1.4 Changes in Real GDP Index, 1990–2004 (1990 = 100)

Source: World Bank (2006).

performance, income and consumption; (ii) an upturn in agriculture due to good weather, high world commodity prices and several agricultural reforms; (iii) an acceleration in foreign direct investment (FDI) inflows, particularly in oil and gas, and in manufacturing to a lesser extent; (iv) better macroeconomic management; and (v) improved political stability (Dowling and Wignaraja, 2006).

The CARs are the major producers and exporters of oil, natural gas and also cotton, gold, and other minerals. Prices for these commodities began increasing in the latter part of the 1990s and have continued upward since then, particularly for oil and natural gas in the last two years and even beginning in 2002. Export earnings from these commodities, and the associated revenues flowing into the government coffers gave CARs the ability to begin addressing social issues, develop and improve infrastructure, and increase economic efficiency.

Since the late 1990s, agriculture was a major driver of growth in the non-oil-exporting CARs — particularly the Kyrgyz Republic and Uzbekistan. This derived from generally favorable weather conditions, high world prices for cotton and wheat, along with several agricultural reforms. Some productivity gains were also achieved from the collective farms producing cotton and wheat. Privatization in Kyrgyz Republic also lifted productivity. However, state control remains pervasive in Uzbekistan for agricultural purchasing, pricing, subsidies, distribution, and provision of inputs.

Average annual FDI inflows in the oil and gas exporting CARs of Kazakhstan, Azerbaijan and Turkmenistan have also increased dramatically as foreign investors have taken stakes in the oil and gas industry. In Kazakhstan FDI doubled to $2.3 billion between 1997 and 2004 and also increased in Azerbaijan and Turkmenistan. FDI in Kazakhstan and Azerbaijan is rapidly approaching foreign investment levels in Central and Eastern Europe and the Baltic States. Substantial raw material deposits along with high international prices, favorable corporation tax rates for foreign investors, a strengthening of supply lines to export, improvements in the energy infrastructure, and low cost technical manpower attracted much of the investment.

Macroeconomic stability — helped by declining inflation — has improved. Inflation rates have fallen significantly throughout the region over the past few years indicating improved macroeconomic management. Average inflation for the CARs as a group declined from 20.4% to 6.9% between 1997 and 2001 and further to 6.3% in 2005.

Improved political stability in the late-1990s was also a significant factor contributing to the region's economic revival. The most notable developments were (i) the ending of the civil war in Tajikistan in 1997 and (ii) the 1994 signing of a ceasefire resolving the conflict between Armenia and Azerbaijan. Tajikistan is remarkable among the post-conflict economies for its speed in forming a functioning government, its focus on implementing an economic development agenda and in seeking aid from multilateral institutions. The ceasefire enabled Azerbaijan to reduce defense expenditures, invest in infrastructure, and concentrate more on attracting FDI into oil and gas. The ceasefire, booming oil, and gas industry, and a high growth environment also stimulated domestic investment in industries closely linked to oil and gas.

One of the keys to more rapid growth in Asia over the past decades is the rising importance of entrepreneurial talent and the spread of new technologies and innovation throughout the region and from industrial countries through foreign direct investment. New theories of economic growth stress the importance of education and new ideas as they contribute to the development and dissemination of new products. Such innovations are responsible for the rapid development of new products and export dynamism of Asian economies. Hummels and Klenow (2005) find that around 60% of export growth takes place through new products rather than through exporting more of the same old goods. Institutional developments are an important input into developing and sustaining innovation and product development. These include competition policy, trade liberalization, industrial licensing, good governance and control of corruption. These are all important components of an innovative economy (see Gill and Kharas, 2006 and Chapter 5 for further discussion).

Growth in Trade

There were several reasons for the recent strong performance in income growth including import demand growth in the United States and Europe, continued structural adjustment, financial market normalization as nonperforming loans have been reduced, and continued rapid growth in industrial production. These developments have been reflected in export growth in developing Asia (see Table 1.3). In the last 2 years, exports have grown over 25% per year with China and India included and a very healthy 16% even without China and India. Both international trade and intra-regional trade have been on the dramatic rise, particularly the latter. In the case of East Asia, its share of intra-regional trade has risen from 42% in 1990 to over 50% of intra-regional trade This is highly comparable to the European Union's 60% share. NAFTA has only 45% share of intra-regional trade and MERCOSUR has only 15%. The story goes that it is due largely to China which has sourced more than half of its imports from East Asia (World Bank, 2006).

This high degree of intra-regional trade between countries of similar factor endowments runs contradictory to the Hecksher–Ohlin trade theory, which postulates that countries should trade more with

Table 1.3 International Trade in Goods — Growth in Value of Exports in US dollars (%)

Country or Region	1988–1997	1998–2007	2004–2005
Developing Asia	15.7	14.2	25.3
Excluding India and China	14.9	8.0	16.2

Source: IMF (2006), Table 23; 2006 and 2007 are forecasts.

countries that have different factor endowments in order to best gain advantages from trade using the classic principle of comparative advantage. Modern international trade theory relaxes many of the classical assumptions of perfect competition by introducing the concept of monopolistic competition and product differentiation. This explains why rich countries at similar levels of income and with similar factor endowments trade more than the countries with very different levels of income and factor endowments. Aside from product differentiation, modern trade theory stresses the importance of scale economies and technology as additional factors in determining the pattern of international trade. Globalization has intensified the importance of new product development and innovation as well as the size of the global market as key determinants of the growth in trade and the ability of countries to compete in world markets.

There is now greater trade in intermediate goods as outsourcing has increased. Yeats (2001) reports that 30% of world trade in manufactures is in intermediate goods. This is particularly true in Southeast Asia and East Asia. Trade in parts and components, particularly in sectors like electrical machinery has been on the rise. The Philippines exports comprise a much larger share of parts and components now (over 50% of exports in 2003) compared to that in 1990 (17.8% of exports). Other Asian economies like Malaysia, Taiwan, Korea, and Thailand face a similar doubling of exports in parts and components over the same period (see Okamoto, 2005 for more details). Greater vertical integration between production processes in different countries in Asia has probably raised total factor productivity (see next section for further discussion on total factor productivity) and has combined with more foreign direct investment to provide further impetus to the surge in exports.

Table 1.4 Growth Rate of Merchandise Exports, 2001–2007

	2001	2002	2003	2004	2005	2006	2007
China	6.8	22.4	34.6	35.4	28.4	20	17
Hong Kong	−5.8	4.9	12.1	15.9	11.2	11	9
Korea	−14	7.9	20.7	30.6	12.1	11	8
Taiwan	−17.3	6.4	10.5	20.7	8.8	9	7.4
Indonesia	−12.3	3.1	8.4	12.6	20.1	12.1	11.9
Malaysia	−10.6	7.2	11	20.9	12	17.8	15.1
Philippines	−16.2	9.9	2.7	9.8	3.7	5.1	8
Singapore	−10.5	5.2	15	24.3	15.7	9.5	6
Thailand	−7.1	4.8	18.2	21.6	15	15.3	12
Vietnam	6.5	7.4	20.4	30.3	20.5	18	15
Bangladesh	12.6	−7.6	9.5	15.9	14	13	12
India	−1.6	20.3	23.3	23.9	15.8	17	17
Nepal	11.7	−20.3	−13.8	14.8	10.3	10.2	12.5
Pakistan	9.1	2.3	20.1	13.5	16	20	13
Sri Lanka	−12.8	−2.4	9.2	12.7	8.9	8	8

Source: ADB (2006); 2006 and 2007 are projections.

As an example of trade growth consider the combined exports of the economies of China, Taiwan, and Korea. The combined exports of these three countries in July 2006 were 2.3 times those in July 2002, an annual growth rate of 24% over the four years. The value of China's exports is now only a few percentage points less than that of the United States and it is likely that China will surpass the United States as the world's biggest exporter by 2007 (see Morgan Stanley's website www.morganstanley.com/gef/team.html for further analysis).

For the decade up to 2007, export growth in developing Asia has been more volatile than income growth. It has been rather susceptible to external shocks. The slowdown in the United States in 2001 had a negative impact on export growth in all three regions. Most countries had negative export in 2001 or 2002 with the exception of China, Pakistan and Vietnam. After recovering, export growth accelerated to double digit rates in many countries, although by 2005, export growth had begun to slow down again (refer to Table 1.4).

Total Factor Productivity (TFP)

While individual estimates of TFP may vary significantly because of different assumptions and time period of estimates, it is now widely accepted that TFP adds to output by developing new production processes that adopt new technology and adds additional output with the same amount of labor and capital. TFP also lifts output by achieving better industrial organization and management, more effective inventory control, greater efficiency in subcontracting, and more efficient marketing and distribution.

Total factor productivity became a more significant contributor to overall growth in developing Asia in the second half of the 1980s and 1990s. The information technology (IT) revolution that began in the early 1990s in the United States contributed to the growth in innovation as enhanced technology came into use. Information technology facilitated a new industrial structure which featured the positioning of production platforms for a variety of inputs and components in many different countries in order to lower costs and increase economic efficiency.

In the view of many observers, TFP and the transfer in technology and increased productivity in East and Southeast Asia through the contribution of such an industrial organizational development were facilitated primarily by multinational firms that began to operate in the region in the 1970s and 1980s and with greater intensity in the second half of the 1980s. Foreign direct investment increased throughout the Asian region in the decade following the Plaza accord of 1985. At the same time, much of the increase in output was the result of rapid growth in the labor force and large increments in investment during the rapid growth phase that ended in 1997 with the Asian financial crisis.

The period directly following the 1985 Plaza Accord resulted in appreciation of the yen and a significant rise in foreign direct investment by Japan in Korea and Southeast Asia, increasing competitiveness and economic efficiency as well as further integrating the industrial sectors of Japan with Indonesia, Korea, Malaysia, Philippines and Thailand. While the impact on overall growth is hard to measure, the inflow of FDI over the next decade did result in an increase in technological transfer and in the contribution of TFP to overall economic growth. Estimates of the contribution of TFP to overall growth in Southeast Asia and Korea are shown in Table 1.5.

Table 1.5 Growth Accounting — Contribution of Capital, Labor, and TFP to Overall Growth

Country and Period	Growth in GDP	Contribution of Capital	Contribution of Labor	Contribution of TFP
Indonesia				
1968–1976	6.7	2.7	1.7	2.2
1976–1986	6.5	3.4	1.9	1.2
1986–1996	7.3	2.6	1.8	2.9
1968–1996	6.9	3.0	1.9	2.1
Korea				
1968–1976	10.3	4.3	2.1	3.9
1976–1986	7.7	3.7	1.8	2.2
1986–1996	8.0	3.3	1.4	3.3
1968–1996	8.6	3.7	1.8	3.1
Malaysia				
1968–1976	7.0	3.1	2.3	1.6
1976–1986	5.6	3.2	2.1	0.3
1986–1996	8.1	2.4	1.9	3.8
1968–1996	6.9	2.9	2.1	1.9
Philippines				
1968–1976	5.5	2.0	2.0	1.6
1976–1986	1.8	2.1	2.0	−2.3
1986–1996	3.6	0.8	1.9	0.9
1968–1996	3.5	1.6	2.0	−0.1
Thailand				
1968–1976	6.7	3.3	2.2	1.3
1976–1986	6.1	2.5	2.0	1.6
1986–1996	9.1	3.2	1.3	4.6
1968–1996	7.4	3.0	1.8	2.6

Source: Berthelemy and Chauvin (2000).
Note: Contribution of capital, labor and TFP may not sum to total growth due to rounding-up. Period of highest contribution of TFP highlighted.

These results suggest that the contributions of TFP to output growth is highest in the decade following the Plaza Accord (1986–1996) in three countries and second highest in the other two. The contribution of TFP is particularly strong in Thailand, Korea and Malaysia, where it contributed 50%, 41% and 46% respectively to overall growth in that decade.

There are other studies of TFP which shed additional light on the contribution of TFP to output growth. The results are dependent on several factors including the share of labor and capital in the production process and how changes in the quality of labor are reflected in terms of the improvements in education and skill. Some of these studies find differences in the contribution of TFP to growth for other time periods. For example, Bosworth and Collins (2003), whose results are recorded in Table 1.6, find no TFP for Indonesia from 1975 to 2000, negative TFP for Philippines, and TFP of between 0.9 and 1.8 for other countries in the region. Weerasinghe and Fane (2005) provide a summary of other estimates for various time periods in Table 1.7.

Considering all these estimates of TFP for Asia together, we conclude that TFP is low in Philippines, around 1% in Malaysia, as much

Table 1.6 Sources of Growth in East Asian Economies (1975–2000) — Average Annual Percentage Change

Country	Output	Contribution of Physical Capital	Contribution of Education	TFP
China	8.8	2.5	0.4	3.9
Indonesia	5.8	2.4	0.5	0.0
South Korea	7.3	3.0	0.7	1.1
Malaysia	6.9	2.2	0.6	0.9
Philippines	3.0	0.8	0.4	–0.9
Singapore	7.7	2.1	0.5	1.8
Thailand	6.5	2.1	0.5	1.4
Taiwan	7.8	2.6	0.4	2.4

Source: Bosworth and Collins (2003).

Note: Last three columns do not add up to output growth column because labor input is not included.

Table 1.7 Estimates of TFP per Year for East Asian Economies; Various Years

Country	World Bank 1960–1980 regressions	Kim and Lau regressions	Young (1960–1987) Growth accounting	Collins and Bosworth (1960–1994) Growth accounting	Weerasinghe (1960–1990) Growth accounting	1968–1996 from Table 6.3
Indonesia	1.5			0.8	0.9	2.1
Korea	3.4	1.0	1.7	1.6	2.0	3.1
Malaysia	1.5			1.0	0.1	1.9
Philippines				–0.4	–0.3	–0.1
Singapore	2.1	1.9	0.2	1.6	–0.1	
Taiwan	3.9	1.6	2.6	2.0	2.6	
Thailand	2.6			1.9	1.8	2.6

Source: Weerasinge and Fane (2005) and Berthelemy and Chauvin (2000).

as 2% in Singapore and Thailand, and between 2% and 3% in Korea and Taiwan. It also appears that TFP increased in the late 1980s and early 1990s, although the evidence for this acceleration is not conclusive. Another explanation for the acceleration in TFP after the Plaza Accord is that earlier estimates of TFP by Young (1995) and popularized by Krugman (1994) were biased downward. Although increases in output were primarily a result of high rates of growth in capital investment and labor force growth there were also few signs of diminishing returns. This suggests that there were also gains in output as a result of TFP.

TFP also increases as labor moves from low productivity jobs in agriculture to higher productivity jobs in industry. Bosworth estimates that the bulk of TFP in Thailand resulted from these sectoral shifts in output (Bosworth, 2006). Although not documented as rigorously as the findings for Thailand, it is possible that similar gains in output from TFP also occurred in other countries in Southeast Asia that experienced substantial shifts out of agriculture, including Indonesia, Malaysia and the Philippines. These TFP gains are possibly reflected in higher TFP estimates for China, Korea, Taiwan, and to a lesser extent Thailand, where industrial sector growth was very strong in the 1980s and 1990s. These shifts were less pronounced in the Philippines and Malaysia, while Singapore had always had a strong industrial sector. Finally, Ferreira *et al.* (2005) remind that TFP measured as a contribution to the percentage growth in output for Asian economies is much higher than the TFP for other developing countries and regions.

While the Plaza Accord was instrumental in attracting Japanese investment to Southeast Asia, other multinational investors were also attracted to the region, particularly to China where policies were being liberalized to attract foreign firms beginning in the late 1970s and continuing in the next two decades (see Chapter 7 for details). In the years following World War II, the United States also began to invest in Taiwan and Korea, partly as a response to the political threats of communism in the region. As interest in Asia increased in the 1980s, the United States invested more heavily in Southeast Asia, beginning in the Philippines, where it had been well established for many years and then in other countries in the region.

Shifting Comparative Advantage

The process of industrialization in Asia has been discussed thoroughly (see Dowling and Valenzuela, 2004 for references). The flying geese paradigm of development first suggested by Akamatsu (1961, 1962) has served as a model for the spread of technology from Japan to the rest of the region. This model explains how Japan served as the leader in technology, beginning with labor-intensive industries and moving over the years to higher skill and capital-intensive industries. The spread of technology then extended to the rest of the region either by foreign direct investment or by imitation of Japanese practices.

However, as the process of technological transfer takes place tensions arise between recipient and investing countries as production becomes more dependent upon highly skilled workers. The investing country has concerns about losing jobs to the recipient country and the recipient country has difficulties in implementing better supply systems and providing skilled workers. As industrialization becomes more sophisticated, tensions increase further as the host country needs to upgrade its technological base and investing countries become more protective of its specialized high technology (Hobday, 1995; Praussello, 2005). Furthermore, as the NIEs have become more developed they have also begun to transfer technology to other countries in the region, particularly Southeast Asia. Despite these limitations and shortcomings, the flying geese model provides some basic and useful insights.

The operation of the flying geese model can be understood in detail by following the shift in production of exports as comparative advantage shifted over time. These shifts are reflected by changing international competitiveness as a result of what Balassa (1965) calls revealed comparative advantage. To follow these shifts we make use of a database constructed by the United Nations (2006, Chapter 3). A commodity trade matrix was constructed drawing on the commodity trade statistics database of the United Nations for two periods — 1962–1980 and

1980–2000. Exports were divided into five categories according to natural resource and technological content use, a classification scheme developed by Lall (2000). The five categories are:

- Primary products (PP), consisting of food, live animals and non-oil minerals such as silver, copper, nickel, aluminum, lead, zinc and tin.
- Natural resource-based manufacturers (NRB) consisting of processed food, wood products, leather, refined petroleum products, dyes, precious stones and organic chemicals. Products can be simple and labor-intensive such as foods and leather or capital-intensive. In any case they depend for the most part on locally available materials.
- Low-technology manufacturers (LT) consisting of textiles, garments, footwear and other leather products, toys, simple metal and plastic products, furniture and glassware. These products have stable and well-known technologies, and cost structures are usually dominated by labor. Products tend to be undifferentiated at the mass produced level.
- Medium-technology manufacturers (MT) consisting of heavy industry products such as automobiles, industrial chemicals, machinery and standard electrical products. These products depend on complex but slowly changing technologies.
- High-technology manufacturers (HT) consisting of more complex electronics including telecommunications, precision instruments, fine chemicals and pharmaceuticals. Products have advanced and rapidly changing technologies, with large research and development spending, and advanced technological infrastructure. High value to weight ratios (as opposed to automobiles) allow many segments of value chain to be located in different countries or regions.

The United Nations looked at the trade specialization pattern over time with reference to these five categories. The diversification in production and exports over time was examined. Diversification is defined as the decline in the share of primary products combined with an increase in the share of manufactured products in total merchandise exports. The category of dynamic products was examined further. These dynamic goods are defined as products which exhibited an increase in market share of global exports in a given period of time. Using these data for all five categories, they compiled a trade specialization indicator (TSI) which shows the change in share of the category of exports as a proportion of total exports for a given country multiplied by the share of the country in total developing country exports of that particular export category.

This TSI, defined in Equation (1.1), is a variation of a measure of revealed comparative advantage. It combines changes in the structure of exports of a particular country with that country's share of world exports. The indicator for a category may be small if changes in the structure of exports of the country have been large but the country has not gained market share relative to other developing countries in that category of products. Countries are put into one of the five classifications of products according to the size of the TSI indicator. For example, if TSI in a country is high for PP this means that primary products were the category where specialization was most significant relative to total merchandise exports of developing countries.

$$\text{TSI} = [(X_{ij}/\Sigma X_{ij})_{t+1} - (X_{ij}/\Sigma X_{ij})_{t}] \times [(X_{ij}/\Sigma X_{ij})_{t+1}] \times 100, \quad (1.1)$$

where X is the value of exports, i = goods, j = country, and t = period.

Σ in the first two terms on the right-hand side of the equation, is over the total number of goods. This term shows the change in share of the commodity in total exports of the country concerned between period t and t+1. Σ in the last term on the right-hand side of the equation is over the total number of developing countries. This term shows the country share of total developing country exports in that particular product category.

Comparison of TSI values for the five different categories of exports for developing Asian economies over the two time periods 1962–1980 and 1980–2000 provides an index of shifting comparative advantage over time for each country and also a ranking of the countries in the region.

These calculations are compiled in Tables 1.8 and 1.9. If a country does not appear in the table, it means that it did not exhibit any changing diversification of exports over the period.

The countries in East Asia and Southeast Asia all developed trade specialization in HT industries in the latter period from 1980 to 2000. Furthermore, the large values for the index in many countries reflect not only a shift in specialization within the country but also a growing share of country exports as a proportion of world exports. Only Hong Kong has a relatively low index, reflecting its failure to

Table 1.8 Trade Specialization Index for 1960–1980 — Category of Specialization and Value of Index

High Technology (HT)	Low Technology (LT)	Medium Technology (LT)	Natural Resource Base (NRB)
Singapore (2.77)	Korea (1.06)	Hong Kong (0.19)	Malaysia (0.80)
Philippines (0.11)			Indonesia (0.94)
			Sri Lanka (0.008)
			Vietnam (0.005)

Source: United Nations (2006), Chapter 3, and Annex, Statistical Table A5.
Note: Specialization index shown in brackets.

Table 1.9 Trade Specialization Index for 1980–2000 — Category of Specialization and Value of Index

High Technology (HT)	Low Technology (LT)	Medium Technology (LT)	Natural Resource Base (NRB)
Korea (1.93)		India (0.11)	
China (1.61)		Indonesia (0.53)	
Taiwan (2.54)		Pakistan (0.26)	
Thailand (1.13)		Sri Lanka (0.17)	
Singapore (3.09)		Vietnam (0.29)	
Hong Kong (0.59)			
Malaysia (3.23)			
Philippines (1.82)			

Source: United Nations (2006), Chapter 3, and Annex, Statistical Table A5.
Note: Value of specialization index shown in brackets.

increase market share in high technology exports. This reflects the growing importance of services in the Hong Kong economy and its role as a marketing and distribution center for Chinese exports. The importance of medium technology in South Asia reflects the growth of manufactured exports to other developing countries and the emphasis that India, Pakistan and Sri Lanka have been placing on these types of manufactured goods which have been promoted by the governments of these countries. Other things equal, we would have expected them to develop comparative advantage in low technology industries.

Another approach to comparative advantage is to look at shifts in import demand in industrial countries and to explore whether Asian exporters were following this pattern. Dowling and Ray (2000) constructed an index of export growth for the Asian economies that is heavily influenced by the changing structure of exports. If changes in export structure are closely aligned with changes in the import demand of industrial countries, then this index will be large and conversely the index will be small if the export structure is not responding. Analyzing the export movement between 1970 and 1995, Dowling and Ray (2000) found that almost all of the countries in East and Southeast Asia moved quickly to take advantage of the growing world demand in the three key industrial categories — electronics, computers and pharmaceuticals. Table 1.10 shows the pattern of export growth in those industries where the share of OECD imports increased between 1970 and 1995. Working with local and Taiwanese companies, Hong Kong did not respond to the shifts in OECD import demand because it was in the process of becoming a service economy. Its share of manufactured exports and industry fell in the 1990s. Indonesia was slow to respond because of a large petroleum economy and also because its industrial sector was slow to evolve and had difficulty making the transition to higher value-added electronics and computer products. Other countries have been making the shift quickly including NIEs outside Hong Kong and Malaysia. Thailand has been less nimble in adapting to shifting demand and China's export base was still primarily focused on labor-intensive products in 1995.

Table 1.10 Share of Asian Manufactured Exports in Sectors Where the Share of OECD Imports Increased, 1970–1995

Country or Region	1970	1985	1995
Indonesia	3.0	9.3	34.2
Malaysia	5.9	36.2	66.2
Thailand	4.2	23.4	53.9
China	19.4	23.9	49.8
Singapore	22.6	40.5	74.5
Taiwan	34.5	50.6	67.8
Korea	29.8	44.7	62.5
Hong Kong	45.6	48.7	49.2
World	43.2	50.6	57.5

Source: Dowling and Ray (2000).
Note: The products considered were computers, electronics, pharmaceuticals, instruments, motor vehicles, chemicals, electrical machinery, other manufacturing and fabricated metal products.

Modern Comparative Advantage

Economic growth has been closely aligned with the speed in which Asian countries were able to pick up demand in fast growing sectors like the three key industrial categories — electronics, computers and pharmaceuticals industry. Interestingly, according to new growth theory, these are exactly sectors which exhibit considerable economies of scale in production. They have elasticity of scale above 1 (World Bank, 2006).

Such economies of scale allow for vertical specialization where production can be spliced into different components and later combined to form a final product, say the assembly of semiconductor chips for use in handphones and computers. East Asia had been among the fastest to take advantage of such a trend. No country monopolizes the production chain but rather each country specializes in niche markets within the sector, and participates in the final assembly of the good through well-developed transport and service networks.

According to Antweiler and Trefler (2002), East Asia exports have been growing most rapidly in electrical machinery, followed by non-electronics machinery, instruments, iron and steel, and pharmaceutical products. Indeed, it has been said the growth of Singapore was not due so much to policy factors but due to the popularity of computer chips, which allowed Singapore to expand exports and raise economic growth. With the recent increase in regional agreements that allows for greater integration in the Asian region (see Chapter 2 for more details), deeper tariff cuts will be pursued to promote trade, and obstacles to FDI will be lessened.

The Role of Investment

Investment played a key role in the growth of Asian economies and this has been highlighted as a key ingredient of the sustained performance that became known as the Asian miracle. As outlined above, structural change and economic dynamism were strongest in East Asia and China. And investment rates also increased more rapidly in these regions. In the forty 3-year periods from 1960 to 2003, average gross fixed capital formation per person increased over 15-fold in East Asia, over 12-fold in China, and a much lower 3-fold in Southeast Asia, and just over 2-fold in South Asia (see United Nations, 2006, Chapter 2). The average investment rates also increased dramatically in East Asia and China and more modestly in Southeast Asia and South Asia (see Table 1.11).

Capital accumulation not only fuels growth through the production function but it also facilitates the rapid implementation of new technology and structural changes which resulted in the growth in industry and services and the decline in agriculture mentioned

Table 1.11 Average Investment Rate for Selected Periods and Regions, 1960–2003

Region	1960s	1970s	1980s	1990–2003
East Asia (without China)	13	21	27.5	33.5
China	22	27	28.5	34
Southeast Asia	17	22.5	2.5	27
South Asia	18.5	18	19	21

Source: United Nations (2006), interpolated from Figure II.3.

earlier. Investment in financial and business services helps to support industrial growth. Investment in machinery and equipment is also more conducive to sustained growth than construction. In the NIEs the shares of manufacturing and financial intermediation, transportation, real estate, and business investment increased as shares of total investment over time at the expense of agriculture. In Korea, for example, gross fixed capital formation in transportation increased from 2% to 12% between 1970 and 2003 while investment in agriculture fell from 14% to 2%. Manufacturing and mining investment increased from 16% to 24%. Investment in Taiwan followed a similar pattern.

Foreign Direct Investment

There is a still unresolved controversy regarding the impact of foreign direct investment on economic growth among academics. This controversy is difficult to resolve because so many other variables such as trade openness, industrial policy, government regulations, and the general external environment are all changing as multinationals begin to invest in a country. One position is that FDI brings with it a host of advantages including dynamic externalities; innovation and technological transfer; production, marketing and distribution expertise (crowding in). These allow countries to increase exports, raise productivity, and serve as role models for domestic firms. Many countries in Asia have lifted export growth substantially with the help of FDI (see Chapter 7 for details of the China experience). Opponents of the importance of FDI stress the displacement of domestic firms and bidding resources away from them (crowding out) and squeezing out domestic suppliers as foreign firms bring more integrated supply chains with them featuring suppliers in other locations.

Whatever the relative merits of these two positions, the historical evidence in Asia suggests that FDI, acceleration in growth, openness to trade, and human resource development have increased in tandem over the past few decades. Where trade regimes have been open and competition encouraged, inward FDI has been strong, growth has accelerated and exports have shifted toward higher value-added industrial products. This has been the experience of Korea and the major countries of Southeast Asia. Where the trade regimes have been closed and international trade and competition not encouraged, growth and development have been slower. This was the case of India and other countries in South Asia in the 1980s and early 1990s. However, changes in policies in the 1990s resulted in a gradual acceleration of growth. In the early years of this century growth in South Asia exceeded that of Southeast Asia. The details of these relationships can be found in the work of a number of authors including Dollar and Cray and many others.

The pattern of FDI inflows are displayed in Table 1.12 and Figure 1.5. China has dominated FDI inflows in the past decade, rising from $14 billion per annum in the early 1990s to over 40 billion by the end of the decade and close to 60 billion by 2006. Greater China, including Taiwan and Hong Kong, has also benefited. There has been a moderation of inflows to Southeast Asia and growth has slowed somewhat. Nevertheless, these countries are also benefiting from the growth of China as trade has increased. Five year averages of FDI from 1980 to the present in Figure 1.5 for the three subregions show a dramatic increase in FDI in East Asia, with FDI tripling between the 1985–1990 and the 1990–1995 period, and then more than doubling again in the next period from 1995–2000. A similar pattern is observed in Southeast Asia although at much lower levels of FDI inflow and lower growth between 1990–1995 and 1995–2000.

FDI in South Asia was close to zero in the early periods and has been growing slowly over time. In the period since 2000, FDI inflows have fallen from their peak levels in the previous 5 years in both East and Southeast Asia, partly as a result of a general slowdown in overseas investments by the United States and European countries in 2000 and 2001 as growth in these countries slowed down (see Table 1.13). Greater China continues to dominate inflows into developing Asia. Richer developing countries with the capacity to absorb high technology FDI and China have benefited the most from outward flows to developing countries from industrial countries. When Singapore is added to the usual greater China countries of Hong Kong, Taiwan and China, the rest of Asia, particularly Southeast Asia, is left with a

Table 1.12 Foreign Direct Investment in Developing Asia (1991–1993 and 2004) — Annual Averages in US billion dollars

	1991–1993	Rank in 1991–1993	1998–2000	Rank in 1998–2000	2004	Rank in 2004
China	14.3	1	41.6	1	54.9	1
Malaysia	5.0	2	3.5	6	4.6	6
Hong Kong	3.9	3	33.8	2	34.0	2
Singapore	3.9	4	11.1	3	16.0	3
Thailand	2.0	5	5.6	5	1.4	8
Indonesia	1.8	6	Not in top 10		1.0	10
Taiwan	1.0	7	2.7	7	n.a	
Philippines	0.9	8	1.6	9	0.5	Not in top 10
Korea	0.8	9	8.0	4	8.2	4
Kazakhstan	0.7	10	Not in top 10		4.1	7
India	Not in top 10		2.4	8	5.3	5
Vietnam	Not in top 10		1.5	10	1.6	8
Pakistan					1.1	9
All of developing Asia	35.4		111.6		72*	
High income	179 (1990)					413

*Not strictly comparable with previous years.

Source: UNCTAD (2006).

Table 1.13 Averages of Foreign Direct Investment (1981–2004) in US billion dollars

Region	1981–1985	1986–1990	1991–1995	1996–2000	2001–2004
East Asia	1	9	30	75	71
South Asia	Negligible	0.3	1	4	5
Southeast Asia	3	7.5	19	27	18

Source: Adapted from UNCTAD (2006).

shrinking share of total FDI flows to Asia that has been growing slowly over the past few years.

FDI inflows have been stimulated by the development of export processing zones EPZs or special economic zones. These zones have grown in importance since they were first introduced in Southeast Asia and China in the 1980s. These zones are discussed further in Chapters 2 and 7.

Production Platforms and Patterns of Technological Transfer

Physical and human infrastructure, FDI, technology transfer, telecommunications, transportation all interact to create a dynamic high productivity atmosphere in several Asian economies over the past several decades. To achieve international competitiveness in particular, product lines, firms and countries have focused on developing regional centers which specialize in particular products or components of products. These highly specialized production platforms are interrelated through efficient transport and telecommunications networks where products are produced and shipped to different assembly locations and scheduled to reduce inventory according to just-in-time arrival scheduling. Often these firms are set up in special economic zones where governments have granted special privileges, reduced red tape, built infrastructure, and worked to create a physical environment that will increase firm efficiency as much as possible. A wide range of technologies and industrial specialties have been developed including simple labor-intensive industries like food processing, wood products, and simple textiles and apparel to highly skilled engineering, chemicals, and pharmaceutical processes. We cover these special economic zones further in Chapter 2. Here we discuss the manner in which countries obtained access to foreign technology and the specific role played by foreign direct investment.

Patterns of technological transfer[3]

In the NIEs there were four distinct patterns of government intervention and openness to foreign investment. Hong Kong and Singapore pursued policies to actively attract foreign direct investment including low taxes, full ownership, access to modern infrastructure, and a well-educated labor force. There was limited innovation by local firms. In Hong Kong, industrial policy was limited and market forces were generally allowed to dictate the pattern of trade and industrial development. In Singapore, the government was more proactive by training the workforce and providing industrial infrastructure. In Korea and Taiwan, governments were less open to foreign direct investment. Rather than encouraging foreign firms to locate in the country, they preferred arrangements where foreign companies hired local subsidiaries to subcontract. These arrangements took one of several forms. Subsidiaries could produce to exact specifications set out by the foreign firm and were accordingly called original equipment manufacturing (OEM). Alternatively, the local subsidiary could be given more input into the production process, where the final product was a combination of foreign and local design inputs called own design manufacturing (ODM). Later on, in the industrialization process, local firms began to produce their own products under their own labels or brands (OBM). Firms in Korea and Taiwan did not generally enter into joint ventures although they did sometimes use licensing arrangements and eventually formed strategic alliances with overseas firms to access technology.

Despite these differences in approach to industrialization in all the NIEs, and eventually the countries in Southeast Asia, four factors underpinned their success: (i) Firms benefited from a stable macroeconomic

---

[3]The next two sections draw on Dowling and Valenzuela (2004, Chapter 5).

environment including low inflation and interest, high saving rates, and prudent macroeconomic policies; (ii) governments followed outward looking export strategies that were beneficial to export-oriented manufacturing; (iii) human resource development strategies supplied a wide range of needed skills through an education system that included technical and vocational schools as well as engineering and management training; (iv) government policies were supportive and stimulating as appropriate. In Korea, large conglomerates (*chaebol*) were formed along the Japanese model (*kareitsu*). In Taiwan, some state-owned enterprises were set up although small and medium scale enterprises were typical. Hong Kong did not adopt an explicit industrial policy but focused more on establishing good relations with China and Taiwan to facilitate the flow of commerce.

In many countries the industrialization drive was coupled with the development of export processing zones (EPZs) or special economic zones (SEZs) to facilitate the growth of exports by increasing the efficiency of export platforms[4]. EPZs in India and China are discussed in greater depth in Chapter 7. Generally SEZs provide tax exemptions on raw materials, duty-free imports of key materials, and cheap, accessible and reliable support infrastructure services including transport (roads, sea, air), power and water. They may also include site rental at reduced rates and tax holidays or other tax breaks. In addition to the SEZs, countries also offer other support to stimulate exports and FDI including duty compensation and duty drawbacks for exporters which exempt imports from import duties and business taxes, bonded warehouses and other promotions (see for example the Thailand example cited by Radelet (1999)).

Experience in electronics

The electronics industry has played a prominent role in the industrialization of Asia. In a world export market of over $800 billion in 2000, the major Asian exporters captured nearly $300 billion (see Table 1.14).

---

[4]EPZ and SEZ are used interchangeably in the text.

Table 1.14 Ranking of Major Electronics Exporters in Asia (in US billion dollars)

County or Region	1997	2000	Global Rank in 2000
World exports	644.1	876.8	
South Korea	33.9	58.7	3
Taiwan	38.6	57.8	4
Malaysia	36.3	52.4	5
China	21.5	43.5	8
Singapore	41.6	41.6	9
Thailand	14.1	21.6	14
Philippines	14.2	9.7	20
Indonesia	2.9	7.3	21
Hong Kong	5.2	4.0	25

Source: Singapore Ministry of Trade and Industry (2002) and Dowling and Valenzuela (2004).

Specializations in particular products differ from country to country. Singapore attracted foreign companies by offering political stability, good infrastructure including telecommunications, transport, and energy as well as liberal ownership terms. By the early 1990s, more than 300 foreign firms had established a presence in Singapore from many companies in Europe, North America and Japan. Singapore developed a focus on disc drives and was the world's largest producer in the 1990s. Recently it has been challenged by China, Taiwan and Korea.

Korea became a leader in the chip industry and in the production of household appliances, particularly TVs and cassettes, and small computers and peripherals. Much of this output was originally undertaken through licenses and OEM subcontracting. Production moved to ODM and there was more research and development spending. The large chaebols, including Samsung, Daewoo and Lucky Goldstar are the major electronics companies and have become an established presence in the global electronics industry. Korea is one of the leaders in the production of dynamic random access memory (DRAM) devices.

Like Korea, Taiwan also discouraged direct foreign investment and focused on subcontracting by local firms under OEM and ODM arrangements. It did not build large conglomerates; instead it relied on small- and medium-sized firms that were able to innovate quickly and hire skilled technicians and engineers from the growing number of graduates from domestic universities. Taiwan became a specialist in the production of desktop and laptop computers as well as disk drives peripheral, monitors and other computer components. Eventually, Taiwanese firms moved to (OBM) and developed a network of suppliers both in Taiwan and elsewhere. Acer is a good example of how Taiwanese companies were able to build a global presence in this way.

Malaysia, the Philippines and Thailand followed the Singapore path of industrialization. They attracted FDI, initially from Japan, and later on from Europe and North America. By the end of the 1990s electronics accounted for the bulk of total exports and manufacturing output. All three countries began by attracting investment in the chip assembly sector and eventually Malaysia diversified into the production of consumer goods including computers and television sets with Japanese investment. The Philippines and Thailand have emphasized lower level simple electronics; the Philippines focused on chips and Thailand on computers, particularly hard-disc drives. By relying on investment by transnational companies, the three economies of Malaysia, Philippines and Thailand have not developed local expertise and extensive linkages with other sectors of the economy.

Production platforms and international division of labor

Global competition and acceleration in the pace of new innovations have created continuous pressure on exporters to respond quickly to changes in taste, new innovations and new technology. This has created added incentives to tighten the linkages between suppliers through the production chain and give them more responsibility for developing products and new technologies. In the lexicon of subcontracting they are moving from OEM to ODM and in some cases to OBM. In the electronics industry outsourcing has increased rapidly and has given rise to consolidation into several large contract manufacturers who are responsible for much of the production of electronics products for the main computer and information technology

Table 1.15 Revenue and Growth of Major Electronics Contract Manufacturers (1994–2001) in US billion dollars

Company	1994	2001	Annual Growth (%)
Flextronics	0.2	12.1	78
Solectron	1.6	18.7	42
Celestica	2.0	10.0	26
Sanmina/SCI	2.4	11.2	25
Jabil Circuit	0.4	4.3	40
Total	6.6	56.4	36

Source: Sturgeon and Lester (2003).

firms such as IBM, Nortel Apple, Hewlett Packard, Sun Microsystems, Cisco Systems and Lucent Technology. These firms have been growing at more than double digit rates for the past decade and have spread to European information technology firms such as Ericsson, Nokia and Alcatel (see Table 1.15).

Each of these large contract manufacturers has established a global network of plants of various sizes and specializations. In Asia they have large-scale investments for high volume production in Southeast Asia and China (see Table 1.16 for Solectrons sites in Asia).

Leading firms such as IBM, Hewlett-Packard, and other US high-technology giants maintain their focus aspects that are essential to being on the cutting edge of technology and maintaining their dynamic comparative advantage. These would include innovation, product development marketing, and advertising that relate directly to brand development. For the other so-called non-core functions, they have come to rely more on specialized suppliers such as contract manufacturers who take care of organizing a whole range of these activities. By divesting themselves of non-core activities lead firms can take advantage of new innovations while spreading risk. This type of vertical de-integration or reverse integration are now being widely employed in the textiles/apparel, motor vehicles and electronics industries. As a part of this process, subcontracting suppliers have become more capable of providing better products, some of them completely designed and finished (ODM) and others made to specification

Table 1.16 Solectrons Sites in Asia

Asian location	HQ	Mfg	Mat	NPI	Serv	Tech
Taipei, Taiwan	X					
Singapore		X	X	X		
Johor, Malaysia		X	X	X		
Penang, Malaysia		X	X	X		X
Suzhou, China		X	X			
Penang, Malaysia		X	X			
Wangaratta, Australia		X				
Singapore			X			
Liverpool, Australia				X		
Bangalore, India						X
Tokyo, Japan	X					X
Kanagawa, Japan			X	X	X	

Source: Sturgeon and Lester (2003).
Note: HQ: Global or regional headquarters; Mfg: Manufacturing facilities; Mat: Materials purchasing and management centers; NPI: New product introduction centers (R&D); Serv: After sales service and repair center; Tech: Technology centers (R&D).

(OEM). According to the surveys carried out by Sturgeon and Lester, fewer firms have been able to proceed to the OBM stage and begin marketing their own products — there are exceptions such as the Korean automobile firms Hyundai, Kia, and the Taiwanese computer companies including Acer.

As globalization strengthened, leading firms are looking for suppliers who also have a global reach. This enables them to provide supplier logistics beyond the local scale and to build an international position, perhaps through mergers and acquisitions. So far Asian firms have not developed this vision or capacity. Only 16% of ASEAN manufacturing firms replying to a World Bank survey employed managers with any foreign experience (reported by Sturgeon and Lester, 2003). Lead firms with a global orientation are also looking for subcontractors that can become partners in developing new products and are able to supply ODM and perhaps OBM rather than just OEM.

Table 1.17 Lead Firm and Contractor Profitability

Firms	Country	Sectors	Five-year Average Profit Growth (%)	Sales Growth 4-year Average (%)
Johnson Electric	Hong Kong	Motors	19	10
Black and Decker	USA	Tools	10	–1
Quanta	Taiwan	Laptop	11	73
Dell	USA	Laptop	9	45
Techtronics	Hong Kong	Tools	7	23
Home Depot	USA	Retailer	8	25
Hon Hai	Taiwan	Cables	11	57
Intel	USA	Microprocessors	34	24
TSMC	Taiwan	Microprocessors	0	28
Phillips	Holland	Electronics	44	6
Hung Hing	Hong Kong	Printing	16	16
Bertelsmann	Germany	Publishing	7	6
Li&Fung	Hong Kong	Trading	3	25
The Limited	USA	Retailing	8	6

Source: Sturgeon and Lester (2003), Table 7.

This is not to say that the marriage between lead firms and Asian suppliers has not been profitable for both the lead firm and the supplier (see Table 1.17). It has been and will continue to be profitable. Nevertheless, there is a challenge for Asian subcontractors to develop a wider focus outside Asia. There are several strategies for doing this which will be discussed in the section on Future prospects.

Shifting Sectoral Income Shares

The sectoral shares of income broken down by agriculture, industry and services are displayed in Figures 1.5–1.7. The typical shift in the pattern of income shares observed in a number of countries is for agriculture to decline monotonically over time and for services to increase monotonically.

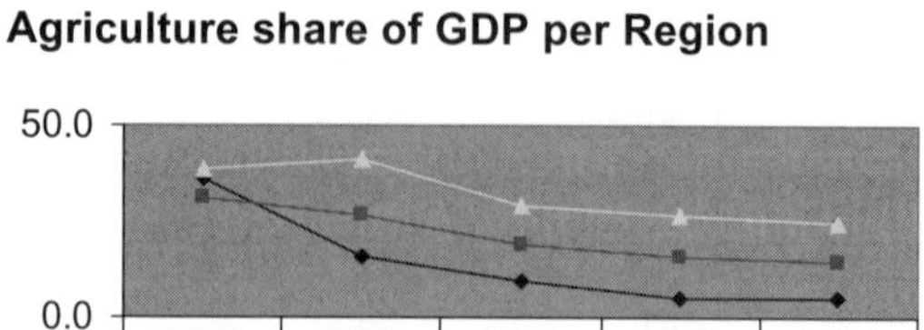

	1970	1980	1995	2000	2004
East	36.0	15.9	9.6	4.9	5.0
Southeast	31.1	26.7	19.2	15.7	14.4
South	38.4	41.3	29.3	26.8	24.7

Figure 1.5 Agriculture Share of GDP (1970–2004) by Region

Source: ADB (2006).

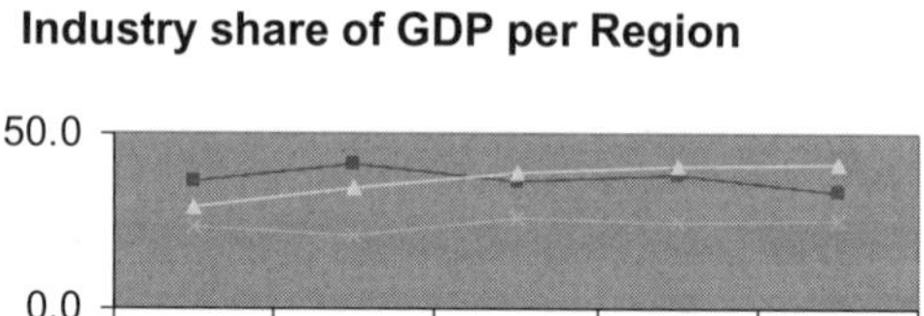

	1970	1980	1995	2000	2004
East	36.6	41.3	36.6	38.2	33.6
Southeast	29.1	34.8	38.9	41.0	41.3
South	23.5	21.1	25.7	24.9	25.3

Figure 1.6 Industry Share of GDP (1970–2004) by Region

Source: ADB (2006).

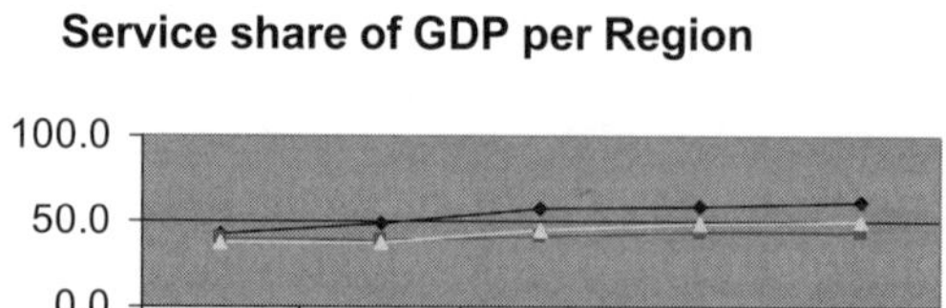

	1970	1980	1995	2000	2004
East	42.9	49.4	57.5	59.3	61.1
Southeast	39.7	38.5	41.8	43.4	44.3
South	38.0	37.6	45.1	48.3	49.9

Figure 1.7 Service Sector Share of GDP (1970–2004) by Region

Source: ADB (2006).

Industry

Industry tends to increase, level off, and then decline. This pattern of structural shifts in shares results from the initial spurt in manufacturing output during the initial phases of the industrialization process. As industry matures and production processes become more complex and additional support services are required, the services sector becomes more and more important. This pattern tends to hold in Asia with a few interesting exceptions. In East Asia the share of the industrial sector peaked in the 1980s. If we consider Singapore as a member of the NIEs then its industrial sector peaked a bit later, sometime in the 1990s. Hong Kong began specializing in services in the 1980s as its role as a financial and management center for southern China grew. By the early years of this century, the industry's share fell to under 10% of the total output. The pattern in China has been somewhat different. The industry sector share has continued to grow and now has nearly two-thirds of output, a much higher proportion than any other country in the region, and perhaps even globally.

In Southeast Asia, industry's share peaked in the 1980s in the Philippines and by the mid-1990s in Malaysia. In the remaining countries of the region, Thailand, Vietnam and Indonesia, industry's share is still increasing. Partly this is a reflection of the lower level of income in the region and the underlying structure of the industrial sector. Thailand, like Malaysia and the Philippines, began to open up its economy in the 1970s producing agricultural products for exports while importing manufactured goods. In the 1980s, Southeast Asia began to develop labor-intensive manufactured products for domestic use and for export. By the end of the 20th century it had become a major exporter of high-technology manufactured goods as reflected by its comparative advantage. Exports of electronics, computer peripherals and parts, automobiles, and agricultural products continue to grow and domestic investment in infrastructure has boosted industrial efficiency and growth in output. Vietnam started later and from a lower base, and is still expanding its industrial sector. Indonesia was also a late starter in the industrialization process and was the most adversely affected by the 1997 financial crisis. As a result, it is still catching up with the rest of the region, although the share of industry has been leveling off at around 44% for the past decade.

In South Asia, the pattern of industrialization followed a different model. Initially, the region followed a Soviet autarchic model with emphasis on heavy industry. A period of slow growth followed and industry languished. By the early 1990s there was some opening up of international trade and a modest quickening of industrial sector output. Industrial output has been rising as a share of income although still very modest when compared with more open economies in Southeast Asia. Industry share of output is less than 25% on average, a full 15% lower than the average in Southeast Asia. Furthermore, industry share of output peaked in 1995 in India, Pakistan and Sri Lanka. This means that industry sector has been below the average for the country for more than a decade and strongly suggest that industry has not yet played a key role in the South Asian growth experience. This is not typical of the growth pattern in Southeast Asia and East Asia where the industrial sector spearheaded growth. Unless further structural reforms are made, it is unlikely that the industrial sector will play a key role in the development of South Asia in the future (see Chapter 7 and future section of this chapter for more details).

Services

Service sector income shares have been increasing in East Asia since the 1970s. Services now dominate the output mix of Hong Kong (91%), Singapore (69%) and Taiwan (70%) serving to complement the high-technology industrial sectors of these mature economies. Korea is also moving toward a higher service sector component as is China. In the latter case the contribution of services is probably still being underestimated. Adjustments from the Soviet accounting system, where services' contribution to income were not recorded, are still being made. Furthermore, there are large informal service sectors in most Asian economies that attract low-skilled workers, where hours are long, wage rates are low, and working conditions are substandard.

In Southeast Asia, service sector shares are clustered closely between 40% and 50% of output. This suggests a similar pattern of service sector growth, composed mostly of high technology support services although tourism also plays a role in most of these economies, particularly in Thailand. Service sector shares have not changed much in the past 6 years aside from Malaysia. Generally, in Southeast Asia service sectors are still considerably smaller than the service sectors in East Asia. This could be because agriculture is still important in all the countries.

In South Asia, the service sector has a surprisingly large fraction of output, larger on average than in Southeast Asia. Recent growth in high technology service in Bangladesh and India may account for some of this strength in services but the overall contribution of this subsector to GDP is minor. A more likely explanation is that internal migrants from rural to urban areas are increasingly being employed in the informal sector and in low-paid entry level jobs in a variety of service industries. This is the direct result of slow growth in industry and an inability of the manufacturing sector to absorb new entrants into the labor force.

Sectoral Employment Trends

Breakdown of employment by sector is more difficult than output breakdowns. Labor market information is often unreliable and incomplete. We do know that shifts in employment lag shifts in output, and as a result a large fraction of the labor force remains in agriculture even after agriculture's output share has declined. This is particularly true in countries where overall growth is slow and the industrial sector has not emerged as the engine of growth.

As an example of these trends, consider countries in Asia where more than 40% of the labor force is still employed in agriculture. Table 1.18 lists these countries. Not surprisingly, most of them are in South Asia. What is surprising is that Thailand is also in this group of countries. Industrialization has resulted in a dramatic shift in output in Thailand but there has not been a commensurate shift in employment.

Table 1.18 Share of Employment in Agriculture, Industry, Services, and Not Adequately Defined — Latest Date between 2000 and 2004

Country	Employment Share of Agriculture (%)	Employment Share of Industry (%)	Employment Share of Services (%)	Not Adequately Defined
Bangladesh	62	12	22	4
China	47	17	13	23
India	66	13	20	0
Indonesia	44	18	36	2
Nepal	75	5	20	0
Pakistan	49	19	32	0
Sri Lanka	42	22	33	3
Thailand	49	18	33	0
Vietnam	69	0	0	31

Source: Estimated from ILO (2005), Figure 3.2.

Compared with the Philippines, the result for Thailand is even more puzzling. The Philippines has been deindustrializing since the 1980s as reflected by a decline in industrial sector share in output. Yet, the Philippines does not appear on the list of countries with more than 40% employed in agriculture. Is this because of faster growth in services in the Philippines than in Thailand? The service sectors in the two countries were the same size relative to total income in 2000. However, the Philippine service grew faster over the 35-year period between 1970 and 2004. Another explanation is that the Philippines has a much higher rate of international migration. Workers that otherwise might have been employed in agriculture have left the country to work in the service sector overseas and this has reduced the agricultural labor force.

The data in Table 1.18 show that despite the growing industrial sector in most countries, agriculture still employs the bulk of the labor force in poorer countries and even in a relatively high-income country like Thailand. All countries in South Asia fall in the over 40% category, and Bangladesh, India and Nepal still have more than 60% of the workforce employed in agriculture. Furthermore, the employment share of industry in South Asia is small. Averaging the five countries' share, only 14% of the workforce is employed in the industrial sector. The picture is not much better in Thailand and Indonesia, where only 18% of the workforce is employed in industry. What is even more startling is the figure for China. In a country where two-thirds of output is now generated by the industrial sector, only 17% of the workforce is employed in the sector. Either the industrial sector workforce is super-efficient and working in a highly capital-intensive environment — two assumptions that are inconsistent with the image of China as a labor-intensive economy — or else there are data inconsistencies, illegal hiring or undocumented workers.

Foreign Aid

The impact of foreign aid on economic growth and poverty reduction has been studied extensively (United Nations, 2006, Chapter 1). When aid is broken down into different components such as bilateral aid and multilateral aid, several tentative conclusions are possible. When aid which is geopolitically motivated and aid which is designed to address human resources and other developmental issues are separated, Reddy and Minoiu (2006) conclude that aid that flows through multilateral aid agencies and a group of Nordic countries, and other developmentally motivated donors (Austria, Canada, Luxembourg and the Netherlands) had a positive impact on development with a time lag. An increase of 1% of GDP in multilateral aid receipts in the 1960s was associated with a half a percent increase in the average growth rate of GDP in the 1990s. Results were for a panel of developing countries where several other variables were included in the regressions including level of per capita income, terms of trade, government corruption, life expectancy, and government consumption among others. The other component of aid, termed geopolitical aid, was negatively associated with growth. Aid from two bilateral donor groups also had a positive effect on growth with a lag. Average growth between 1980 and 2000 was over 1% higher for countries that received an additional 1% of GDP as aid transfers from these Nordic countries in the 1970s and 1980s. These results were obtained from regression analysis for a number of countries, not just in Asia.

In recent years, aid in Asia has been confined to China, South Asia, a few countries in the Mekong region, Philippines and Indonesia, and Central Asia. These aid flows are summarized in Table 1.19. Flows have been generally falling since 1998 both absolutely and as a percentage of Gross National Income (GNI). In a few countries, aid makes a highly significant contribution to the investment flow when measured as a percentage of gross capital formation. In Cambodia, Kyrgyz Republic, Lao PDR and Tajikistan, they comprise more than 50% of gross capital formation in 2003 and an even higher proportion in 1998. In these countries aid is indispensable, providing a crucial complement to domestic resource mobilization, which is often small and inadequate.

In the rest of the region, foreign aid is a minor component of capital formation. Richer countries now go to commercial sources for funds or borrow from multilateral institutions at market rates of interest rather than relying on concessional aid from multilateral or bilateral donors.

Issues and Prospects for Growth and Structural Change

Income growth

In this section, we explore the prospects for future growth and structural change. There are many possible approaches from simple growth, to back of the envelope, to computable general equilibrium to elaborate econometric models. We adopt a simple growth framework with structural parameters estimated from historical experience. We begin with some simple growth models and growth accounting.

Growth models such as the Harrod (1939), the Solow model (1956), and the new growth theories of Romer (1986), Mankiw *et al.* (1992), and Lucas (1988) highlight the importance of the rates of saving, investment and the capital stock, the capital output ratio, labor force growth, and technological change and innovation as the primary factors determining long-term growth rates.

Table 1.19 Aid Flows — 1998 and 2003

Country	1998	2003	1998	2003	1998	2003	1998	2003
	Aid in million $	Aid in million $	Aid per capita in US$	Aid per capita in US$	Aid as % of GNI	Aid as % of GNI	Aid as % of gross capital formation	Aid as % of gross capital formation
Bangladesh	1158	1393	9	10	2.5	2.5	12.1	11.5
Cambodia	337	508	28	38	11.1	12.5	91	54
China	2456	1325	2	1	0.3	0.1	0.7	0.2
Kazakhstan	223	268	14	18	1.0	1.0	6.4	3.4
Kyrgyz	240	198	50	39	15.3	10.7	94	64
India	1610	942	2	1	0.4	0.2	1.8	0.7
Indonesia	1266	1743	6	8	1.4	0.9	7.9	5.2
Lao PDR	276	299	55	53	22.1	14.3	88	69
Malaysia	208	109	9	4	0.3	0.1	1.1	0.5
Nepal	402	467	18	19	8.3	8.0	33.3	30.9
Pakistan	1053	1068	8	7	1.7	1.3	9.6	7.8
Philippines	618	737	8	9	0.9	0.9	4.7	4.9
Sri Lanka	425	672	24	35	2.7	3.7	10.7	16.5
Tajikistan	161	144	26	23	12.7	9.9	78.9	51.4
Turkmenistan	24	27	5	6	0.9	0.4	2.0	1.7
Uzbekistan	158	194	7	8	1.1	2.0	5.1	9.8
Vietnam	1177	1769	15	22	4.4	4.5	14.9	12.8

Source: World Bank (2006).

If we refer to a production function there are several components of growth. In a Solow model, growth in income equals the rate of growth of education adjusted labor force (h) times labor's share in income (ls) plus the rate of growth of capital times capital's share in income ($1 - ls$) plus a residual (a) which we call total factor productivity, (TFP).

$$y = (ls)\, h + (1 - ls)\, k + a, \tag{1.2}$$

where y is the rate of change in income, h is the rate of change in education adjusted labor input, k is the rate of growth in capital, ls is labor's share in income, ($1 - ls$) is capital's share in income, and a is TFP. If we assume that the capital to output ratio is fixed in the short run, then we can substitute y for k on the right-hand side of Equation (1.2) and rearrange so that

$$y = h + a/ls. \tag{1.3}$$

This is the simple growth model framework suggested by Bosworth (2006), where income growth (y) is a function of the growth of the labor force adjusted for improved quality by higher education and better health (h), the share of labor in total income (ls), and a measure of output growth resulting from innovation and technological change designated as TFP (a).

By looking at the historical values for h, a, and ls we can establish a benchmark rate of growth that seems plausible for the future. This growth in income has to be matched by enough investment on the supply side to sustain growth. Consider a simple Harrod–Domar model where income growth is a function of the saving rate (s), the capital output ratio (k), and the rate of depreciation (d)

$$y = (s/k) - d. \tag{1.4}$$

Looking at the Asian economies for the last couple of decades, making a few assumptions about a, d, and using average historical data, a range of future income growth trajectories can be derived. These hypothetical growth scenarios can then be compared with projections made by forecasters such as consensus economics, the IMF, the ADB, and Deutsche Bank Research.

From this analysis, a broader consensus forecast emerges and is discussed. To begin, consider the values of *h*, *a*, and *ls* for Asia. Unadjusted labor force estimates are shown in Table 1.20 for the period from 1980 to 1995 and 1995 to 2005.

Estimates of labor force growth (*h*) are displayed in the final two columns of Table 1.20. Labor's share in income is more difficult to measure. Bosworth (2006) uses an estimate of 56% from production function estimates of labor and capital share. Many others, including

Table 1.20 Labor Force (in millions)

Country	1980	1995	2005	Implied Annual Change — 1980–1995 (%) (*h*)	Annual Changes from 1990–2003 (*h*)
China	539	709	784	2.0	1.1
Hong Kong	2	3	3.5	3.0	1.9
South Korea	16	22	23	2.0	1.8
Indonesia	59	89		3.2	2.4
Malaysia	5	8		3.3	2.9
Philippines	19	28		2.9	2.8
Singapore	1.2	1.7		0.3	2.3
Thailand	24	34	37	2.6	1.1
Vietnam	26	37		2.6	1.9
Bangladesh	41	60		2.9	2.1
India	300	398		2.0	2.1
Nepal	7	10		2.7	2.6
Pakistan	29	46	52	3.7	2.7
Sri Lanka	5	8	10	3.7	2.0
Japan (for comparison	57	66	65	1.0	

Source: 1980 and 1995 from World Development Indicators (WDI) and quoted by ILO; 2005 from ILO database of total active population between 15 and 65 and spliced with 1995 labor force estimate from WDI. Data for 1990–2003 are taken from world development indicators, Table 2.2.

Sarel (1995) and Young (1995) have used higher values of 65% in estimating the value of TFP.

By looking at discrepancies in different estimates of TFP, Weerasinghe and Fane (2005) suggest that Singapore and Malaysia may have lower labor shares. In the exercise below, we use the 65% share of labor in income assumed by Young and many others. An upward adjustment in labor force growth of 1% is made in East and Southeast Asia to reflect upgrading of skills and education. A lower figure of 0.5% is used for South Asia under the assumption that skill and educational upgrading are generally lower in South Asia than in East and Southeast Asia. TFP estimates vary widely and we have elected to combine the results displayed in Table 1.20 and summarized there. TFP is low in Philippines (0.3%) and Indonesia (0.5%), around 1% in Malaysia, as much as 2% in Singapore and Thailand, and between 2% and 3% in Korea and Taiwan. There are few reliable estimates for TFP in South Asia. We assume it to be 1%.

Putting these estimates of TFP, *h*, and *ls* together, we obtain the following estimates of income growth from Equation (1.3), which are displayed in Table 1.21.

These estimates of growth potential or long-term growth from the demand side have to be balanced against the saving and investment rates. Will there be enough productive capacity to produce these growth rates? Looking at Equation (1.4) we need the depreciation rate of capital, the saving rate, and the capital/output ratio. A depreciation rate of 5% is often used and the average saving rate for the past decade is shown in Table 1.22. The capital/output ratio will have a tendency to increase over time as the capital stock grows and diminishing returns set in. This will be offset to some extent by technical progress. Estimates of the capital/output ratio for Asian countries are difficult because the capital stock is essentially unknown. For argument's sake we will compute several different estimates with different capital/output ratios and growth rates in income to see what that implies for the rate of saving and investment needed to sustain growth rates shown in Table 1.21. Recall

$$s = yk + d \tag{1.5}$$

Table 1.21 Estimated Long-Run Growth Using Equation (1.3)

Country	*ls* (Labor Share in Income)	*h* (Augmented Labor Force Growth)	*a* (TFP)	*y* (Income Growth)
China	0.65	2.1	3.9	8.1
Indonesia	0.65	3.4	0.5	4.2
Korea	0.60	2.8	2.0	6.1
Hong Kong	0.65	2.9	1.5	5.2
Malaysia	0.65	3.9	1.0	5.4
Philippines	0.65	3.8	0.3	4.2
Singapore	0.60	3.3	2.0	6.6
Taiwan	0.65	2.8*	2.0	5.9
Thailand	0.65	2.1	2.0	5.2
Bangladesh	0.65	3.1	1.0	4.6
India	0.65	2.6	2.0	4.1
Nepal	0.65	3.1	0.5	4.5
Pakistan	0.65	3.2	1.0	4.7
Sri Lanka	0.65	2.5	0.5	3.3

*Assumed to be same as Korea's augmented labor force growth *h* is derived as final column in Table 6.13 plus 1%.
Source: Table 1.6, Table 1.7 and Table 1.20.

These estimates are displayed in Table 1.22.

How do these figures for the saving rates compared with historical saving rates? Saving rates for the NIEs of Korea, Singapore and Taiwan have been averaging between 30% and 35% for the past two decades according to the IMF (see Dooley *et al.*, 2005, Charts 2 and 3) while saving rates for emerging Asia outside of the NIEs have been rising from around 25% in the late 1980s to nearly 40% in 2004. Much of this is because of increased saving in China although savings in other countries have also been steady or rising. Savings rates for selected countries are also displayed in Table 1.23.

Comparing the final column of Table 1.21 with Table 1.23 we see that actual domestic saving rates are usually higher than the saving rates implied by fitting historical values to Equation (1.3) in East Asia and Southeast Asia. If the capital to output ratio was lower than the ratio of 4 assumed in the second column of Table 1.21, the implied

Table 1.22 Estimated Saving and Investment Rates Implied from Growth Rates in Table 1.21

Country	y from Table 1.14	k	d (in %)	$S = yk + d$ (in %)
China	8.1	4	5	36.4
Indonesia	4.2	4	5	21.8
Korea	6.1	4	5	29.4
Malaysia	5.2	4	5	25.8
Philippines	5.4	4	5	26.6
Singapore	4.2	4	5	21.8
Taiwan	6.6	4	5	31.4
Thailand	5.9	4	5	27.6
Bangladesh	5.2	4	5	25.8
India	4.6	4	5	23.4
Nepal	4.1	4	5	21.4
Pakistan	4.5	4	5	23.0
Sri Lanka	4.7	4	5	23.8

Source: Table 1.14.

saving rate would be even lower[5]. This information suggests that resource mobilization has not been a problem in these two regions of Asia. This is confirmed by the build up in foreign exchange reserves over the past few decades and the increase in outward foreign investment by Singapore, China, and other countries in recent years.

For South Asia the results are not as clear. Saving rates are not as high as they are in Southeast Asia and East Asia. Comparisons between investment rates and implied saving rates from Equation (1.3) show a particularly large shortfall in Pakistan. The volume of domestic saving may have been a constraint on growth in South Asia and foreign aid has served to fill some of the gap. More needs to be done to raise saving rates in the region and to transfer surpluses from the richer Asian economies. See Chapter 2 for details of some possibilities for doing this.

Given that saving has not been a constraint on growth in Southeast Asia and East Asia, and also assuming that constraints in South Asia will be removed by resource transfers, we can compare the final column in

[5]IMF estimates of capital output ratios for Asian economies reported in IMF (2006, Table 2.6) suggest a ratio closer to the range of 2.5–3.0.

Table 1.23 Saving and Investment Rates

Country	Domestic Saving Rate as Percent of GDP (1993)	Gross Domestic Investment Rate as Percent of GDP (Average of 2001–2005)	Saving Rate Implied by Equation (1.3)
China		38.0	36.4
Indonesia	31	20.8	21.8
Korea	35	29.7	29.4
Malaysia	38	22.3	25.8
Philippines		17.2	26.6
Singapore	47	20.8	21.8
Taiwan		19.0	31.4
Thailand	36	26.3	27.6
Bangladesh		23.6	25.8
India		26.4	23.4
Nepal		25.2	21.4
Pakistan		17.0	23.0
Sri Lanka		23.7	23.8

Source: IMF (2006) and ADB (2006). Domestic saving rates in South Asia are proxied by gross domestic investment, which will be higher than saving because of positive capital inflows.

Table 1.21 with the actual and projected growth rates in Table 1.24 for the decade ending in 2010. These are displayed in Table 1.25.

The residuals are systematically positive in South Asia. The model underpredicts actual growth. In East Asia, the model systematically overpredicted growth except in the case of China. In Southeast Asia there were no apparent systematic errors. In two of the three financial crises countries, Korea and Thailand, the model overpredicted growth. In Thailand, lower than expected growth could also be the result of the adverse impact of the 2001 recession in the United States, some residue from the 1997 financial crisis, and political uncertainties in recent years. In Taiwan, structural adjustment to the growing strength of China, combined with political uncertainties has kept growth rates subdued.

Growth was much stronger in India and to a lesser extent in the other South Asian countries, perhaps because TFP was higher than the assumed 0.5%. If we raise TFP to 1% this would raise growth by

Table 1.24 Income Growth in Developing Asia (2001–2010)

	2001	2002	2003	2004	2005	2006	2007	2008	2009	2010	Average
China	8.3	9.1	10	10.1	9.9	9.5	8.8	8.8	8.9	8.7	9.21
Hong Kong	0.6	1.8	3.2	8.6	7.3	5.5	5	5.5	6	5.5	4.9
Korea	3.8	7	3.1	4.6	4	5.1	4.9	5	5	4.8	4.73
Taiwan	–2.2	4.2	3.4	6.1	4.1	4.4	4	4	4.5	4.4	3.69
Singapore	–2.3	4	2.9	8.7	6.4	6.1	4.6	5.8	6	5.8	4.8
Indonesia	3.8	4.3	5	4.9	5.6	5.4	6	5.5	6	5.5	5.2
Malaysia	0.3	4.4	5.4	7.1	5.3	5.5	5.8	5.6	5.7	5.6	5.07
Philippines	1.8	4.4	4.5	6	5.1	5	5.3	5	5.4	5	4.75
Thailand	2.2	5.3	7	6.2	4.5	4.7	5.5	5.1	5.3	5	5.08
Bangladesh	5.3	4.4	5.3	6.3	5.6	6.5	6	5.7	5.9	5.6	5.66
India	5.8	3.8	8.5	7.5	8.1	7.6	7.8	8	8	7.7	7.28
Pakistan	1.8	3.1	4.8	6.4	8.4	6.5	7.3	7	7.2	6.9	5.94
Sri Lanka	–1.5	4	5.9	5.5	5.7	5.3	5.2	5.5	5.7	5.5	4.68

Source: Figures 1.1–1.3, and author's projections.

Table 1.25 Comparison of Historical Growth with Model Growth Rates

Country	(2) Average Growth — 2001 to 2010	(3) Income Growth from Table 1.21	(2) – (3). When Value is Positive, Actual Growth Exceeds "Long-Run" Projected Growth
China	9.2	8.1	1.1
Hong Kong	4.9	5.2	–0.3
Korea	4.7	6.1	–1.4
Taiwan	3.7	5.9	–2.2
Singapore	4.8	6.6	–1.8
Indonesia	5.2	4.2	1.0
Malaysia	5.1	5.4	–0.3
Philippines	4.7	4.2	0.5
Thailand	5.1	5.2	–0.1
Bangladesh	5.7	4.6	1.1
India	7.3	4.1	3.2
Pakistan	5.9	4.7	1.2
Sri Lanka	4.7	3.3	1.4

Source: Tables 1.21 and 1.24.

about 0.8% per annum for all of South Asia. Yet it would still not remove the underprediction completely. TFP may be even higher or adjusted labor force growth should be higher.

Consensus Economics prepares forecasts for the Asian region based on inputs from a number of different forecasters including commercial banks, financial houses and government agencies. These inputs are averaged to give a consensus view of prospects for these economies for East Asia and Southeast Asia. Comparisons of country forecasts for 2006–2009 by Consensus Economics are compared with Table 1.24 values and displayed in Table 1.26. (Consensus Economics does not publish forecasts for Bangladesh, Pakistan or Sri Lanka.)

Consensus Economics forecasts are somewhat lower than our forecasts for China, Hong Kong and Singapore. For other countries the forecasts are close, although Consensus Economics forecasts are also a little lower for the Philippines and India.

Table 1.26 Growth Forecast Comparisons

Country	2006 Table 1.24	2006 Consensus Economics	2007 Table 1.24	2007 Consensus Economics	2008 Table 1.24	2008 Consensus Economics	2009 Table 1.24	2009 Consensus Economics
China	9.5	7.9	8.8	7.8	8.8	8.3	8.9	7.8
Hong Kong	5.5	4.5	5	4.2	5.5	4.5	6	4.6
Korea	5.1	4.7	4.9	4.9	5	4.9	5	5
Taiwan	4.4	4.3	4	4.5	4	4.5	4.5	4.5
Singapore	6.1	4.9	4.6	4.8	5.8	4.6	6	4.3
Indonesia	5.4	5.5	6	5.8	5.5	5.4	6	5.4
Malaysia	5.5	5.5	5.8	5.7	5.6	5.6	5.7	5.7
Philippines	5	5	5.3	4.6	5	4.8	5.4	4.8
Thailand	4.7	5.7	5.5	5.0	5.1	4.9	5.3	5.2
Bangladesh	6.5		6		5.7		5.9	
India	7.6	6.9	7.8	7.3	8	7.2	8	7.3
Pakistan	6.5		7.3		7		7.2	
Sri Lanka	5.3		5.2		5.5		5.7	

Source: Table 1.24 and Consensus Economics (2006) projections made in July.

Factors contributing to growth

The exercise conducted in the previous section is designed to raise the readers' awareness of the major factors contributing to growth. Analysis of these factors can give us clearer insight into Asian growth prospects for the next decade and a half. Over time, labor force growth declines and TFP also slows as the leverage from shifting from agriculture to industry and services dissipates. As the size of industry and services growth diminishes along with the growth in the labor force, there is a natural reduction in the sustainable long-term rate of growth in output and output per capita. There will also be diminishing returns to scale as the amount of capital per worker continues to grow. To offset these tendencies, economies have to become more efficient in the use of resources, technology and manpower. We look at a few aspects of these challenges to raise output and productivity.

Investment

The rate of investment is crucial. Historically, Asia has been the highest saving region in the world and this is projected to continue. However, some developments may slow the rate of investment. Consumer credit is a growing feature of the Asian economies. In industrial countries the spread of consumer credit has been negatively correlated with saving and there is a possibility that this could develop in developing Asia. However, Japan is a case where saving rates remain high despite the widespread access to credit. Investment is particularly critical for funding of physical infrastructure in South Asia to increase economic efficiency and international competitiveness and to facilitate the ongoing transformation of industry into a higher technology orientation.

Total factor productivity

Total factor productivity in all of its different aspects is important. TFP reflects innovation and technical transfer as well as the movement to higher value-added goods and gains in labor productivity. Countries that continue to transform their mix of products from low-value agricultural products and labor-intensive manufacturers to higher value industrial products and services benefit from the additional output generated and also from TFP that raises the skill levels of the entire labor force. In Southeast Asia, for example, United Nations (2006, Chapter 2) shows that nearly all of the annual growth in labor productivity from 1991 to 2004 in Southeast Asia came from gains from relocation of labor from low productivity jobs in agriculture or services to high productivity manufacturing or service sector jobs. Fewer reallocation gains were made in China and the rest of East Asia, and also in South Asia. Labor productivity in Asia outpaced all other developing regions during this period. According to some observers, there has been a trend toward a slowing of TFP growth in developing Asia over the past decade (Bosworth, 2006).

Labor force growth

The demographic transition has taken place at different rates in the different regions. Fertility and population growth are still relatively high in South Asia and this will help the region raise growth levels if productive jobs can be found for new entrants into the labor force. Increases in labor force participation rates for women will also increase the labor force in countries where women's opportunities continue to expand. In other countries, tradition and custom could constrain the entry of women into the labor force (see Chapter 4 for additional discussion).

Education and health

The augmented labor force variable h is a crucial variable in the growth equations reviewed in the previous section. The rate of increase in educational attainment and improvements in health and sanitation are equally important as absolute increases in the size of the labor force. In East Asia and Southeast Asia, raising the highly skilled and professional components is critical. In many of these countries, universal primary and secondary education have been achieved and it is time to widen access to higher and professional education, particularly in business and engineering. In the poorer countries emphasis should be on secondary and technical training as well as tertiary education in business and engineering. These skills also have an important positive impact on poverty (see Chapter 4 for further discussion).

Governance and corruption

Corruption and good governance do not enter directly into the growth models. However, they enter indirectly by raising or lowering the economic efficiency of the economy through the capital to output ratio and the productivity of labor. Honesty and trust bring greater efficiency and stronger growth with them while inefficiency and corruption result in misallocation of resources and economic inefficiencies. We discuss these factors in greater depth in Chapter 5.

Foreign direct investment and export promotion

We saw in the review of the past that FDI and export processing zones played a crucial role in the growth experience of the Asian region. They will continue to be crucial in the future. China is continuing to develop processing zones along the Yangtze River, and India is also developing export processing as a strategy (see Chapter 7). Other countries in the region, including Singapore, Vietnam, and Bangladesh are also focusing more on developing export zones that can compete internationally. As these zones spread their influence and increase industrial efficiency, entire economies will become more capable of manufacturing products for international sale.

Growth Projections to 2020

What will the growth profile for the region look like for the decade from 2010 to 2020? Current population growth provides a rough indication of comparative labor force growth between countries in 20 years. Table 1.27 displays gross population growth rates from 1950 to 1975, 1975 to 2000, and 2000 to 2025 for developing Asia. (see East–West Center, 2002, Appendix Table 2). Growth in the latter period from 2000 to 2025 is substantially lower than in the earlier periods.

Table 1.27 Population Growth, 1975–2000 and 2000–2025 (%)

Country	1950–1975	1975–2000	2000–2025	2000–2025 Implied Annual Population Growth Rate (in %)
China	67	37	15	0.58
Hong Kong	122	56	26	0.98
Korea	73	39	16	0.63
Indonesia	69	57	28	1.15
Malaysia	100	81	70	2.25
Philippines	110	80	41	1.40
Singapore	121	77	24	0.90
Thailand	109	52	23	0.88
Bangladesh	80	82	53	1.80
India	73	62	34	1.16
Pakistan	77	100	77	2.40
Sri Lanka	80	49	19	0.75
USA			19*	1.00

*US figure is for 2000–2020 and source is US population census.
Source: East–West Center (2002).

Actual employment growth in the forecast period from 2010 to 2010 will depend upon a host of other factors as well as lagged population growth, including mortality in the labor force, labor force participation rates for men and women, years of education, and changes in retirement age. What is important to note, however, is the decline in the number of new entrants into the labor force. Those numbers are closely related to population growth a couple of decades earlier; and those numbers are declining. Nevertheless, population growth toward the end of the 20th century and in the early years of the 21st century will largely determine the rate of labor force growth between 2011 and 2020.

The working age populations of the three largest countries in Asia are shown in Figure 1.8. It is clear that Japan's working age population peaked in the late 1960s and leveled off until the 1990s when it began to decline rapidly. China will reach its peak soon and

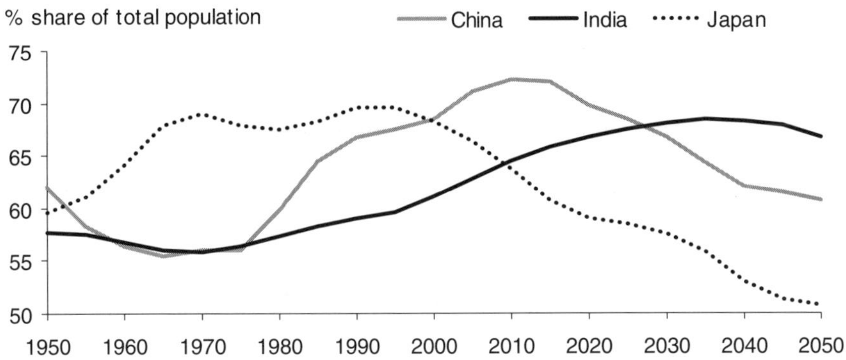

Figure 1.8 Working Age Populations in China, India and Japan

Source: UN Population Statistics.

Table 1.28 Change in Working Age Population (15–59) in China by Decade by Millions of People

1975–1985	163
1985–1995	124
1995–2005	107
2005–2015	33
2015–2025	–29
2025–2035	–77

then begin to decline while India's will not peak until 2030 or thereabouts.

Economic growth projections for the decade from 2011 to 2020 are displayed in Table 1.29 and compared with the growth in the previous decade and with the ADB (1997) forecast. In China, a moderation in growth in the 2011–2020 period is projected. This results from slower growth in the labor force (Table 1.28) and a reduction in the level of TFP. If TFP is 3% and the augmented labor supply, *ls* falls to 1.5% as additions to the labor force fall dramatically[6] and

[6] The decade addition to the labor force from 2005 to 2015 displayed in Table 1.28 is less than one-half of 1% of the Chinese labor force of around 740 million in 2005.

Table 1.29 Projected Growth from 2011–2020 Compared with the Previous Decade

(1)	(2)	(3)	(4)	(5)	(6)
Country	2001–2010 Growth in GDP	2001–2010 Growth in GDP per capita	2011–2020 Growth in GDP	2011–2020 Growth in GDP per capita	1995–2025 Growth in per capita income in ADB (1997)
China	9.2	8.6	7.0	6.4	6.0
Hong Kong	4.9	3.9	4.0	3.0	2.1
Korea	4.7	4.1	4.0	3.4	3.5
Taiwan	3.7	3.1	3.5	2.9	3.1
Singapore	4.8	3.9	4.0	3.1	2.4
Indonesia	5.2	4.2	5.0	4.2	5.0
Malaysia	5.1	2.9	5.0	2.8	3.9
Philippines	4.8	3.4	4.6	3.2	5.3
Thailand	5.1	4.2	5.0	4.1	3.8
Bangladesh	5.7	3.9	6.0	4.2	3.9
India	7.3	6.2	7.6	6.5	5.5
Pakistan	5.9	3.5	5.9	3.5	4.4
Sri Lanka	4.7	4.0	4.8	4.1	3.9
Azerbaijan	10	9	8.2	7.2	na
Kazakhstan	7.2	6.2	6	5	na
Turkmenistan	9.1	8.1	5.2	4.2	na
Kyrgyz Republic	6.1	5.1	5.9	4.9	na
Tajikistan	7.2	6.1	5.9	4.9	na
Uzbekistan	5.7	4.7	5.8	4.8	na

Source: Author's projections, historical growth to 2005 from ADB (2006) and Table 1.13. For Central Asia, Dowling and Wignaraja (2006) and author's projections from 2015–2020.

labor's share remains at 65%, then $y = h + a/ls$ is $1.5 + 3.0/0.65 = 1.5 + 4.6 = 6.1\%$. Even if TFP remains very high and the augmented labor supply grows a bit faster, growth will still slow to around 7%.

For the other countries in East Asia and Singapore, growth will also slow but not as rapidly as in China. The major reason for the slowdown in East Asia and Singapore is more moderate labor force growth and diminishing returns to capital as TFP slows. Also growth in greater China (including Hong Kong, Taiwan and Singapore) will be adversely affected by slower performance in mainland China.

In Southeast Asia growth is projected to fall slightly, although not as much as East Asia. Much will depend upon how effectively Southeast Asian economies respond to the challenges of industrial growth presented by China. Somewhat slower labor force growth will also have a negative impact on economic prospects although the full impact of the demographic transition on these economies will not occur until the later part of the forecast period and beyond. We also have to recognize that the first decade of the new millennium started with very slow or negative growth in these countries as they were adversely affected by the downturn in the United States. Keeping this in mind the forecast for 2011 to 2020 is only slightly lower than the earlier period growth.

Economic prospects in South Asia depend more upon policy adjustments than in the other regions. Liberalization measures have not yet been fully implemented in South Asia and bureaucratic restrictions are still in place in many countries. India has yet to come to grips with restrictions in labor movements, particularly hiring and firing, that have been an impediment to rapid growth in manufacturing. These issues are taken up in some depth in Chapter 7.

The labor force in South Asia will continue to grow rapidly through the decade. The challenge will be to find gainful employment for new entrants and to upgrade skills to take advantage of new opportunities for industrial sector growth in labor-intensive and some skill-intensive occupations. Infrastructure bottlenecks will also restrain output growth unless more resources are made available to improve and increase the efficiency of transportation, power, water, telecommunications, and other services critical for the growth of industry. These issues are discussed in more depth in Chapter 2. Keeping these factors in mind and assuming that bottlenecks will be removed and appropriate policies undertaken, the region should be able to grow at the same or even higher rates than in the previous decade.

In the Central Asian Republics, growth will remain robust for most of the forecast assuming that needed policy reforms are introduced. In the three oil producing economies (Azerbaijan, Kazakhstan and Turkmenistan), an implementation schedule for WTO membership is being developed and trade reforms will continue. Progress will be made in large-scale privatization, corporate governance, and financial sector liberalization. In the three non-oil economies (Kyrgyz Republic, Tajikistan, and Uzbekistan), exports will be promoted, financial sector reforms carried out, particularly of state-owned banks, and privatization of the productive sectors will deepen. In the non-oil CARs, progress in reform will be especially visible in the agricultural sector including removal of quantity and price controls, dismantling of public monopolies, increased investment in rehabilitation of irrigation systems, and rationalization of agricultural subsidies (Dowling and Wignaraja, 2006).

There will be a coherent and cohesive development of the CARs as a region through a number of initiatives in regional cooperation including transport, trade, and energy. These are described more fully in Chapter 2. An integrated rail and road network system will be implemented. Customs codes will be harmonized, border procedures will be streamlined, and best practices information and inspection systems will be adopted. The prevalence of non-tariff measures on regional trade will be significantly reduced as a result of improvements in institutions. An efficient regional energy market and system will emerge through implementation of structural reforms, tariff reforms, and significant investments in regional energy infrastructure. This will be accompanied by greater transparency and more rational pricing in line with market conditions.

Comparing the differences in projected growth rates by subregion we notice a strong tendency toward convergence in per capita income. For the 2011–2020 average growth in East Asia outside of China is just over 3% (column 5 in Table 1.29), in Southeast Asia the average growth in per capita income is about 3.6. Rates in South Asia are a full percentage point higher at 4.6. In the Central Asian Republics the energy boom in the three oil producers and continued market reforms is projected to lift per capita income at the rate of 5.2, the highest of all the regions. These differential growth rate projections in per capita income reflect the maturation of the East Asia economies which will have standards of living comparable to the United States by 2020. They also reflect the gains that will continue to be made by transition economies in Central Asia as well as China and Vietnam as they move from central planning to the market economy.

Export Growth

Export growth is generally dependent upon the rate of growth in income in export markets and the degree of additional market penetration made possible by better products and lower prices. Exports are also more volatile than income, particularly if the import demand elasticity of destination countries is high. From historical experience Asian exports are particularly prone to downturns in the US market.

Keeping these factors in mind, we have constructed a simple model relating projected income growth with export growth over the past few decades and derived an export growth elasticity with respect to income growth. This is a reduced form estimate and not a behavioral model of export behavior

$$\text{Log (export growth)} = a + b \log \text{(income growth)} + \Sigma c_i \text{ (country dummy variables)}, \quad (1.6)$$

where b is the constant elasticity of export growth with respect to income growth. The estimates of b from (1.6) are displayed in Table 1.30 for a pooled regression of several Asian economies. The elasticity is large and significant, reflecting a strong response of exports to changes in income.

These results suggest that exports and economic growth are positively related. Most regression coefficients are positive. However, only a few are statistically significant. To gain more degrees of freedom a pooled regression with intercept dummy variables was performed as suggested in Equation (1.6). These results showed that the export elasticity was positive and significant at the 1% level. A one percent increase in income leads to a 0.566% increase in exports.

Table 1.30 Elasticity of Export Growth with Respect to Income Growth, 1985–2004

Country	Elasticity of Export Growth with Respect to Income Growth
China	–0.06
Hong Kong	0.36*
Korea	0.12
Singapore	1.77
Taiwan	–0.01
Indonesia	0.82
Malaysia	0.26
Philippines	0.47
Thailand	0.20*
Bangladesh	0.88
India	–0.01
Pakistan	1.25
Sri Lanka	1.13*

Source: Data from World Development Indicators CD-ROM 2002 and WDI Online.
Note: Significant *t*-statistics are denoted with an asterisk.

Standard of Living in 2020

In the ADB forecast for the future contained in its forward looking publication published in 1997 (ADB, 1997) a table showed the GDP per person relative to the United States in 2025. It might be useful to undertake a similar exercise a decade later. We can do this in one of two ways — either by projecting income per capita in US dollars using current exchange rates or by projecting growth in GDP using purchasing power parity. Both sets of figures are displayed in Table 1.31 for 2005. Since purchasing power parity is supposed to give a better indication of relative standards of living since all goods enter into the calculation, we make projections based on this way of measuring income. However, since purchasing power changes over time the projections are only a rough guess as to the relative standards of living in 2020.

For the projections we take the growth rates from Tables 1.29 and 1.21. We begin the forecast from the last actual figure in 2005. The forecast for US growth depends upon the time period used. GDP growth

Table 1.31 GDP in 2005 in PPP and Exchange Rate Terms from ADB (1997)

Country	GDP per capita in PPP terms US$	GDP in exchange rate US$
China	6800	1256
Hong Kong	32,900	24,515
Korea	20,400	16,538
Taiwan	27,600	15,531
Singapore	28,100	26,777
Indonesia	3600	1112
Malaysia	12,100	4812
Philippines	5100	1091
Thailand	8300	2794
Bangladesh	2100	na
India	3300	693
Pakistan	2400	na
Sri Lanka	4300	na

Source: ADB (1997) and CIA Factbook (2006).

has slowed down over the past decades, from 3.46 in 1959 to 2005, 3.01 in 1980 to 2005, and 2.97 in 1990 to 2005 (United States Government, 2006, Table B5). Assuming that growth will slow somewhat further over the next 15 years, we assume US growth of 2.9% per annum from 2006 to 2020. Using PPP for the United States of $41,800, this gives a PPP value of $64,181 for the United States in 2020. In per capita terms we assume a population growth rate of 0.75% based on Census Bureau estimates of population growth between 2010 and 2020. This results in a PPP per capita in the United States of 54,827 in 2020.

According to our projections, the Asian region will grow faster than the United States. By 2020, Hong Kong, Singapore and Taiwan will have achieved higher per capita income than the US level in 2005. Other countries have also improved in their performance relative to the United States, as displayed in Table 1.32. The catch up to the United States is not as dramatic as the forecasts contained in ADB (1997). Taking into account that we have 10 more years of data, this result is somewhat surprising, particularly since our per capita growth rates are somewhat higher than the ADB (1997) forecasts for East Asia.

Table 1.32 Income per Capita in 2005, 2020 and 2025

(1)	(2)	(3)	(4)	(5)	(6)	(7)	(8)
Country	2005 PPP Income per capita	2005 PPP Income per capita as % of US per capita income in 2005	2020 PPP Income per capita	2020 PPP Income per capita as % of US per capita income in 2020	2025 PPP Income per capita	2025 PPP Income per capita as % of US per capita income in 2025	2025 Income per capita as % of US per capita income from ADB (1997), Table 2.11
United States	41,800		54,827		60,016		
China	6800	16	18,874	34	25,738	43	38
HongKong	32,900	79	57,550	105	68,022	113	117
Korea	20,400	49	35,278	64	41,697	69	83
Taiwan	27,600	66	43,540	79	49,987	83	88
Singapore	28,100	67	48,063	88	55,990	93	107
Indonesia	3,600	9	6602	12	7994	13	36
Malaysia	12,100	29	18,786	34	21,515	36	71
Philippines	5100	12	8396	15	9829	16	29
Thailand	8300	20	15,252	28	18,646	31	47
Bangladesh	2100	5	3919	7	4814	8	17

(*Continued*)

Table 1.32 (*Continued*)

(1)	(2)	(3)	(4)	(5)	(6)	(7)	(8)
Country	2005 PPP Income per capita	2005 PPP Income per capita as % of US per capita income in 2005	2020 PPP Income per capita	2020 PPP Income per capita as % of US per capita income in 2020	2025 PPP Income per capita	2025 PPP Income per capita as % of US per capita income in 2025	2025 Income per capita as % of US per capita income from ADB (1997), Table 2.11
India	3300	8	8575	16	11,749	20	24
Pakistan	2400	6	5911	11	7021	12	18
Sri Lanka	4300	10	8082	15	9880	16	25
Azerbaijan	4800	11	15,242	28	19,177	32	Na
Kazakhstan	8200	20	18,281	33	21,713	36	Na
Kyrgystan	2100	5	4384	8	5257	9	Na
Tajikistan	1200	3	2652	5	3242	5	Na
Turkmenistan	8000	19	14,969	27	17,865	30	Na
Uzbekistan	1800	4	3724	7	4596	8	Na

Source: Author's projections and ADB (1997).

Part of the difference between our forecasts and the ADB (1997) forecasts is due to the financial crisis in Southeast Asia which occurred just after ADB (1997) was published. This crisis moderated performance for several years and is also evident from the forecasts of 2020 and 2025 as well. Differences may also be the result of differences in the forecasts of the US growth rate and the rate of population growth.

In any event, both sets of results suggested show some convergence toward the US growth rate. This can be seen by comparing the ratio of PPP per capita income in 2005 and 2020 as displayed in columns (3) and (5) in Table 1.32.

Policy Challenges and Responses

The growth assumptions from Equations (1.2) and (1.3), and the data assembled to compare saving rates with growth rates were designed to explore the plausibility of continued growth without significant bottlenecks. In order to achieve sustained growth in per capita income of the magnitudes displayed in Table 1.29, continued reforms and proactive government policies in a number of areas will be required. These aspects are explored in more detail in the remainder of this book.

The trajectory of growth will be subject to a number of challenges. Briefly, aside from China, East Asia faces similar challenges to those confronting industrial countries in Europe, North America and Japan. These include slower population growth and greater burden to support an aging population; technological challenges to maintain competitiveness in a global environment; maintaining growth at a sustainable pace while containing inflation, and avoiding recessions. Capital accumulation has been high for the past few decades. As a result, future opportunities for productivity growth from shifts from low productivity to high technology jobs are limited.

In China, regional income imbalances between rural and urban areas have grown rapidly and need to be addressed along with environmental protection. Restrictions on migration have been relaxed to some extent, but more needs to be done to encourage those willing to migrate. At the same time the government has to develop policies that stimulate the western provinces and rural areas in general. EPZs along the upper reaches of the major rivers are a move in this direction. Government plans are in the works to raise subsidies for farmers and for education in rural areas. Manufacturing has grown rapidly and has created imbalances that need to be addressed. The government needs to develop the service sector and introduce policies that help to support industry. Insurance, banking, telecommunications, and other services will expand. Slower growth in industry and investment will be accompanied by faster growth in consumption and greater emphasis in the quality rather than the quantity of investment. Macroeconomic management is complicated by the role of local governments, which are responsible for the bulk of investment and are not easily controlled by the central government. Ultimately, any bad debts that arise from poor investments by local governments are the responsibility of the central government. This creates a moral hazard problem and a risk of overheating that has not yet been effectively dealt with. Independent local governments that also control spending make fiscal policy difficult to implement at the country level, leaving monetary policy as the only effective macroeconomic management tool.

In Southeast Asia, Thailand and Malaysia have begun to pull away from the rest of the region. By 2020, their per capita income will be nearly as high as Korea's in 2000. They could begin to qualify as high income developed countries by 2020. Thailand will have to deal with growing income inequality created by the rapid growth of the Bangkok metropolitan area and the continued slower growth of the northeast and southern provinces, and the political uncertainties that surfaced in 2006 may continue to slow growth. Malaysia has to deepen its services sector by encouraging greater foreign participation and reducing restrictions on foreign participation in professional service, banking and telecommunications (ADB, 2006). Continued upgrading of human resources for the electronic age are critical. Indonesia and Philippines will face a different set of challenges. Governance issues are a paramount concern in both countries. Bureaucratic procedures and corruption are reducing efficiency and constraining the investment environment and political volatility has deterred investors. These must be addressed if growth momentum is to be maintained. Failure to complete the demographic transition has been moderated by continued out-migration and increasing remittances in the Philippines. This is not a long-term solution and greater efforts need to be made to restrict population growth. Government revenue needs to be increased to meet growing infrastructure bottlenecks, particularly in the power sector, and to improve efficiency and international competitiveness. In Indonesia, manufacturing sector growth has lagged, particularly since the financial crisis. Particular attention is needed to attract new investments in the oil and gas sector and to improve the general investment climate, including a better legal framework and reform of customs and the tax structure. Infrastructure, particularly in transportation, has to be upgraded to increase competitiveness. Throughout Southeast Asia, internationally competitive manufacturing will depend to a great extent on the efficiency of export platforms and the ability to meet the challenges of China and India, either by establishing niche markets or integrating into the supply chains of these larger economies. Many of these issues are examined in Chapters 2, 4 and 5.

In South Asia, India has the challenge of boosting industrial sector growth and reducing poverty. In 2005, agriculture employed nearly two-thirds of the labor force while providing less than 25% of GDP. These ratios will change over time but not quickly enough unless bottlenecks to industrial growth can be removed. It is particularly important to increase labor market flexibility, streamline bureaucratic procedures, and promote efficient production platforms such as SEZs to facilitate exports of manufactured goods. Infrastructure investment is also crucial for lowering the costs of power, transport and communications. In Bangladesh, fiscal effort has to be improved to address a number of development challenges. FDI inflows that have already increased in recent years have to be further encouraged by appropriate policies and complementary investment to increase efficiency. This should be part of a general program to increase spending on necessary infrastructure in power, telecommunications, and ports that are required to increase efficiency. Governance and corruption issues must also be addressed to strengthen the business environment and attract both domestic and foreign investment. In Pakistan, improved relations with India are important to maintaining an attractive investment environment and economic stability. Textiles and clothing are a key sector comprising about 70% of total exports and 40% of industrial employment. To maintain export share after the opening up of the global market at the end of the quota system will require further increases in labor productivity and efficiency. Diversification to other product lines is also possible. Deregulation and privatization in banking and services should continue to lift efficiency and facilitate investment. In Sri Lanka a resumption of civil conflict is the biggest risk. The forecast assumes this will not happen. Infrastructure spending is critical for future improvements in efficiency and ability to compete in global markets. The policy environment is being improved to provide incentives for investment which would rise in tandem with improved land transport and a cheaper, more reliable power source. Exports of tea and clothing could be threatened by new regulations imposed by the European Union. Competition from other producers could also threaten clothing exports, which currently accounts for almost 50% of export earnings and the bulk of employment in the manufacturing sector. EPZs are being upgraded to improve efficiency to meet these competitive threats. Many of these issues are explored further in Chapters 2 and 7.

New growth theories advance that internationally competitive firms (exporters) are driving industrial growth, and hence the importance of innovation in the manufacturing sector (Gill and Kharas, 2006). So it depends on which country can innovate most effectively. Korea has fostered its own local brand like Samsung and Daewoo, and Singapore has focused on electronics, which proved to be beneficial for achieving rapid economic growth. However, some criticize that such growth in Singapore has been devoid of local content, depending mostly on foreign MNCs.

It remains to be seen how Singapore and other Asian countries develop their own talents, continue to innovate, and compete globally. In any event, a premium will be placed on innovation, better quality new products and lower prices from greater economies of scale, and higher production volumes. China has insisted on joint ventures and alliances in the past so as to imitate technology and foster new innovations. Due to its inherent advantage of a large domestic market MNCs are generally more willing to accept greater risks in China than in smaller countries that cannot offer the same domestic market advantages. But that does not mean that the transfer of technology is necessarily smoother in China. Whether MNCs will indeed transfer high quality technology to China firms is a question yet to be answered as is the question of whether the local labor force has the capacity to absorb skills. South Asian countries like India face a number of challenges, including continued infrastructure bottlenecks and labor market rigidities. Many of these issues are discussed further in Chapters 2 and 7.

Risks to the Forecast

There are two kinds of risks to the forecast. The first risk is that the assumptions regarding domestic reforms and policy initiatives will not be realized. The forecasts assume that the reform process will continue where needed in the region. A major risk is that these reforms will stutter or stop. If this happens, growth trajectories to 2020 will have to be adjusted downward. Adjustments will be greater where policy implementation is critical — primarily in South Asia, China, Philippines and Indonesia — and less critical in the richer economies that are already in the high income categories. In these countries the major risks are that domestic macropolicies will be misdirected, resulting in either excessive inflation or recession.

There are also external risks to the global environment. Linkages between developing Asia and industrial countries through foreign trade are strong. Analysis of variations in growth in developing Asia suggests that recessions or slowdowns in the United States and developing Asia are closely related. The last recession in the United States in 2001 was the major cause for slower growth in developing Asia. We have not factored a slowdown in growth in the United States into the forecast. Our growth projections for the 2005 to 2010 are based on forecasts of international financial institutions such as the IMF, World Bank and the ADB, central banks, planning and finance ministries in developing Asian and commercial banks and financial houses. Some of these forecasts were compiled by consensus economics. These forecasts were generally made under the assumption of a smooth future trajectory of income growth in the global economy. For the 2010–2020 decade, our forecasts were based on assumed longer term structural changes in the Asian economies including trends in labor force growth, technological change, and foreign trade. Relationships between the United States and Europe and developing Asia are examined in Chapter 3.

Bibliography

Akamatsu, K. (1961). A Theory of Unbalanced Growth in the World Economy. *Weltwirtschaftliches Archiv*, 86(2), 196–215.

Akamatsu, K. (1962). A Historical Pattern of Economic Growth in Developing Countries. *The Developing Economies*, 1(1), 3–25.

Antweiler, W. and D. Trefler (2002). Increasing Returns and All That: A View from Trade. *American Economic Review*, 92(1), 93–119.

Asian Development Bank (ADB) (1997). *Emerging Asia*. Manila: ADB.

Asian Development Bank (ADB) (2006). *Asian Development Outlook 2006*. Manila: ADB.

Balassa, B. (1965). Trade Liberalization and 'Revealed' Comparative Advantage. *The Manchester School of Economic and Social Sciences*, 33, 99–123.

Berthelemy, J.-C. and S. Chauvin (2000). *Structural Changes in Asia and Growth Prospects After the Crisis.* Centre d'studes prospectives et d'information internationals.

Bosworth, B. P. (2006). Economic Growth in Thailand: The Macroeconomic Context. *Brookings Institution Paper*, 15 June.

Bosworth, B. P. and S. M. Collins (2003). The Empirics of Growth: An Update. *Brookings Papers on Economic Activity*, 2003(2), 113–179.

Central Intelligence Agency (CIA) (2006). *World Factbook 2006.* Washington, DC: CIA.

Cleary, T. (1995). *The Tao of Organization: The I Ching for Group Dynamics.* Shambhala Publications.

Consensus Economics, Inc. (2006). http://www.consensuseconomics.com/ [Retrieved 25 October 2006].

Deutsche Bank (2005). Global Markets Research: Economic Research, 7 May.

Dooley, M., D. Folkerts-Landau and P. M. Garber (2005). Savings Gluts and Interest Rates: The Missing Link to Europe. NBER Working Paper No. 111520.

Dowling, J. M. and D. Ray (2000). The Structure and Composition of International Trade in Asia: Historical Trends and Future Prospects. *Journal of Asian Economics*, 11(3), 301–318.

Dowling, J. M. and R. Valenzuela (2004). *Economic Development in Asia*. Singapore: Thomson Learning.

Dowling, J. M. and G. Wignaraja (2006). Central Asia After Fifteen Years of Transition: Growth, Regional Cooperation, and Policy Choices. Working Paper Series on Regional Economic Integration No. 3, Office of Regional Economic Integration, Asian Development Bank.

East-West Center (2002). *The Future of Population in Asia*. Honolulu, Hawaii: East-West Center.

Ferreira, P. C., S. Pessoa and F. A. Veloso (2005). East Asian Miracles on the Tyranny of Numbers. *World Perspectives*, March.

Gill, I. and H. Kharas (2006). *An East Asian Renaissance: Ideas for Economic Growth*. Washington, DC: World Bank.

Harrod, R. F. (1939). An Essay in Dynamic Theory. *Economic Journal*, 49, 14–33.

Hobday, M. (1995). *Innovation in East Asia: The Challenge to Japan*. Cheltenham: Edward Elgar.

Hummels, D. and P. J. Klenow (2005). The Variety and Quality of a Nation's Exports. *American Economic Review*, 95(3), 704–723.

International Labor Organization (ILO) (2005). *World Employment 2004–2005*. Geneva: ILO.

International Monetary Fund (IMF) (2006). *World Economic Outlook 2006*. Washington, DC: IMF.

Korea (Various Years). National Statistics Office.

Krugman, P. (1994). The Myth of Asia's Miracle. *Foreign Affairs*, 73(6), 62–78.

Lall, S. (2000). Selective Industrial Trade Policies in Developing Countries: Theoretical and Empirical Issues. QEH Working Paper No. 48, Oxford, UK.

Lucas, R. (1988). On the Mechanics of Economic Development. *Journal of Monetary Economics*, 22(1), 3–42.

Mankiw, N., G. D. Romer and D. N. Weil (1992). A Contribution to the Empirics of Economic Growth. *Quarterly Journal of Economics*, 107(2), 407–437.

Morgan Stanley. http://www.morganstanley.com/gef/team.html.

Naya, S. F. (2002). *The Asian Development Experience: Overcoming Crises and Adjusting to Change*. Manila: Asian Development Bank.

Okamoto, Y. (2005). Emergence of the "Intra-mediate Trade": Implications for the Asia-Pacific Region. Paper presented at PAFTAD 30 on 19–21 February 2005, organized by East-West Center and Rosenberg Institute of Global Finance, Brandeis University. http://www.eastwestcenter.org/stored/misc/paftad_30_okamoto.pdf [Retrieved 12 December 2006].

Praussello, F. (2005). Country Case Studies on Unequal Technology Transfer. Working Paper 12, Universita di Genova, Dipartimento di scienze Ecooniche e Fianaziarie.

Quibria, M. G. (2002). Growth and Poverty: Lessons from the Asian Miracle Revisited. ADB Institute Research Paper 33. Tokyo: ADB Institute.

Radelet, S. (1999). Manufactured Exports, Export Platforms, and Economic Growth. CAER Discussion Paper No. 43, November.

Reddy, S. G. and C. Minoiu (2006). Development Aid and Economic Growth: A Positive Long-Run Relation. Working Papers 29. United Nations: Department of Economics and Social Affairs.

Romer, P. M. (1986). Increasing Returns and Long-Run Growth. *Journal of Political Economy*, 94(5), 1002–1037.

Sarel, M. (1995). Growth in East Asia: What We Can and What We Cannot Infer from it. IMF working paper. Also reprinted in P. Anderson, J. Dwyer and D. Gruen (eds.), *Productivity and Growth: Proceedings of a Conference Held at the Coombs Centre for Financial Studies.* Sydney: Reserve Bank of Australia.

Singapore Ministry of Trade and Industry (2002). Declining Global Market Shares of Singapore's Electronic Products: Is it a Concern? Singapore Ministry of Trade and Industry, *Occasional Paper*, 28 February 2002. Available from http://www.mti.gov.sg/public/PDF/CMT/NWS_2001Annual_Electronics.pdf?sid=92&cid=1035.

Solow, R. M. (1956). A Contribution to the Theory of Economic Growth. *Quarterly Journal of Economics*, 70(1), 65–94. Reprinted in Stiglitz and Uzawa (1969).

Stiglitz, J. and H. Uzawa (eds.) (1969). *Readings in the Modern Theory of Economic Growth.* Cambridge, MA: MIT Press.

Sturgeon, T. J. and R. K. Lester (2003). *The New Global Supply-Base: New Challenges for Local Suppliers in East Asia.* Cambridge, MA: MIT Industrial Performance Center.

UNCTAD (2006). Foreign Direct Investment Database. http://www.unctad.org/ [Retrieved 25 October 2006].

United Nations (2006). *World Economic and Social Survey.* New York: United Nations.

United States Government (2006). *Economic Report of the President 2006.* Washington, DC: United States Government Printing Office.

Weerasinghe, N. and G. Fane (2005). Accounting for Discrepancies among Estimates of TFP Growth in East Asia. *Economic Papers*, Economic Society of Australia.

World Bank (1993). *The East Asian Miracle: Economic Growth and Public Policy.* New York, NY: Oxford University Press.

World Bank (2006). *World Development Indicators Online.* Washington, DC: World Bank.

Yeats, A. J. (2001). How Big is Global Production Sharing? In S. Arndt and H. Kierzkowskyi (eds.), *Fragmentation: New Production Patterns in the World Economy.* Oxford: Oxford University Press.

Young, A. (1995). The Tyranny of Numbers: Confronting the Statistical Realities of the East Asian Growth Experience. *Quarterly Journal of Economics*, 110, 641–680.

Chapter 2

Regional Cooperation and the Global Community

We cannot enter into alliances until we are acquainted with the designs of our neighbors.

— Sun Tzu (554–496 BC), The Art of War

Introduction

Asia began to play an important role in international trade and the interplay of forces within the global community when Japan opened up to foreign trade and began its export push to the United States and Europe in the 1970s and 1980s. Other countries in Asia also began to adopt export promotion policies, and as a result, external trade in the entire East and Southeast Asian region accelerated in the last three decades of the 20th century.

Three watershed developments helped to define the scope and pace of these developments in international trade and intraregional trade. The first was the realignment of the Japanese exchange rate with the US dollar and European currencies as a result of the Plaza Accord agreement in 1985. By realigning the yen against the US dollar and European currencies, particularly the German Mark, the Plaza Accord resulted in a dramatic change in Japanese competitiveness. Over the next few years Japan began to transfer its productive capacity in many labor intensive industries to countries in Southeast Asia, which were also opening up their economies to international trade and direct foreign investment. Many firms in the apparel and textiles sector, toys, labor intensive electronics, and information technology products began to develop export processing facilities in Malaysia, Indonesia, the Philippines and Thailand.

The second watershed occurred in China in 1978 when the regime under Deng Xiaoping instituted a series of reforms which introduced market forces into the agricultural sector and the small-scale industrial sector — the household responsibility system and town and village enterprises. These reforms were joined by the development of export processing zones which mushroomed in the 1980s and 1990s (see Chapter 7 for details). The blossoming of China and the synergies with the overseas Chinese in Hong Kong, Singapore and Taiwan created a rapid growth of intraregional trade in what came to be called Greater China, usually taken to mean China, Hong Kong and Taiwan and also sometimes adding Singapore.

The third watershed was the financial crisis of 1997 and 1998 which exposed many weaknesses in the financial architecture of developing Asia and highlighted the process of financial contagion that overtook the entire region in a matter of a few months.

By the early years of this century, the Asian region had become a closely integrated and interlinked web of trade and financial relations rivaling the European Union in the extent and complexity of its interrelationships. In fact the amount of intraregional trade relative to total trade in Asia is now approaching the level of intraregional trade in the European Union and higher than NAFTA. Further, the share of intratrade in Asia has grown much more rapidly in Asia than in the other two regions since 1980 (see Table 2.1). One important factor has been the rapid growth of China as a processing center for final assembly of a wide range of products as well as the general growth dynamism of the Asian region as a whole.

FDI has played a critical growth in the process of export led growth and to the growth in intraregional trade. This FDI has come from Japan, Europe and North America and more recently from the NIEs — Hong Kong, Singapore, Taiwan and Korea and has been mostly directed to Southeast Asia and China. Estimates reported in ADB (2006a, Table 2.2) suggest that 60% of China's manufactured exports are generated by foreign-owned firms while in Malaysia and Singapore, the shares were even higher at 73% and 86% respectively. Similar findings of higher export propensities of foreign-affiliated firms

Table 2.1 Intraregional Trade — Total Intraregional Trade as Percent of Total Trade

Country/Region	1980	1990	2003
China	29	21	32
Korea	29	29	40
Japan	21	21	36
ASEAN	18	19	24
ASEAN +3	29	29	39
South Asia	3	3	6
Pan Asia	33	42	53
EU	57	65	66
NAFTA	33	37	45

Notes: ASEAN + 3 includes ASEAN, China, Korea and Japan; Pan Asia includes ASEAN + 3 + Hong Kong, Taiwan and South Asia.
Source: IMF (2005); Rana (2005).

are reported in Indonesia, Thailand and Vietnam where foreign-owned firms exported at least 50% of manufactured export production (ADB, 2006a). Foreign-owned firms typically have operations in several countries and these serve as export platforms for products at different stages of production which are then shipped for assembly or further processing to other locations in Asia. One of the strongest recent trends is for China to be the final point of assembly and for export.

This system of international production sharing which involves the fragmentation of vertical supply chains to take advantage of location-specific advantages has, almost by definition, involved greater intraregional trade, since each of the countries in Southeast Asia and East Asia began to specialize in a few components in the supply chain such as electronic chips and components for televisions and computers (see Dobson and Chia, 1997).

As an example, consider the four major brand name laptop computer companies and their outsourcing business as displayed in Table 2.2. Both Dell and HP outsource all of their production, while Toshiba and IBM outsource smaller proportions. Taiwan and Korea are the major outsourcing countries and in Taiwan there are only a few companies

Table 2.2 Proportion of Outsourced Manufacturing of Major Notebook Companies (2002)

Company	Proportion of outsourcing (%)	Outsourcing to Taiwan (%)	Outsourcing to Korea (%)	Taiwanese suppliers
HP	100	87	13	Inventec, Arima, Quanta
Dell	100	95	5	Quanta, Compal, Wistron
Toshiba	24	24	0	Compal, Inventec
IBM	40	25	15	Wistron, Quanta

Source: Yang (2006).

that supply all four brand name manufacturers. Furthermore, many of the Taiwanese companies have begun to move some of their production facilities to China, where land and labor costs are lower.

Asia's role in the global trading system has also increased dramatically in the past several decades. This is primarily the result of the dynamism and export-led orientation of East and Southeast Asia. Overall, the results have been impressive. From 1950 to 1997, Asian countries grew by nearly 6% a year on average, nearly twice as fast as the rest of the world and more than 1½ times faster than the United States (see Figure 2.1 and also Chapter 1). Asian countries now account for about one-fourth of the global economy, in terms of both size and recent economic growth.

Trade was critical to success. During 1960–2005, Asia's share in global trade increased from 11% to 26% and this upward trend is likely to continue.

The region is also playing a more important role in the world trading system and the policy agenda of the WTO. The global community has been negotiating the next round of trade liberalization (the Doha round) for the last few years. As of the drafting of this chapter the negotiations have come to a halt. A failure to agree on tariff reductions for agricultural products has been the main sticking point and the reason given to the news media for the cessation of talks. China, India and the other countries in Asia have a big stake in the successful completion of this round of trade negotiations. China

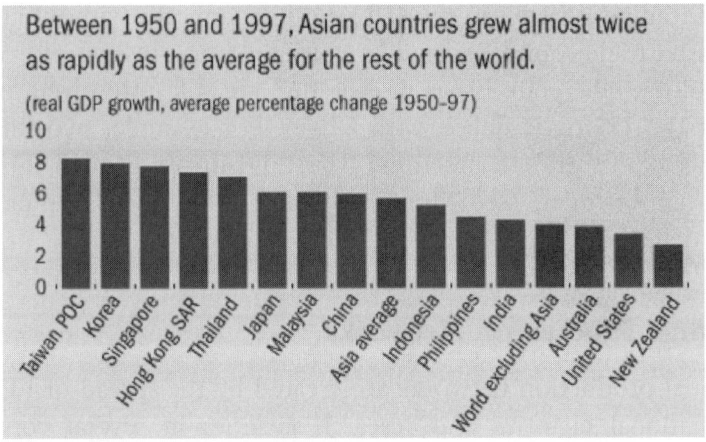

Figure 2.1 Fast Growers

Source: Maddison (2003).

has already benefited from the end of the Multi-Fibre Agreement, though the EU and the United States have imposed some trade barriers to keep the exports of China from growing too fast (see Chapter 7 for more details). Other issues such as intellectual property and trade-related issues are also important to Asia. In this chapter we will not focus on these matters or the WTO. Instead we will be looking at the evolution of regional trading arrangements and the rise of preferential trading arrangements in Asia.

It is important to note that even though trade discussions at Doha dragged on and eventually failed after 5 years of negotiations, there have been gains in international trade. Global trade volumes — calculated as the average of export and import growth — rose by 6.6% per annum over the 2002–2005 period; that pace was about 50% faster than the 4.3% average world GDP growth over the same time frame (see Morgan Stanley, 2006). And the ratio of exports to world GDP rose by 4.0% points from 24.1% in 2001 to 28.1% in 2005 — the strongest four-year increase since the early 1970s. The growth in global exports accounted for fully 40% of the cumulative increase in world GDP over the past four years and that is important to remember. It could be that trade barriers were already low enough that the global trading system can more forward dynamically without another successful trade round, assisted by continued progress in information technology. There are issues in Asia that are still unresolved, in particular the US trade deficit with China. For example, two-thirds of China's export growth has come from FDI-driven growth fueled by western investment (see Morgan Stanley, 2006 and Chapter 3 for more details relating to the US trade deficit issue).

Regional Production Networks

Regional production networks are created by the spread of FDI and multinational factories and research facilities in several countries. Regional production networks also facilitate the research, development and production process and lower costs. Technology clusters are a key component of regional production networks as is subcontracting and establishment of specialized production platforms for specific components. Companies separate different stages of production and develop a flow of product from purchase of raw materials to final assembly and shipment based on geographic location and cost factors. These are further integrated with research and development and new product development centers. For example Seagate, a hard disc drives manufacturer, has product development centers in the United States and Singapore, assembly sites in China and Malaysia, Singapore and Thailand. Wafer fabrication takes place in Northern Ireland and the United States and further assembly takes place in Thailand, Malaysia and Singapore (see Yusuf and Evenett, 2002). If the external environment changes or developments suggest a more efficient production approach, shifts are made between different production and research platforms. Clusters take advantage of synergies with other producers in similar industries and efficiencies are created by lumping power, water, and transport and sewage infrastructure for many producers. Clusters exist in several locations in southern China in export processing zones in Shenzen, the Pudong area near Shanghai, Fujian and Xiamen. These clusters are closely related to companies in Hong Kong which handle marketing, finance and distribution. There is a cluster of industries in the regional growth triangle of Singapore,

Batam Island in Indonesia, and southern Malaysia. Production platforms have been developed in Indonesia and Malaysia, and finance and marketing take place in Singapore. There is a cluster of electronics firms in Penang, Malaysia and an industrial cluster has been developing on the east coast of Thailand. Clusters relate to each other as seen from the Seagate example and together help to define a regional production network.

Increases in the efficiency of the international transportation network have been a major factor in the development of regional production networks. Countries specialize in the production of particular components which are shipped to another country for further processing and final assembly for export or domestic sale as part of a regional production network. Container ships are now five times bigger than they were 30 years ago and travel faster. Information technology now allows companies to track shipments and improve the reliability and speed of movement of goods. Transiting time through ports has fallen as handling facilities have improved, more cranes have been installed to handle the increase in container traffic, and bureaucratic procedures have been streamlined. These developments have increased efficiency and dramatically reduced logistical costs per unit.

FDI by multinational firms (Transnational Companies or TNC) is a key ingredient in the formation and growth of clusters and the build-up of regional production networks. The top 50 TNCs worldwide have a very strong presence in East Asia and Southeast Asia. UNCTAD has developed a dataset which shows TNCs active in Asia and the share of total foreign assets these companies hold in Asia and other regions. Southeast and East Asia have the greatest number of companies on the top 50 list (33 out of 50) and the largest share of total foreign assets (73%) well ahead of Latin America at 12 of 50 and 22% of total foreign assets (see Table 4.9 in Sakakibara and Yamakawa, 2003).

UNCTAD also constructed a transnationality index (available at http://www.unctad.org/templates/Page.asp?intItemID=3277&lang=1) which shows how countries, as opposed to firms, are involved in international production. Globally, Hong Kong ranked first, Singapore was fourth and Malaysia eight. As Asia has evolved as a manufacturing center and developed more regional production networks, its share of manufactured goods in total exports has increased dramatically, doubling from 10.8% of world trade in manufactures not based on natural resources to 22.6% between 1985 and 2000. All of these evidences support the contention that the East Asian and Southeast Asian region comprise a broadly defined regional production network.

Institutional Initiatives Toward Regional Cooperation in Asia

ASEAN and SAARC. Two important regional bodies designed to promote trade and investment on a regional basis were formed in the 1960s and 1980s respectively[1] — the Association of Southeast Asian Nations (ASEAN) in 1967 and South Asian Association for Regional Cooperation (SAARC) in 1985. ASEAN was established in 1967 in Bangkok by Indonesia, Malaysia, Philippines, Singapore and Thailand. Brunei Darussalam joined in 1984, Vietnam in 1995, Laos and Myanmar in 1997, and Cambodia in 1999. The aims of ASEAN are: to accelerate the economic growth, social progress, and cultural development in the region and to promote regional peace and stability. ASEAN stresses mutual respect for independence and sovereignty of nations as well as political and economic cooperation.

[1]Other regional organizations include the Bangkok agreement involving Bangladesh, India, LaoPDR, Korea, and Sri Lanka and later joined by China; Closer Economic Relations Trade Agreement (CER) between Australia and New Zealand; Asia Pacific Economic Cooperation (APEC) involving Asian countries as well as the United States and several Latin American countries; the Economic Cooperation Organization (ECO) involving the central Asian Republics; South Pacific Regional Trade and Regional Trade and Economic Cooperation Agreement (SPRTECA)which extends preferences throughout the Pacific to a number of different island economies, and Shanghai Cooperation Organization (SCO) which includes China, Russia, and four of the Central Asian republics. APEC includes many Asian countries as well as the United States, Canada, and some Latin American countries. Member countries are Australia, Brunei Darussalam, Canada, Chile, People's Republic of China, Hong Kong, China, Indonesia, Japan, Republic of Korea, Malaysia, Mexico, New Zealand, Papua New Guinea, Peru, The Republic of the Philippines, The Russian Federation, Singapore, Chinese Taipei, Thailand, United States of America and Vietnam.

A number of declarations and accords have been signed to further these objectives (see ASEAN Secretariat, 2006). In recent years greater emphasis has been laid on economic cooperation, particularly trade cooperation including the Framework Agreement on Enhancing Economic Cooperation in 1992, which began a move toward an ASEAN Free Trade Area or AFTA which aimed to eliminate tariff and non-tariff barriers among the member nation states. In 1995, ASEAN agreed to accelerate the timetable for the realization of AFTA from the original 15-year timeframe to 10 years. Additional agreements were reached to implement this free trade vision including telecommunication links and a gas pipeline project. ASEAN has also adopted a common effective protection tariff system (CEPT). The CEPT is the mechanism by which tariffs on goods traded within the ASEAN region, which meet a 40% ASEAN content requirement, were reduced to 0–5% by the year 2002/2003 for Southeast Asian nations, by 2006 for Vietnam, 2008 for Laos and Myanmar, and 2010 for Cambodia. The tariff reductions are scheduled for "fast" and "normal" tracks. Tariffs on goods in the fast track were largely reduced to 0–5% by 2000. Tariffs on goods in the normal track were reduced to this level by 2002, or 2003 for a small number of products. Currently, about 81% of ASEAN's tariffs are covered by either the fast or normal track. Exclusions are allowed for temporary reasons, for sensitive agricultural products, and for general reasons. The content requirement excludes many products that are exported where the value-added in the ASEAN countries is less than 40%.

SAARC is an association of seven countries of South Asia: Bangladesh, Bhutan, India, Maldives, Nepal, Pakistan and Sri Lanka. Afghanistan has been accepted as a new member, but the membership procedures have not yet been completed. SAARC encourages cooperation in agriculture, rural development, science and technology, culture, health, population control, narcotics control, and anti-terrorism. In 1993, SAARC countries signed an agreement to gradually lower tariffs within the region through SAARC Preferential Trading Arrangement (SAPTA). In 2002, SAARC countries framed the South Asia Free Trade Agreement which sought to establish a zone in South Asia in the beginning of 2006.

How successful have ASEAN and SAARC been in raising the share of intraregional trade in total trade? It is difficult to make a definitive statement since we do not know what would have happened in their absence. Gravity models have been used to estimate how much they have contributed to the growth in trade. The gravity model suggests that trade between two countries will depend on the geographic distance between them and the relative size of their economies. Using such a model research suggests that AFTA had no significant impact on intrabloc trade while SAPTA had a positive effect. In the case of SAPTA it should be recognized that despite being a member in SAPTA, India and Pakistan are trading at a 70% lower rate than would be expected given their relative size and geographic proximity [see ADB (2002), Part 3; and Frankel and Wei (1998)]. Furthermore, Table 2.1 shows that the share of intra-SAPTA exports was only 6%, much smaller than intra-trade within ASEAN.

In addition to the gravity models, several computable general equilibrium models have also been constructed to investigate the impact of PTAs on trade. Robinson and Thierfelder (1999) concluded that PTAs create trade rather than divert it. Several other studies suggest that larger PTAs like the EU and NAFTA have positive welfare effects, particularly if trade preference is liberalized in a non-discriminatory manner by including those outside the agreement. These models also suggest that PTAs have trade creation effects that dominate trade diversion. Results of gravity and CGE modeling suggest that there are negative impacts on non-members and smaller countries that are not part of a larger PTA often suffer. Furthermore, there is the risk that PTAs do not confer the same benefits of general reciprocity of most favored nation (MFN) treatment which extends trade concessions to all trading partners, and may therefore obscure the transmission of trade externalities to many countries both inside and outside the agreement.

Compared to other regions the size and scope of PTAs in Asia has been limited. ASEAN is the most visible, most active, and includes only a small component of East Asian, Southeast Asia, South Asia, and Central Asia region. Therefore, Asia has developed primarily as a region where trade barriers have been lowered by reciprocal agreement in accord with MFN principles without much formal structure.

Furthermore, the impression that AFTA and SAPTA may not have made much difference in the pattern of trade is confirmed by a glance at Table 2.1. This is particularly true of SAPTA, which has been handicapped by the ongoing conflicts between Pakistan and India. Nevertheless, in order to sidestep this problem and move forward with regional integration within South Asia and to build bridges with Southeast Asia, an alternative grouping, Bay of Bengal Initiative for Multisectoral Technical and Economic Cooperation (BIMSTEC) which drops Pakistan and adds Thailand and Myanmar to SAARC has been formed (see Bhattacharya and Biswa, 2006).

Turning our focus to ASEAN, much has been written about the challenges it faces in negotiating tariff reductions and dealing with differences among its different members with different living standards, geography and economic specialization. Trade with ASEAN has accelerated in recent years and AFTA and ASEAN may have played an important role in this growth despite the findings of the gravity model exercise. Trade within ASEAN has grown rapidly, particularly between 1990 and 2004, although not as much as the intra-trade share of the other Asian countries and groupings in Table 2.1. For example, intra-trade within ASEAN + 3 has risen more rapidly than the trade within ASEAN, about 35%, from 29% to 39% of total trade versus 25% for ASEAN (19% to 24% of total trade) between 1990 and 2003. Other than a rise in the amount of goods traded, tourism within ASEAN has increased rather dramatically, facilitated by the ASEAN passport (see http://www.aseansec.org/64.htm for further details on ASEAN).

However, lists of excluded items to which preferential tariffs are applied, tend to leave out sectors where domestic firms are heavily involved. For example, motor vehicles and agriculture are still excluded from the items subject to free trade by 2010 (see ADB, 2002, Part 3; and Office of the United States Trade Representative, 2000).

In evaluating the success of ASEAN it is important to consider the criteria used for this evaluation process. ASEAN's primary purpose for most of its history was to reduce tariffs and to make the region more competitive in international markets. Increasing intraregional trade within ASEAN was never an objective. However, once this primary objective was generally achieved, ASEAN has not been able to develop another objective such as achieving deeper regional integration. Furthermore, ASEAN has not become an integrated region able to attract foreign investment and lower production costs. The market is still small and fragmented and business costs are raised by the continued existence of different product standards and expensive bureaucratic procedures. This is particularly true when cost estimates are compared with doing business in China (see reference to study by McKinsey and Company in Sakakibara and Yamakawa, 2003).

Following the Asian financial crisis in 1997, ASEAN invited broadening economic and technical cooperation including a financial assistance package of loans and grants from Japan. It is possible that this new and expanded configuration of countries will be able to build on the ASEAN experience to develop a more deeply integrated regional institution.

ASEAN asked China, Japan and Korea to attend their meeting at the end of 1997, and this was the beginning of ASEAN + 3. Subsequently, ASEAN and these three countries have entered into negotiations to lower trade barriers through preferential trade agreements and have developed ways to further the economic and social development of ASEAN. ASEAN + 3 has a much larger share of world exports than ASEAN alone and this share has been growing much faster than ASEAN's share (see Table 2.3). While ASEAN has raised its share of world exports from less than 1% in 1970 to 1.6% by 2000, a more than threefold increase, its share of world exports is still tiny. ASEAN + 3 on the other hand has more than doubled its share of

Table 2.3 Merchandise Exports within Regional Groups as a Share of World Exports

Region	1970	1980	1990	2000
APEC	20.9	19.1	27.1	35.8
ASEAN	0.5	0.7	0.9	1.6
ASEAN + 3	na	3.7	4.6	7.0
European Union	27.3	24.4	29.5	22.3
NAFTA	7.9	5.5	6.8	10.7

Source: IMF (2005); Sakakibara and Yamakawa (2003).

world exports since 1980 — to 7%, only a few points lower than NAFTA, although still far below the EU.

Integrating ASEAN into more broadly defined East Asian economic community including Japan, Korea and China has proved difficult. There is the implicit concern within ASEAN that its interests will be diluted or subverted by a wider regional arrangement, because China and Japan are much bigger economies than ASEAN. Nevertheless several possible alternatives exist for creating a more broad-based East Asian economic community (see Soesastro, 2001, 2003). One is through the establishment of an East Asia Free Trade Area that would include all ASEAN + 3 economies. Another is through the formation of a free trade area in Northeast Asia, comprising Japan, Korea and China, that is subsequently amalgamated with the ASEAN Free Trade Area. A third is through the amalgamation of three separate ASEAN + 1 agreements — the ASEAN–China, ASEAN–Japan, and ASEAN–Korea agreements. None of these alternatives aside from FTA between ASEAN and Korea and ASEAN and Japan has been pursued yet. A fourth, discussed later in this chapter, is based on the notion of monetary regionalism and would focus on financial cooperation and integration.

Before closing this section a word about APEC is called for. APEC does not negotiate reductions in trade barriers and does not have a substantive role in other trade negotiations. It does undertake research and has a number of committees which study various aspects of regional integration in Asia. APEC also has made a series of proposals for reform of financial systems and trade liberalization and reform in Asia. In recent years it has turned its attention to issues of international terrorism and general security concerns. APEC can also serve as a forum for discussing matters of concern to Asian countries in a narrow forum than the WTO or the United Nations. APEC also sponsors the Pacific Economic Cooperation Council (PECC) that prepares various reports of developments in Asia.

Bilateral trading arrangements

Although ASEAN has been somewhat effective in developing trade linkages within Southeast Asia, overall international trade within the

Asian region has expanded rapidly without a strong unifying framework such as the EU or NAFTA. With the Doha round of international trade negotiations foundering, many countries in Asia have become more frustrated with WTO and the prospects for resolving ongoing trade issues such as agriculture and intellectual property rights. Rather than trying to develop a counterpart to the EU, which would require harmonizing relations within the entire Asian region including South Asia, East Asia, and Southeast Asia, many countries in the region have turned to bilateral trade agreements as a way to solidify trading relationships with their neighbors. Preferential trade agreements (PTAs) can facilitate trading agreements between "like-minded" countries without waiting for the cumbersome WTO negotiations to be finally resolved. Rather than applying tariffs to all trading partners, in line with most favored nation principles, such agreements establish rules of origin which govern those goods subject to the bilateral agreement. In the last decade a large number of such agreements have been signed, both between countries and between regional groupings. The scope of these agreements varies from exchange of minimal tariff preferences to deeper agreements that go beyond tariff reductions to include trade in services and economic cooperation. Some agreements are within Asia and others are with countries outside the region. The latter have grown substantially in the last few years as India and China negotiate pacts that guarantee a smooth flow of minerals and other strategic materials required for rapid industrialization.

The Asian Development Bank (ADB, 2006a, Section 3) estimates that nearly 100 such agreements have been signed in the past decade and 55 more are being discussed. Worldwide, the number of agreement in force by the end of 2006 is estimated at 300. Competition to sign agreements and the power that comes with having a large traded goods sector complicates the analysis of trade preferences and can freeze out small countries with little leverage. Hub countries such as China, with many preferential trade agreements will be able to exercise market power relative to spoke countries including small economies such as Nepal in South Asia and Lao PDR in Southeast Asia that have few agreements. For example, hub countries can require spoke countries to use primary or intermediate products produced in the hub in order to qualify for tariff preference. In any case the bewildering set of agreements, local content requirements, and tariff regimes further complicates international trade. Opportunities for corruption also multiply as customs officials can choose to interpret rules to lower tariff burdens. To simplify the process of harmonization of tariffs among the different PTAs there are proposals to use the strong hubs like the EU, NAFTA, or large countries like China to set content regulations.

This does not address the problems of how to provide an overall umbrella that will harmonize regulations among all the different hubs. Furthermore, the research on the effectiveness of PTAs reviewed above suggests that small countries negotiating with other small countries or groups of countries will probably not achieve a significant reduction in overall trade barriers or increased economic efficiency. Many of them are between countries that are parts of larger groupings already — in Central Asia, between Kazakhstan and Uzbekistan, and Tajikistan and Kazakhstan; or in Southeast Asia between Korea and Singapore, and Thailand and Peru. Many of the recent agreements and planned agreements are bilaterals between Singapore and other countries in Asia or Latin America; between Japan and other Asian countries; between Taiwan and Latin American countries and the PRC and a variety of countries. The question arises — would it not be better to concentrate on more widely focused PTAs that expand existing arrangements such as ASEAN + 3 to include other countries in the region such as South Asia, rather than narrow bilateral agreements?

There may be other reasons for the rapid growth in bilateral PTAs. These include trade facilitation through developing and expanding opportunities for harmonizing standards, regulations, and financial relationships. PTAs also provide a way to reduce military and political tensions by increasing trust and providing a ready forum for discussing political, social, and economic issues. PTAs could have also contributed to the development of cross-border growth triangles and sub-regional cooperation. There are several such organizations including Indonesia, Malaysia, Singapore Growth Triangle (IMS-GT); Indonesia, Malaysia, Thailand Growth Triangle (IMT-GT); Brunei Darussalam, Indonesia, Malaysia, Philippines East ASEAN

Growth Area (BIMP-EAGA); and the Ayeyawady–ChaoPraya–Mekong Economic Cooperation Strategy (ACMECS), sometimes called the Greater Mekong subregion (see Anwar, 1994 and Pomfret, 1996). PTAs can also serve as self-reinforcing mechanisms to continue the process of economic reforms and facilitate the transfer of technology and the inflow of FDI, particularly if some of the members are more highly developed and willing to share technology with the smaller or less-developed members.

Regional Cooperation in Transportation

Transportation infrastructure has grown dramatically in Asia over the past few decades. Transportation is an important component of the cost of doing business and the income elasticity of demand for freight and passenger transport is high, between 1.5 and 2 for most Asian economies (see UNESCAP, 2006). Logistics costs are typically more than 20% of sales and transport comprises more than half of this total (see UNESCAP, 2005). Landlocked countries face transport costs that are 50% higher than those countries with direct sea access.

Sea and air transport have grown even faster than the overall transportation infrastructure as the region became an important exporter. For example, developing Asia accounts for about a quarter of global GDP compared with over 60% of global container traffic. In the 1990s, container traffic through the ports of Asia increased 23-fold in China, 5-fold in Malaysia, and 4-fold in Korea, Indonesia, Bangladesh and India. Over the same period, aircraft departures doubled in the Asian region, increased 11-fold in China, and 6-fold in Korea (UNESCAP, 2005). Road and rail traffic between countries have increased much more slowly.

The internal transportation network in Asia, mostly consisting of roads, but also inland waterways and rail systems, has been extended, allowing greater access to rural residents in poor areas. By 2002, 90% of China's population lived within 2 km of an all weather road. Figures are lower for India (50%) and Thailand (43%), yet there has been substantial improvement in the past decade. Better and more reliable road access allows farmers to bring goods to market quickly at less cost and can further develop the potential for involving more remote areas. In China there are plans to extend the Yangtze river transportation system into the western provinces, linking with the road system to bring goods to market in Wuhan or other cities.

Regional cooperation in transportation began many years ago with the development of the Asian Highways Network and the Trans-Asian Railway Network. The Asian Highway project was initiated in 1959 with the aim of promoting the development of international road transport in the region. During the first phase of the project (1960–1970) considerable progress was achieved. After the war in Vietnam, regional political and economic changes spurred new momentum for the Asian Highway Project. It became one of the three pillars of Asian Land Transport Infrastructure Development (ALTID) project, endorsed by ESCAP Commission at its 48th session in 1992, comprising Asian Highway, Trans-Asian Railway, and facilitation of land transport projects. The intergovernmental agreement on the Asian Highway Network was adopted in 2003 and begun in 2005 (see Figures 2.2 and 2.3 for the extent of the systems).

Both of these transportation systems have worked to increase bilateral trade and regional cooperation and social development. Harmonization of gauge differences and filling in of missing links have begun on the rail system and containers are being shipped between Western Europe and Northeast Asia with a significant reduction in elapsed transport time relative to sea transport. While sea container traffic between Europe and Asia is still much larger than traffic by rail, there is scope for more rail and rail combined with sea linkages with Europe and the Middle East.

There have also been a number of sub-regional transportation initiatives which have been taking place within the broader framework of the regional rail and road developments. We mentioned some of these organizations in footnote 1. Of particular note are the regional initiatives in the Central Asian Republics and in the Mekong countries. In the Mekong countries there are three road development projects linking the countries of the sub-region along the

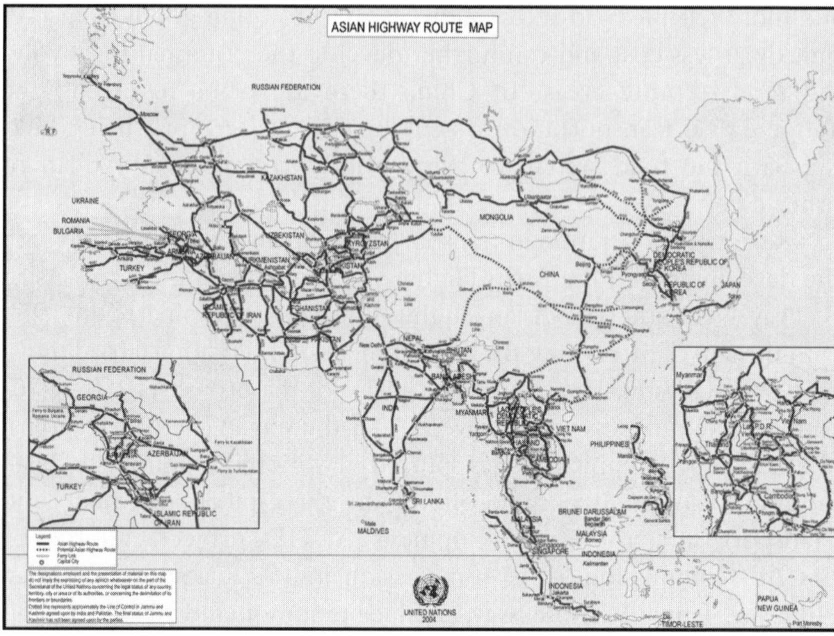

Figure 2.2 Asian Highway Network

Source: UNESCAP (2002).

north–south corridor (Lao PDR, Myanmar, Thailand, China and Vietnam), the east–west corridor (Lao PDR, Myanmar, Thailand and Vietnam), and the southern corridor (Cambodia, Thailand and Vietnam). The southern corridor will also connect major cities, promoting cross-border trade and investment. All of these projects are ongoing.

In the Central Asian Republics there are ongoing road and rail projects and planned linkages with Europe and South Asia and also with China. An integrated rail and road network system will be implemented. Customs codes will be harmonized, border procedures will be streamlined, and best practices information and inspection systems will be adopted. The prevalence of restrictive non-tariff measures on regional trade will be significantly reduced as a result of improvements in institutions. An efficient regional energy market and system will emerge through implementation of structural reforms, tariff reforms,

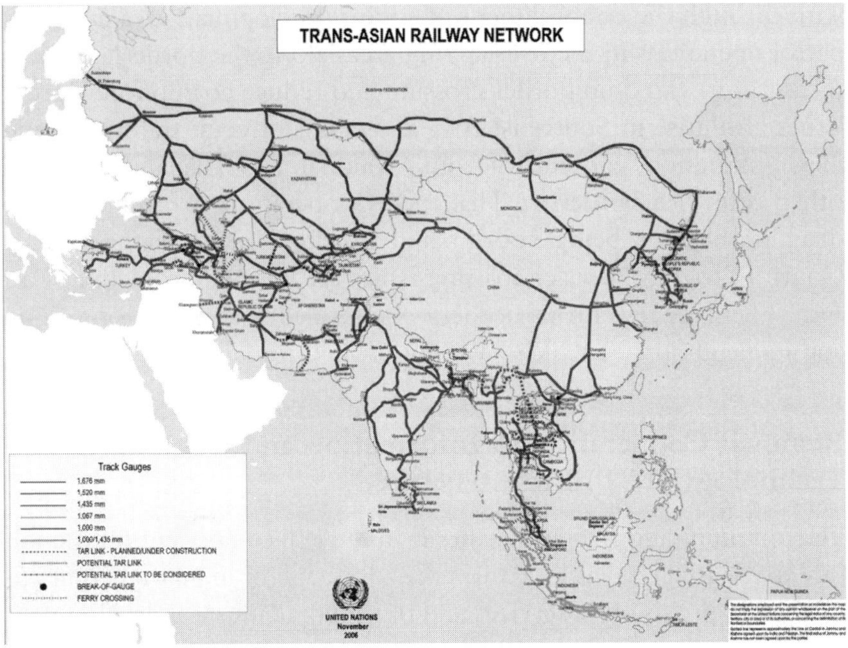

Figure 2.3 Trans-Asian Rail Network

Source: UNESCAP (2006).

and significant investments in regional energy infrastructure. This will be accompanied by greater transparency and more rational pricing in line with market conditions. These projects are crucial for the development of international trade with countries outside the region since all of these countries are landlocked and one (Uzbekistan) is double landlocked. Negotiations for harmonization of border crossings and tariff issues are sticking points at many of the borders, particularly between Kazakhstan and Uzbekistan. See Dowling and Wignaraja 2006 for further details.

Border crossings and differences in technological specifications of train tracks can add costly delays. In the case of Thailand and Malaysia concern over infiltration of separatist guerillas from Malaysia into Thailand have created numerous delays and conflicts at the border. Resolving these issues has been difficult although they are being dealt with through the establishment of a joint development area and joint patrol operations in the overlapping areas along the border as well as measures to speed up border crossing and reduce corruption. Border issues also arise in Southeast Asia as a result of drug trafficking and illegal migration, particularly from Myanmar into Thailand as well as other countries bordering Thailand. Cooperation to deal with the drug problem is being coordinated by the United Nations Drug Control Program. Border crossing issues are less of a problem in East Asia. There are fewer road borders — essentially between Hong Kong and its neighbors — which have not created significant difficulties.

Regional Cooperation in Information and Communication Technology (ICT) Infrastructure

Telecommunications infrastructure is an integral component of technology that has been developed to take advantage of improved competitiveness in provision of services and to support industrial development in Asia. A variety of investments have been made in the past two decades to increase telecommunications connectivity and flexibility. They have provided support for the expansion of the internet and other information-based networks that have increased productivity in the industrial sector and contributed to subcontracting and the development of global supply chains and the break up of the vertical integration of production systems. As a result, telecommunication developments have gone hand in hand with industrialization. Industrial countries in Asia — the NIEs and Japan have well-developed connectivity and technology infrastructure. Poorer countries including countries in South Asia and parts of Southeast Asia have much lower levels (see Table 2.4). Notice the big gap after Taiwan. This hierarchy has led observers to suggest that a digital divide has arisen between rich and poor countries that is contributing to the divergence in living standards between these two sets of countries.

Broadly speaking, the ICT infrastructure includes telephone mainlines, mobile telephones, and internet connectivity and accessibility. There are two possible technologies — land-based telephone and cable access using land-based cables and relays and satellite transmission.

Table 2.4 Connectivity and Technology Infrastructure Scores in Asia

Country	Connectivity and technology infrastructure score
Hong Kong	8.1
Singapore	7.6
Korea	7.2
Japan	6.9
Taiwan	6.7
Malaysia	4.1
Thailand	3.0
China	2.5
India, Indonesia, and Sri Lanka	1.4
Pakistan and Vietnam	1.2

Note: Rankings are from 0–10, 10 being the highest and 0 the lowest.
Source: Economist Intelligence Unit (2005).

Broadband access using advanced cable technology allows many users to use the same cable connection. The systems can be linked together to provide land-based telephone reception as well as mobile phone and internet connectivity. Each government has developed a strategy for implementing a mixture of these technologies and a regulatory apparatus with government and private sector involvement in different proportions. There is an array of technical and economic issues that has to be addressed in setting up a national telecommunications system. We will not address these issues in this section. Rather, we focus on ways that regional cooperation can be used to provide additional service, increase efficiency and reduce costs in expanding the ICT network. Telecommunications technology is changing rapidly — becoming faster and more efficient all the time. Asia has embraced the spread of this technology and the participation of the private sector, particularly in the new areas of cellular and internet service. In the Asian region as a whole 89% of service providers were in the private sector, compared with the 69% of cellular service and 51% of fixed line local and long distance service by 2003 (International Telecommunications Union as quoted in UNESCAP, 2006).

Because they are able to afford access to many users at the same time, broadband technologies are being investigated and implemented in many countries (see Agrawala and Sehra, 2002). This is being done with the help of a number of international agencies through regional cooperation and sharing of information, technologies, hardware and software. They include APEC, ASEAN, UNDP, Asian-Pacific Telecommunity (APT), and International Telecommunications Union (ITU). Bilateral aid is also being extended such as Japanese and ASEAN cooperation.

Individually, or collectively, these governments and agencies are addressing issues such as the role of the government in stimulating and encouraging expansion of the broadband network, developing regulations to stimulate competition and expansion of the infrastructure base, addressing the needs of rural users, establishing broadband for businesses, involving the private sector in broadband deployment, and developing local internet web content.

Investment in ICT infrastructure has been increasing over the past few years. Japan has made the largest investments, followed by China, ASEAN and Korea (Table 2.5). Investment growth in China has been somewhat faster than in the other countries and regions. Per capita investment in Japan is much higher than that in Korea and the gap is even greater between Japan and ASEAN + 1 and China.

ICT density reflects the level of spending per capita although the gap is not so large for telecommunications, particularly since the introduction of mobile phones. Telephone density in China is around

Table 2.5 Investment in ICT (2002–2005)

Country/group of countries	Investment (in $US billion)	Population (in million)
Japan	9.4	127.0
China	2.9	1300
ASEAN + 1	2.3	1500
Korea	1.3	47.5

Note: ASEAN + 1 includes ASEAN plus India.
Source: Estimated from data supplied by International Data Corporation and reported in UNESCAP (2006).

24 per 100 persons for both fixed and mobile access compared with only 4 for India and 46 (fixed) and 72 (mobile) for Japan (ITU, 2004).

Internet connectivity is also being used for distance education and linking more remote areas with urban centers of learning. The internet is also a useful vehicle for sharing new technology and information with rural agriculture centers and small businesses that would otherwise be unreachable on a timely basis. ICT is also a powerful tool for increasing the information available to women in rural areas although it requires a commensurate increase in women's education.

The choice of technology be it satellite, land-based fiber optics, or some combination of the two such as local wireless technology (WI-FI) depends upon the needs of the local communities and the general population, the goals of the government, and the relative cost of the different technologies.

The experience of the United States and to a lesser extent in Europe in the 1990s demonstrates the added output that can be generated by adapting new processes and systems that have been made possible by using new information technology and the internet. Total factor productivity increased as a result and the beneficial effects of this technological revolution were reflected in lower inflation and levels of unemployment even as growth accelerated. Similar beneficial productivity impacts of ICT are being felt in the industrial sectors of Asian economies as well.

Total factor productivity has increased since the mid-1980s as more foreign investment has been attracted to the region. Foreign firms have brought with them improvements in marketing, distribution, inventory control, and production platforms that have raised output and labor productivity. These technological innovations have been made possible largely through the use of computers to aid in production scheduling and delivery, inventory management focused on just on time arrival, improvements in efficiency, and reductions in waste. Regional production networks have been developed and created as a result (see above). These production networks are connected by vertically integrated production platforms in different countries and by information and technology relations that coordinate research, design, production services, manufacturing, component and raw material supplies, marketing, distribution, and after-sales service (Yusuf and Evenett, 2002).

Energy Infrastructure

Rapid growth in economic activity in the Asian region has been accompanied by an increase in energy consumption. Per capita consumption of commercial energy is still low compared with industrial countries, although it is growing rapidly. Consumption is estimated to have grown by 5% per annum between 1980 and 2002, slightly lower than the growth in income but still much higher than the growth rate in energy consumption both globally and in industrial countries (see UNESCAP, 2006). Commercial energy consumption per capita is still low, less than 1 ton of oil equivalent, per annum, particularly in the poorer countries in the region where consumption is less than 0.5 ton of oil equivalent.

Coal is the main source of energy used for the production of electricity followed by petroleum (see Chapter 6 for more details). Alternative fuels such as solar, geothermal, and nuclear and hydropower comprise a small component of total energy consumption and production. The Asian region has a good share of the estimated world reserves of coal, natural gas, and oil. China and India have 10% and 12% share respectively of the world's reserves of coal, and Kazakhstan has over 3.3% of global reserves of oil and coal, and 1.7% of natural gas reserves. Uzbekistan and Turkmenistan together account for nearly 3% of the global reserves of natural gas.

Regional cooperation in the energy sector began several decades ago and has expanded to include a number of different sectors and countries. Trans-boundary power sharing has been investigated by UNESCAP (1987). Projects have been implemented where Nepal sells hydropower to India, and an integrated power grid for members of ASEAN and there are several gas and petroleum pipelines that run between different countries in the region. International gas pipelines currently under construction include an offshore pipeline from the Thailand Malaysia Joint Development Area and another bringing gas from Indonesia's Natuna Gas field to Singapore. There are also petroleum pipelines running from Kazakhstan to Russia and through Dagestan and Chechnya to Novorossisk, and a second route west to the Georgian port of Supsa.

A number of gas and petroleum pipelines are under construction and planned from the Central Asian Republics to Russia, Europe and China. Several are being held up by political uncertainties and others by lack of financing or disputes between the countries involved. For petroleum, Russia favors a northern route. Kazakhstan would expand its existing pipelines to link them with the Russian network of pipelines. Azerbaijan would build a pipeline from Baku to Novorossisk. The most direct, and cheapest, route is to south, across Iran to the Persian Gulf. Iran already has an extensive pipeline system, and the Gulf is a good exit to Asian markets. However, US sanctions on Iran block this option. Azerbaijan, Turkey, Georgia, and the United States favor a western route. According to one variation, oil and gas would flow to the Georgian port of Supsa. From there, it would be shipped through the Black Sea and the Bosporus to Europe. Despite the staggering costs it would take to construct, China is exploring an oil pipeline across Kazakhstan to China. The American oil company Unocal has proposed the construction of oil and gas pipelines from Turkmenistan through Afghanistan to Pakistan and later to India. Afghanistan's long war has prevented this project from moving forward. If some degree of stability returns to Afghanistan, the project may be resurrected. High oil costs and the expanding capacity of Kazakhstan and Turkmenistan in the Central Asian Republics suggest that at least one or two of these pipelines would be built within the next few years (see Worldpress.org, 2006 for a map of the region and further details).

Now, there is some trade in liquefied natural gas (LNG) within the Asian region but it is small and is not expected to be able to compete with natural gas pipelines. For natural gas one proposal is to deliver Russian natural gas from Sakhalin Island to Japan through a sub-sea pipeline. Another would pipe Siberian gas to consumers in northern China. There are also plans to link gas networks in Indonesia, Malaysia and Thailand, and Vietnam with China. Internally, Philippines is building a pipeline from offshore natural gas reserves in Palawan to Manila, and China is building a pipeline to Shanghai from the Xihu Trough.

Regional Cooperation in International Migration

International migration is a subject that has received considerable attention in the past few decades. Initially, the two oil shocks in the 1970s and the rapid growth of oil-rich economies in the Middle East stimulated labor migration of unskilled and semi-skilled labor from Asia. Subsequently, rapid growth combined with slower labor force growth in the richer countries of East Asia stimulated migration from poorer countries in the Asian region. There are two types of migration — through legal channels and approved migration agencies and irregular migration, which is a blanket term used to describe illegal or undocumented migration. There are also two kinds of migrants, those who hold professional accreditation or who are otherwise highly skilled and the unskilled or semiskilled. These two groups operate in separate markets and have distinctly different issues to be dealt with.

Official migration figures have been compiled from a number of sources (see Table 2.6). Annual flows amount to over 2 million migrants annually in 1998. While the number of migrants returning to their home country is not known, many of the seasonal migrants stay for prolonged periods of time up to 2 years and longer. The bulk of

Table 2.6 Official Reported Outflows of Migrant Workers (in thousands) — Selected Years

Country	1980	1990	1998
India	236	144	355
Indonesia	16	86	412
Philippines	215	599	562
Thailand	22	63	175
Pakistan	130	116	104
Bangladesh	30	104	268
Sri Lanka	8	43	158

Source: Wickramesekera (2002).

Table 2.7 Stock of Foreign Workers in Asian Economies (in thousands)

Country	1996	2000
Taiwan		326
Hong Kong	164	217
Japan	610	710
Korea	210	286
Singapore		612
Indonesia	25	17
Malaysia	745	800
Philippines	4	
Thailand	1034	1103
China	80	
Vietnam		30

Source: OECD (2003).

these migrants are unskilled and semiskilled workers. The remainder of this section will focus on these migrants unless otherwise indicated.

The number of foreign workers residing in Asian countries is estimated to be more than 6 million (see Table 2.7). These figures do not include migrants living in other countries outside of Asia. Furthermore, these figures are probably underestimated, particularly for Malaysia and Thailand, where the number of irregular migrants is large (see Wickramesekera, 2002).

Estimates of irregular migration suggest that around 40% of total migration is irregular and this percentage is probably much higher in Thailand and Malaysia where porous borders, ethnic similarities, and sharp differences in income create strong incentives to migrate illegally.

The IMF study by Burgess and Haksar (2005) estimates that migrants from the Philippines have nearly matched the additions to its labor force. The total number of Philippine nationals living overseas is about 25% of the country's labor force, totaling almost 8 million people. Migration rates for other countries where English is widely spoken may be somewhat comparable although not as large proportionally. Burgess and Haksar note that lower middle income countries show remittances of only 1.4% of GDP compared with the Philippines' 9%.

Regional cooperation for migration policies in Asia involves both bilateral and regional issues. In South Asia, the bulk of migrants are to the Middle East (over 90% from Bangladesh, India, Pakistan and Sri Lanka — see Waddington, 2003 and www.livelihoods.org). Ignoring the policies of recipient countries in these countries for the moment, we can examine migration issues and policies of sending and receiving countries in Asia.

Aside from migration of unskilled and semiskilled workers there are also policies regarding migration of professionals and students. The costs of education in the sending country and the skill of the migrant weighed against the possible benefits of remittance income and return migration when the work visa has expired. India and the Philippines, in particular, have lost a lot of professionals. Migrants from the Philippines have been mainly in the medical fields (mostly nurses and some doctors), and those in information technology and the health professions from India. One approach to dealing with these issues is to encourage temporary instead of permanent migration. Several industrial countries have developed visas that encourage this migration route (Khadria, 2006 and see also Chapter 1 for additional discussion of skilled migration).

Issues in sending countries

Migration policies range from unregulated regimes where the migration decisions and flows are left to the market to more regulated and state-managed policies. In the case of regulation, the state establishes a legal framework that controls and licenses employment agencies that recruit workers for overseas positions. These employment agents liaise with firms and agencies in receiving countries and the markets do the rest. This is the typical system which is in place in most Asian countries including Bangladesh, India, Indonesia and the Philippines, and in Asia generally 90% of job placements are by private agencies. In some countries, state-managed firms are set up to recruit workers and subsidies can be given to encourage particular kinds of migration. In Asia, only Korea and Pakistan follow this model, which allows for some private sector participation. In China and Vietnam, the state controls the entire migration process (see Waddington, 2003 for details). In Southeast Asia the majority of the migrants are women, primarily nurses, domestic workers, or those working in the hospitality or entertainment business.

The key to a successful migration policy is flexibility, allowing competition to flourish and protection of the rights and well-being of migrants. Predeparture training and provision of information are important components of the latter, including who to contact to report contract violations, physical abuse, or other problems or issues. These policies can be the subject for bilateral and multilateral talks between supplying and recipient countries in the region and in other regions. Experience in such negotiations is not extensive. Abella and Lonnroth (1995) report that there are virtually no agreements on labor migration between suppliers of migratory workers and receiving countries. Furthermore, internationally agreed standards under the UN International Covenant on Economic, Social and Cultural Rights have been difficult to be enforced on migrants. Host countries are often not interested and sending country governments have little leverage. To bridge this gap the UN adopted a special treaty in 2003 specifically protecting migrants' rights and privileges.

Issues in receiving countries

Receiving countries have a different set of issues, focusing primarily on domestic labor market conditions, the problems of irregular migrants, law and order, and protecting local customs and heritage. Some migration of unskilled and semiskilled workers is a necessity to fill entry level jobs that local workers are reluctant to apply for. In some cases these workers are deported after the jobs are completed, as in the construction of the Hong Kong airport or the recent deportation of Indonesian workers as a component of Hong Kong's strategy for diversifying its labor pool. In many countries the focus is on controlling the flow of irregular migrants. Policies to eradicate these flows have been generally unsuccessful in the long run although they can have a short-term impact. The Thai policy of registering all irregular migrants without the fear of deportation is a first step toward understanding the extent and scope of migration issues. Closer liaison with the governments of sending countries is also being undertaken.

Remittances

Remittance income is an important component of the balance sheet of sending countries. It is a large item in the balance of payments and contributes to the level of domestic consumption and saving among the poor. In the Philippines remittances of over $ 6 billion annually are estimated to be around 9% of GDP. In Pakistan the corresponding figures are $ 3 billion and over 5% of GDP. In India remittances are close to $ 12 billion per year, though a much smaller fraction of GDP (Burgess and Haksar, 2005; Waddington, 2003), and in Pakistan and Bangladesh remittances average $ 4 billion and $ 3.3 billion respectively. In other countries remittances play a smaller role in the economy (see Table 2.8).

Foreign earnings can be remitted through formal channels in the banking system or informally through friends and family or non-bank financial intermediaries. The mechanisms for facilitating remittances are many and varied (see ADB, 2006b). To facilitate the flow of remittances, countries have tried to bring down the cost of remittances by encouraging flows through the banking system and also by encouraging workers to open bank accounts which will help to facilitate these transactions. In India and Pakistan interest

Table 2.8 Workers' Remittances in $ billion

Country	1999	2004
Bangladesh	1.6	3.4
India	11.1	23.0
Pakistan	0.1	4.2
Sri Lanka	1.0	1.3
Philippines	2.0	6.0
Indonesia	na	1.0
Thailand	na	1.0

Source: World Bank (2005); Burgess and Haksar (2005).

rates on migrant deposits have been raised and foreign currency bonds issued and deposits in non-resident Indian (NRI) accounts are exempt from wealth tax. In some cases more favorable exchange rates are offered as well as foreign currency accounts in domestic banks.

Evidence from Southeast Asia suggests that the bulk of remittance income goes into consumption, debt reduction, and sometimes education for household members in the home country. Some are saved. Surveys show that 50% of Filipino families saved some of the remittances compared with 80% of Malaysian families receiving remittances. There has been less emphasis on developing financial products to attract remittances from low-income migrants. Saith (1999) recommends schemes that help returning migrants invest their money wisely in savings accounts, in small-scale businesses, or in the capital market. In any event, some financial counseling of migrants both before and after return is advisable and is ongoing in some countries.

Regional Cooperation in Banking and Finance

The 1997/1998 financial crisis in Asia demonstrated the importance of regional interactions in determining the pace of financial and economic growth and development in Asia. Weakness in Thailand's financial sector and foreign exchange market failed to control short-term overseas borrowing of US dollars and a speculative bubble in the real estate and stock market. This resulted in increased pressure in the foreign exchange market and an eventual collapse in the value of the Thai baht. The crisis spread quickly to other countries as other currencies came under pressure to devalue. Indonesia and Korea were the most adversely affected. This international financial crisis resulted in a downturn in economic activity in the region which lasted a couple of years.

Coming out of the crisis, economies in Southeast and East Asia began to reassess the strengths and weaknesses of their financial and banking systems. A number of reform measures were introduced to strengthen the sector including: allowing greater flexibility and competition among banks, including foreign banks; requiring greater transparency of operations and introducing stronger procedures for screening loan applications; more flexible exchange rate regimes; reducing the ties between banks and industry; writing off non-performing loans (see Dowling and Valenzuela, 2004, Chapter 3; Goldstein, 1998).

The rapid spread of the crisis beyond Thailand showed the importance of establishing regional cooperation measures to prevent a similar crisis in the future. Regional cooperation also developed as a way to improve the efficiency and reliability of the financial sectors in the Asian region. To create greater international financial stability, Asian economies have begun to build up foreign exchange reserves (see Table 2.9). By 2005 foreign exchange reserves in the region had grown to $ 1814 billion from $ 766 billion in 2001. The dramatic build up in China's reserves and large current account surpluses have also resulted in pressure for currency appreciation, bringing the importance of exchange rate policies in the region into sharp relief.

Regional cooperation beyond the immediate reaction to the financial crisis has manifested in greater regional financial cooperation, financial surveillance, and financial services including bond market development. We look at each of these as follows.

Table 2.9 Gross International Reserves (in US billion dollars)

Country	2001	2005
China	212	819
Hong Kong	111	124
Korea	102	210
Taiwan	122	253
India	51	137
Indonesia	28	35
Malaysia	31	70
Singapore	76	117
Thailand	33	52

Source: ADB (2006a), Table A20, rounded to nearest US billion dollars.

Regional financial cooperation

Banking and financial systems in Asia have diverse architectures and institutional arrangements. Although they are generally dominated by a domestic banking system with weakly developed bond markets, there are differences in the amount of government regulation, sophistication of commercial banks, and degree of competition. To further develop regional cooperation and pave the road for deepening of financial systems, domestic financial systems have begun to liberalize by allowing competition and greater international cooperation.

Following the Asian financial crisis, finance and central bank representatives from the Asian region met with officials from the World Bank, ADB and the IMF in Manila to create a framework to enhance financial stability in the region, the so-called Manila Framework Agreement. It suggested a mechanism for regional surveillance, cooperation to strengthen financial systems and regulations, strengthening IMF capacity to respond to crises, and financial measures to supplement IMF's capacity. Many of these suggestions have been subsequently adopted. Initiatives have begun to monitor capital flows within the region by ASEAN + 3. Training and development programs have also been developed to upgrade skills of commercial bankers in cooperation with the IMF. A regional system of clearing and settlement as well as regional credit guarantee and credit rating agencies are also being considered. Harmonizing legal and regulatory systems is also being considered. Swap agreements have also been made following the Chiang Mai initiatives after a meeting of regional finance ministers of ASEAN + 3 economies in Chiang Mai, Thailand in 2001[2]. A number of swap and repurchase arrangements have been made already in 2001 and 2002 and more may be forthcoming. A list of existing bilateral swap agreements is shown in Table 2.10. Japan has taken a leading role under the Miyazawa initiative. The swap is arranged so that a country under speculative attack on its currency

[2]Prior to the Chiang Mai Agreement, ASEAN finance ministers established a surveillance program to supplement the global surveillance undertaken by the IMF. It was essentially superseded by the Chiang Mai Agreement.

Table 2.10 Swap Agreements in Asia

Agreement	Date signed	Amount in US$ billion
Japan–Korea	2001	7
Japan–Thailand	2001	3
Japan–Philippines	2001	3
Japan–Malaysia	2001	3.5
Japan–PRC	2002	3
Japan–Singapore	2002	Under negotiation
China–Malaysia	2002	1.5
China–Philippines	2002	3
China–Korea	2002	2
China–Thailand	2001	2
Korea–Malaysia	2002	Under negotiation
Korea–Philippines	2002	Under negotiation
Korea–Thailand	2002	1

Source: UNESCAP (2002, Chapter 8).

can borrow, usually in US dollars, from another country and use the funds to buy its own currency and stabilize the exchange rate. Swaps of 90 days duration can be renewed up to seven times with interest tied to LIBOR. There are plans to augment these arrangements by a regional financial arrangement similar to the special drawing rights from the IMF organized by Asian central banks. Central banks would earmark a small proportion of reserves to a pool available for drawing with some leverage. Such a region-wide system would provide much more liquidity than the existing patchwork of bilateral arrangements (see UNESCAP, 2004 for further suggestions). Swaps and financial integration can be augmented by clearing and settlement arrangements and credit rating and guarantee mechanisms. These are discussed further in the section on future outlook for financial cooperation.

There are some concerns about these swap agreements and their impact on risk and their effectiveness. There may be a moral hazard introduced since these swap agreements imply a commitment to support currencies in the case of a liquidity crisis. It is similar to the possible moral hazard created when the IMF implicitly agreed to support financial institutions in Asia prior to the financial crisis, perhaps contributing to the crisis. The risk of moral hazard could be reduced by imposing conditionality or tying swaps to some predetermined standards for financial and macroeconomic probity. These swap agreements also impose credit risk on the suppliers of the hard currency in the swap. Risks might be lowered if the swaps were tied to IMF conditionality assuming the IMF conditionality would serve to stabilize currencies. In any event, Standard and Poor's has indicated that if these kinds of swaps did not involve IMF conditionality then credit ratings in the region could be adversely affected (see Sakakibara and Yamakawa, 2003). It may also be that Chiang Mai type initiatives are no longer needed since the Asian crisis economies have moved to more flexible exchange rates and currencies are less likely to be put under pressure by speculators.

Financial surveillance

In order to avoid a repeat of the 1997 financial crisis, several systems have been put in place to detect emerging macroeconomic financial and corporate vulnerabilities. These efforts include collecting and analyzing a variety of financial and macroeconomic variables on a periodic basis and training researchers and bank officials in how to use financial software developed to implement this early warning system. A series of workshops have been developed for this purpose and also to disseminate these findings and implement appropriate action. A unit has been proposed to monitor (i) liquidity positions, (ii) implementation of common prudential standards, (iii) policy implementation, and (iv) reform and coordination of policies (Rana, 2002).

Financial services and bond market development

Asia continues to rely heavily on the banking system for financial intermediation. Bank lending is the most popular form of debt finance by business firms and the bond and commercial paper markets are generally thin and not well developed. Closely held companies who are self-financed are also much more common in Asia than in other regions and high saving rates contribute to the slow growth in alternative debt instruments. Observers believe that continued heavy reliance on bank finance could raise the risk of financial weakness and contagion as experienced during the Asian financial crisis, particularly if the banking system and its customers do not become sophisticated and develop alternative financial instruments (Legot *et al.*, 2004).

Since the Asian financial crisis there has been some movement toward more corporate bond issues as bank lending decreased, mostly in Korea but also in Thailand and Malaysia. A number of initiatives have been undertaken to strengthen bond market development in Asia through regional cooperation. The Asian Bond Fund was launched in 2003 by a group of Asian Central Bankers. The initial offering of this Fund is $1 billion to be invested in a basket of dollar denominated bonds issued by Asian developing countries. The second round is a much larger fund involving private sector co-investors and includes bonds denominated in regional currencies. These funds are designed to serve as a catalyst for domestic financial reform in East Asia by encouraging East Asian countries to issue bonds for the Fund to buy. The potential challenges in developing an Asian bond market and the impact the Asian Bond Fund will have on this market including the nexus between central bank monetary policy and ownership of the Asian Bond Fund are discussed in the section dealing with future issues in the later part of this chapter.

There have been several institutional developments in the past decade that are designed to promote greater monetary and financial integration in South Asia. In 1998 SAARC heads of state agreed in principle to establish a Network of Central Bank Governors and Finance Secretaries of the SAARC region dubbed SAARCFINANCE. This was implemented later in the year and the SAARCFINANCE Network was established. The basic objective of the Network is to discuss macroeconomic policy issues and establish closer links among the members through staff visits and regular exchange of information, consider harmonization of banking legislations and practices, work toward a more efficient payments system and achieve better monetary and exchange rate cooperation. Additionally the Network will forge closer cooperation on macroeconomic policies, train staff for the central banks, ministries of finances and other financial institutions of SAARC members, and promote research on economic and financial issues. The secretariat of the Network is held on a rotating basis and the Chair of SAARCFINANCE moves with the SAARC Chair. Each central bank has established a cell in its research department to coordinate the activities of SAARCFINANCE. SAARCFINANCE meets twice a year concurrently with the IMF/World Bank annual and spring meetings. There are also staff exchanges and seminars and workshops conducted on SAARC issues (see also Dasgupta and Maskay, 2003). There has also been further integration between South Asia and East and Southeast Asia (see Rana and Dowling, 2006, Chapter 5).

Regional Cooperation and the Environment

There are a number of areas where regional cooperation can be helpful in resolving environmental problems. In this section we consider water resources and air pollution.

Water resources[3]

International agencies are often able to resolve water rights issues involving several countries when other efforts have failed. The Aral Sea is an example. The Aral Sea has shrunk dramatically in the past four decades. The flow of water from the Amu and Syr rivers was diverted for irrigation and over the years the Sea has shrunk to a fraction of its size in the 1960s when it was the fifth largest lake in the world. The Aral Sea basin stretches over five Central Asian republics of Kazakhstan, Kyrgyz Republic, Tajikistan, Turkmenistan and Uzbekistan. UNEP and the World Bank began working on environmental issues in the Aral Sea basin in the late 1980s and early 1990s and made recommendations to stabilize the environment of the sea, rehabilitate the disaster zone around it, and undertake comprehensive management of

[3]The first three paragraphs of this section draw heavily on Schiff and Winters (2002).

its waters. The five CARs agreed and there were a series of meetings and agreements, one of which established the Inter-Disciplinary Coordinating Water Management Commission, which directly manages water resources and is composed of members from the five CARs. So far it seems to be managing yearly water allocations for each country quite well.

In another example of regional cooperation, Thailand and Laos agreed to share the hydropower resources of Laos' Nam Ngum River — a Mekong River tributary — despite poor relations and mistrust between the two countries. This was made possible through the auspices of the Mekong Secretariat. The project was funded by a variety of international donors and it now provides cheap power to Thailand and provides Laos an opportunity to earn substantial foreign exchange.

The international community also helped overcome seemingly intractable problems between India and Pakistan over the waters of the Indus River. India and Pakistan were unable to agree on a division of the Indus River Basin waters after the 1947 partition, and tensions between them rose a year later when India diverted water used to feed Pakistan's irrigation system. After much negotiation and civil works to help divide the flow from several tributaries, the Indus Water Treaty was signed dividing the Indus river and its five tributaries, and a fund administered by the World Bank is used to fund additional civil works.

Other water sharing agreements have been signed by India and its neighbors. In 1996 India and Bangladesh also signed a 30-year water agreement which determines the amount of water released by India to Bangladesh at the Farakka Barrage, primarily during the dry season. The agreement is reviewed every 5 years. In 1997 India and Nepal signed an agreement governing the flow of water and electricity from the Mahakali River. One remaining contentious water issue remains between China and India which centers on the Bramaputra, which flows through Tibet to India and Bangladesh. There are plans by China to divert the river to tap its hydro-potential. Observers contend that better cooperation between Bangladesh and India on the

Bramaputra could tap its potential for commerce and transport by bringing tea and oil from Assam to Bangladesh.

Air pollution

Air pollution is discussed thoroughly in Chapter 6. The costs of air pollution are high and fall primarily on those living in the polluting country. There are, however, widespread cases of spillovers to neighboring countries. Acid rain and other pollution associated with burning of fossil fuels are the most widely observed and discussed. In Asia, China is the biggest offender. Acid rain generated in China falls on Korea and Japan, while mercury and other trace elements contained in coal ash are being deposited as far away as the west coast of the United States. Pressure on the Chinese authorities by the affected countries has been exerted through diplomatic channels and by working with international bodies including the UN and the World Bank to deal with the problem. The general problem of emissions from coal-fired generating plants in China, particularly in the north, is also relevant for global warming and this has put additional pressure on China to take remedial action. So far nothing has been done to put systematic pressure on China through the UN or other international fora. Japan has helped by offering assistance to install scrubbers on coal-fired plants to cut down sulfur emissions and this should also help to reduce the amount of mercury pollution.

Aside from noting the build up of mercury levels in fish and birds eggs, the US government has not done much to deal with the reduction in mercury levels. A recent conference on the spread of air pollution to neighboring countries suggested dealing with the problem through existing frameworks dealing with the environment and regional cooperation. While the impact of air pollution on plants, farmland and human health on those in close proximity to the source of pollution has been studied, there has been little systematic study of the science of this inter-country and inter-continent pollution and its effect on plants, fish, animals and people (for further discussion on air pollution and global warming, see Chapters 6 and 7).

Issues and Prospects for Regional Cooperation

Institutional initiatives toward regional cooperation in Asia

South Asia and Southeast Asia have developed regional cooperation initiatives independently, primarily because of regional, ethnic, and governmental differences in approach to economic growth and development. In the past few years, however, there has been a growing interest in regional cooperation bridging these two regions. There are two existing institutions that provide a partial bridge of that gap. They are the Bangkok Agreement involving Bangladesh, India, Lao PDR, Korea, Sri Lanka and China; and the Bay of Bengal Initiative for Multisectoral Technical and Economic Cooperation (BIMSTEC) which drops Pakistan and adds Thailand and Myanmar to SAARC. Both these are ad hoc groupings which were begun without any specific purpose of reaching regional cooperation more widely.

Given current interest in trading arrangements, the growth of PTAs and the initiatives of multilateral arrangements such as ASEAN and SAARC, not to mention APEC, it may be useful to consider a meeting between the members of SAARC and ASEAN + 3 to discuss a broader grouping of Asian economies which would cover the entire Asian region, including all the countries of South Asia, East Asia and Southeast Asia. Hong Kong and Taiwan could be included as part of greater China group, if political opposition to the inclusion of Taiwan can be overcome.

What would be the benefits from adding yet another Asian organization to the list that seems to keep growing? One advantage is that it would be all inclusive and represent more than half the population of the globe. Its economic power would dominate other developing countries and regions. It would also afford an opportunity to begin formal talks to lower trade barriers and encourage further trade between the members of ASEAN and SAARC on a multilateral rather than bilateral basis. It would take away some of the complications resulting from a variety of PTAs with different local content requirements and other special problems that arise from the growth of these bilateral arrangements. It would also provide an effective counterweight to the EU and NAFTA. Additional details would have to be worked out, including the role of ASEAN + 3 and BIMSTEC. Preliminary meetings could be hosted by a regional multilateral institution interested in promoting free trade in the region, such as the ADB or ESCAP.

Energy

The Asian region is expected to be an important source of new energy resources as well as a growing consumer of these resources. The efficient balance of demand and supply in the region can be facilitated by greater regional cooperation. Regional cooperation in petroleum and natural gas pipeline development has great potential. Among the Central Asian Republics, three countries (Kazakhstan, Turkmenistan and Uzbekistan) are major energy exporters of both oil and natural gas. Development in the CARs will require resolution of a variety of issues within the region as well as negotiations with countries outside the region. Regional cooperation in the CARs will continue, with emphasis on resolving pricing and border issues in the region and also with countries outside the region, primarily Russia, and also China, Pakistan, Iran, Ukraine and Turkey. The Central Asian Republics will play a key role in future energy development in the region and will grow as a major exporter to Russia, Europe, and potentially China. There are a number of challenges including pricing and construction issues as well as risks of political instability and internal security as the potential for energy exports from the CARs is further expanded.

Turning to other countries in the region, both India and China have vast coal reserves that they will continue to exploit. However, coal technology will have to be upgraded to address environmental concerns and in any event will have to be augmented by imports of other fuels to supplement domestic resources. These energy-thirsty giants will continue to look for secure, reliable, and reasonably priced energy resources, primarily oil and natural gas, in other countries. They have already struck deals with several oil-producing countries in Africa and Latin America outside the OPEC cartel and will also be importing from Russia.

Similarly, the energy-rich countries in Southeast Asia, particularly Indonesia and to a lesser extent Philippines, Malaysia and Vietnam, will be under pressure to increase production, not only to meet domestic demands, but also to export. Recent changes in the regulatory environment in Indonesia suggest that exploration will resume after a long hiatus. This could bring Indonesia back to the position of being a new exporter of energy in a few years.

Aside from regional cooperation to facilitate the trade in energy products, there are opportunities for sharing technology and policies for improving energy efficiency and reducing environmental impacts. Pricing policies can be introduced that reflect not only the private cost of extracting resources but also the environmental damage created by various kinds of emissions. Many of these issues are discussed in more detail in Chapters 6 and 7. Nevertheless, it is important that regional cooperation issues, particularly when they involve spillovers such as acid rain, pollution of rivers and global warming, be discussed by all the parties involved. ESCAP, ASEAN, ADB, SAARC, and other regional bodies need to take the lead in spearheading this agenda.

Information Technology

The benefits of information technology gained by the richer countries of the region in the past two decades can be extended into the future only if the region continues to implement new developments that will further increase productivity and raise the living standards. There are many opportunities in distance education, health services, software development, and the service industry. Resources from the private sector are being allocated to develop these capabilities. The governments of the region can offer support by adopting legislation that facilitates the development and dissemination of ICT innovation.

One challenge for future development of information technology is for the richer countries in the region to share this technology with the poorer countries. Narrowing the digital divide can take place in many different ways, both between countries and within countries. ICT can be used as a business tool to facilitate international trade. There are trade networks that share information such as the shipping network set up by the Sri Lanka Ports Authority and an associated center for transport standards and several networks set up by China (see UNESCAP, 2004). ICT can be developed as a valuable tool for small- and medium-scale industries with limited budgets for advertising and marketing to access global markets and also to link suppliers and buyers in a global network. ICT can also be used as a teaching tool to train teachers, extend distance learning, and streamline the delivery of public services by standardizing forms for procurement, registration, etc. Many of these ICT arrangements can be extended from country to country by regional cooperation, either on a bilateral basis or through regional organizations such as ADB, ESCAP, ASEAN, APEC and SAARC as well as local and regional NGOs. The digital divide can be closed in this way. It will require additional budget allocations as well as decisions regarding what hardware to use and the linkages required to reach appropriate target groups. If the primary objective is reaching the poor there will be a different set of priorities than if the rich are catered to. Generally, the richer segments of society will be able to access ICT much more easily than the poor.

By setting up satellite-based multipurpose community telecenters in a host of rural locations many more poor people can be reached, than by alternative means. Such centers can provide knowledge and information to local communities and facilitate the development of local businesses and further entrepreneurship and responsive government. Development of these kinds of centers can be supported by multilateral agencies and regional development organizations. UNESCAP (2004) argues that the convergences of ICT services with the Internet can attract small businesses into niche markets if service can be extended to rural areas and regulations can be streamlined to reduce costs of doing business. One problem has been economies of scale. Infrastructure costs can be high although benefits can accrue through wider application of technology and economies of scale in a variety of areas including transportation, health information, education, weather forecasting, agricultural extension, and many others. Since the marginal costs of additional use are essentially zero, it is important to think of all the possible uses that can be made of a telecenter and other ICT investments. ESCAP, UNDP, UNESCO and other organizations have already developed programs which countries can access. These telecenters can also build and strengthen communities and create awareness of business opportunities.

There are also issues of privacy, regulation, and piracy along with the content of the Internet that has to be dealt with as ICT becomes more widely available. A number of meetings have been held dealing with ICT and Internet issues including the Fifth Regional Inter-agency Working Group Meeting on ICT in 2003 (see www.unescap.org/oes/state/st030813.pdf) which drafted recommendations to create an information society to assist in achieving Millennium Development Goals (MDG)[4].

Resources have to be allocated if these goals are to be achieved. Governments in the poorer countries have not been able to gather resources to address these issues and as a result the digital divide is increasing. New measures have to be taken including assistance from the richer countries in the region to offer technical support and resources to put more satellite systems into use in the poorer parts of the region, particularly South Asia. Women and the poor should be targeted in conjunction with education programs to raise literacy. Otherwise these resources could go to waste.

There are a number of issues regarding allocation of frequencies for communication in Asia that are of technical nature and beyond the scope of this chapter. There have been many new satellites launched in the past few years to handle the increased traffic from the explosion in cell phone use and the internet. There are also new technologies that are expanding the spectrum of possible frequencies. It is difficult to predict the outcome of all these developments both on the supply and the demand side. Asian governments should be allocating some technical resources to follow up on these developments and assessing their role in implementing new technological developments and, in the case of the richer countries in the region, in developing these technologies.

[4]See Chapter 4 for more discussion on Millennium Development Goals and the Millennium Development Report.

Migration

The pace of migration from poorer to richer countries in the Asian region is expected to continue at current levels and may even increase. In the medium term, economic growth prospects for the richer countries are good and there will be an increased demand for unskilled and semiskilled labor in construction, services, and labor-intensive manufacturing. Furthermore, population growth from natural increase in the richer countries is expected to continue to decline creating additional labor shortages that will have to be filled by migration. For example, projections for Japan over the next few decades under one scenario show immigration growing from 2 million per annum in the next few years to 10 million by 2030 if the labor force is to remain at current levels (McDonald and Kippen, 2000). Within the Asian region, Korea is the only other country which will be experiencing a decline in fertility over the next few decades. Since our focus is on Asia, we have not looked at Europe and North America as far as future migration goes. However, there will likely be further immigration pressure as domestic fertility in these two regions also declines and labor force either grows slowly or declines. At the same time the pool of labor in the poorer countries, such as Pakistan, India and Philippines is not expected to diminish due to sustained upward pressure on population, even though growth is projected to be strong in these countries.

For migration issues, the region needs to establish a more consistent framework for the sending and recipient countries to work with, possibly including migration as a component of trade agreements. Another is for consultants to develop a template for best practice in sending and recipient countries to avoid problems such as abusive practices, excessive work hours, no rest days or time off, and fees that are imposed after the migrant has arrived in the country. There should be systematic monitoring of impacts of local labor markets in recipient countries and reporting problems of abuse and exploitation in a timely manner. There are often conflicting interests between sending and recipient countries. However, if friendly dialogue can be established between the sending and recipient countries there is opportunity to create win-win situations where both countries can benefit. Countries that are both recipients and senders such as Malaysia and Thailand can receive double benefits from such negotiations.

Remittance income will continue to be an important source of foreign exchange as well as income for thousands of poor families in Indonesia, the Philippines and South Asia. In order to facilitate the repatriation of foreign income by migrants, there is a need to introduce a flexible and modern system of transferring money in all recipient countries. This would include bank to bank transfers and ATM linkage, linking remittances to other bank products, and services including investment advice and competition between different agencies to achieve the best service for migrants and to ensure upgrading of remittance services with better IT technology for making transfers as they become available. There has to be continued reduction in fees for transfers through automation and modern banking facilities. All of the countries in the region already have an elaborate network of state, local, and foreign banks to implement these transfers. There is also a need for better education of migrants in terms of options for spending remittance income. Liaison with banks and other financial intermediaries should also be established to ensure that returning migrants and their families receiving remittances are aware of an array of investment opportunities beyond augmenting consumption, investing in their local village or building a new house.

In any event, the governments have a set of policy objectives for migration and the remittances that are associated with them.

These objectives should include:

- Providing information on living conditions, logistics, and other important information about the recipient country to migrants before they leave their home country;
- Keeping a registry of migrants in the sending country including age, gender, any medical problems, place of residence, whom to contact in case of emergency, and other vital information;
- Reducing domestic unemployment through migration;
- Generating foreign exchange income;
- Improving the wages and working conditions of employed nationals working overseas;

- Reducing the cost of emigration by reducing taxes on migrants and curbing monopoly practices by recruiters;
- Helping migrants save and invest remittances wisely;
- Expanding and diversifying the number of potential destinations for migrants;
- Improving the skill mix of migrants;
- Monitoring the impact of migration on the local labor market in the sending country;
- Focusing on issues concerning female migrants, particularly in the entertainment and household services areas, where the bulk of unskilled and semiskilled women are employed.

Dialogue with countries in other regions should also be undertaken, particularly in the Middle East, where many countries are recipients of migration from Asia, South Asia in particular. These discussions would focus on issues of particular relevance to sending countries like working conditions, exploitation, and levels of remuneration.

We have considered the topic of migration of highly skilled workers and professionals only briefly. There are many issues involved here, particularly questions of brain drain that involve recipient countries in North America and Europe. We discussed some issues in the review of past migration activities. To build on current policies that are encouraging short-term migration of professionals, it is important to consider issues of the local labor markets in sending countries as well as conditions in recipient countries. Student migration, school fees, and work permits beyond the completion of schooling are important issues to be considered, particularly when the skills of these graduates are badly needed in the sending country.

There can also be domestic issues within sending and recipient countries which have limited direct relevance for regional cooperation in Asia, the resolution of which would make a significant impact on the migration and remittance pattern. For example, the immigration of nurses from the Philippines and of doctors from India and the Philippines to the United States is a migration issue that needs to be addressed within the United States. The supply of local doctors and nurses in the United States has been falling over time and some outside observers believe that there is no reason why the United States should not be training more health professionals from its own stock of highly qualified college graduates. These and other migration issues relating to the United States and Europe are discussed further in Chapter 3.

Banking and Finance

For economic growth to continue at a robust pace, the Asian economies must continue to maintain prudent monetary and fiscal policy with low inflation. Stable and predictable monetary and fiscal aggregates are needed to underpin growth and create an environment that is favorable to technical change, domestic and foreign investment, and economic progress in reducing poverty and uplifting living standards. The financial system must also evolve by developing a deeper market for equities to complement the banking system. A dynamic bond market is a key component of this deepening process. Commercial bank liabilities are largely short-term deposits and these are ideally matched by short-term loans. Because of this need to match the maturity of loans and investments, banks are not in a position to finance many long-term investments. A viable and robust bond market provides long-term investment opportunities and reduces the chance of a maturity mismatch that can arise when banks completely dominate the financial sector. Such a mismatch, compounded by exchange rate risk, was a major reason for the Asian financial crisis. An active bond market protects against such a crisis and strengthens the financial system while providing long-term investment opportunities for businesses. Equity markets provide similar protection. Domestic bond and equity markets also reduce the need to rely on foreign capital for long-term investment and can serve to narrow the interest rate spread between the domestic foreign markets for investment funds. Less reliance on foreign funding and the potential mismatch between short-term foreign exchange maturities and longer term domestic maturities can also be mitigated by robust local bond and equity markets.

One of the cornerstones of a viable and mature domestic bond market is the market for government bonds. Such a market will present a range of maturities and establishes a yield curve as a benchmark for commercial paper and bonds as well as creating an environment that encourages longer term domestic investment commitments. Looking at the Asian region, only Hong Kong and Singapore currently have such well-developed government bond markets, even though neither government issues bonds to finance fiscal deficits. Nevertheless, both Hong Kong and Singapore have government bond auctions at regular intervals and as a result there is a benchmark yield curve for government bonds that serves as a reference for commercial transactions. Generally bond markets in other countries in the region are less well developed. Fabella and Madhur (2003) suggest the following hierarchy of bond market development in East and Southeast Asia based on size, secondary market turnover (proxy for liquidity), market infrastructure, regulatory framework, investor base, and openness to foreign investment. From highest (i) to lowest (iv) the countries are arranged as follows:

(i) Hong Kong, China and Singapore
(ii) Korea, Malaysia and Taiwan
(iii) China, Philippines and Thailand
(iv) Indonesia

Other countries can learn from Hong Kong and Singapore in developing a competitive government bond market. Once such a market has been established it can serve as a vehicle for deficit financing. However, this should not be the reason for establishing the government bond market. South Asia, which was not part of this exercise, might be in categories (iii) (India, Bangladesh and Sri Lanka) and (iv) (Nepal).

Further development of corporate bond markets in Asia is being hampered for a number of reasons aside from a poorly developed government bond market. These include lack of transparency and legal protection of investors, no regional prudential standards, transactions taxes that discourage bond trades in the secondary market, small investor base, and illiquid secondary markets. There is also opposition by central banks that fear their control over monetary policy will be threatened and there is no regional clearing house for settlements or widespread electronic clearing and transfer mechanisms. There are also a few institutional investors, and investor confidence is eroded by an underdeveloped legal framework and lack of information about bond participants. Nevertheless, there is some evidence that in Malaysia, Korea and Thailand, progress has been made as market infrastructure has been strengthened, benchmark yields established, and foreign participation expanded (Park, 2001; Sakakibara and Yamakawa, 2003).

The Miyazawa Initiative was initiated by Japan in 1998 to assist Asian countries affected by the financial crisis and to contribute to the stability of international financial markets. As part of this initiative, particular emphasis was placed on the development of bond markets. This has been interpreted as a support for an Asian bond market. The backbone of such a development has been Japanese guarantees of Asian sovereign debt issues. These guarantees have strengthened local markets and could provide a springboard for the development of a regional bond market. However, there has been little support for such a regional bond market within the region outside of Japan. The requirements for the establishment of such a market are similar to those of a viable local bond market and would include a regional bond-trading center and a communications network and a regional credit rating agency and clearing house (Sakakibara and Yamakawa, 2003, p. 79).

As noted above bond market development must be accompanied by other banking reforms and these measures have been discussed elsewhere (for example, in Chapter 1). The point of emphasizing bond market development in this section on regional cooperation is to point out the importance of cooperation with Singapore and Hong Kong to develop the bond market and also to discuss how bond markets can be used to finance new infrastructure and rehabilitation of existing facilities in Asian economies. These projects are estimated by the World Bank to cost over $ 228 billion annually between 2006 and 2010 (see UNESCAP, 2006 for a summary). Infrastructure needs of the different sub-regions within Asia would vary (Fay and Yepes, 2003).

South Asia would have substantial investment needs in electricity generation ($ 11 billion per annum) and roads ($ 6 billion) and an overall total of around $ 28 billion for the 6-year period from 2005–2010. East Asia including Southeast Asia, China and the NIEs would require more than three times this investment, with electricity generation ($ 25 billion), telephone ($ 17 billion) and roads ($ 12 billion) taking the bulk of investment.

Some of this funding will come from domestic tax revenues and some from the private sector and international donors. There will, nevertheless, be a significant shortfall in funding. Public spending for infrastructure is not expected to pick up much from existing levels of 27 billion annually nor is private sector spending of 20 billion. Thus, the shortfall is expected to be about $180 billion annually. This is in the range of the annual surplus of saving over investment in the region over the past few years. China's foreign exchange reserves build up alone was over $ 600 billion between 2001 and 2005 (see Table 2.9). When the surpluses of Japan are added to those of the rest of the region, the amount swells to more than 200 billion over this forecast period.

One possibility for filling this gap would be to use some of the accumulated reserves of the richer Asian economies to fund the infrastructure needs of the poorer countries. This could also be a way to further develop the Asian bond market. Poorer Asian governments could issue long-term bonds to finance infrastructure and guarantees for these loans could be provided by governments of wealthier countries and/or by international donors.

Such an enormous enterprise brings to mind the Marshall plan where reconstruction of infrastructure in Europe and Japan was funded by the United States following the end of the World War II or the initiatives introduced by United States Secretary of the Treasury James Brady (so called Brady bonds) which helped to rescue Latin American countries in the 1980s. Brady bonds were US dollar-denominated bonds issued by Latin American countries and backed by US Treasury bonds. Brady bonds helped to reduce the debt held by Latin American countries that were defaulting on loans. Because the Brady bonds were backed by US bonds, repayment of principal was insured. They were and continue to be widely traded in international markets.

In Asia, infrastructure investment in the richer countries can probably be financed from domestic sources while the poorer countries would require larger transfers. These could be facilitated by locally issued bonds offered at attractive rates and guaranteed by an international agency such as the proposed Asian investment bank. The richer countries could back the bonds issued by poorer countries to fund their infrastructure needs and these bonds would be traded internationally. Such a program would boost the liquidity of the Asian bond market as well as serve as an aid to poorer countries of the region to finance their infrastructure needs. Because of the guarantee of the richer or surplus countries (Japan, Korea, Taiwan, Singapore, or even China) interest rates and risk would be lowered.

An Asian Brady bond market is only one of many possibilities if the Asian bond market does mature and attract buyers. It will have to compete with the global market where there are many countries and firms competing for funds. Without further deregulation and greater competition and transparency this may not happen. The Hong Kong economists Paul Legot, Douglas Arner and Liu Qiao (2004) have a number of suggestions for replicating the Eurobond market success. They suggest that ASEAN + 3 members: (i) adopt essential legal, fiscal and systemic regulations to increase competition, increase market participation and harmonize markets; (ii) develop a collaborative regional public debt market for domestic and major currency issues monitored by a regional body and housed in an established financial center such as Tokyo, Singapore or Hong Kong; and (iii) set up a new regional body to support securitization and spread risk.

Legot *et al.* spell out the details of how these proposals can be implemented in their paper. The proposed infrastructure funding suggestions mentioned above could easily be incorporated into such a framework. Another suggestion is to develop a regional development bank in the mold of the European Investment Bank, the European Union's (EU's) long-term financing institution owned by the member countries. The Asian counterpart would be an independent government-owned Asian investment bank that would promote regional capital markets and serve as an intermediary to reduce risk and provide security for the infrastructure needs of the region. This bank would be owned by Asian central banks or governments and would be capitalized with funds from surplus countries in the region.

Development of bond markets has to go hand in hand with further strengthening of banking systems in Asia. There are a number of proposals for regional cooperation to achieve such an objective. Barry Eichengreen (2001) suggests a forum based on ASEAN + 3 platform called the Asian Policy Forum (APF) that would (i) provide technical assistance on prudential supervision to national governments; (ii) administer training programs for auditors and inspectors; (iii) provide clearing and settlements to member central banks; and (iv) establish financial standards. This would be a more forceful and proactive approach than ASEAN is currently using. This APF would coordinate with other countries and regions outside of Asia and harmonize Asian standards with global standards developed by industrial countries, such as BIS and Basle Accords. Proposals such as the Eichengreen (2001) and Legot *et al.* (2004) suggestions need to be considered by member governments in Asia if they are serious about developing stronger bond markets in the region.

Transportation

Regional cooperation in transportation can be effectively improved through the three major regional organizations ASEAN, SAARC and ECO. ASEAN has made the most progress in effectively linking the transport network and integrating these developments with telecommunications. The key agreement is the ASEAN Plan of Action in Transport and Communications (see www.aseansec.org/7373.htm) which started in the mid-1990s. Many of its initiatives have already begun and will be carried forward in the future. They include integration of different transportation modes to provide door-to-door delivery of goods to reduce marketing and distribution costs, developing interconnectivity in telecommunications to facilitate trade and commerce both with ASEAN and between ASEAN and other countries, harmonizing land transport to facilitate border crossings, developing rules for handling hazardous materials, and developing human resources to increase operational efficiency and capacity utilization in transport and communication. The ASEAN highway network has been developed and will continue to evolve. It stresses safety, uniform design standards and compatibility between countries. Further development of the ASEAN highway network will be integrated with the rest of the Asian region through Asian highway and rail systems.

The Central Asian Republics face many challenges in developing a transportation network for the region. Aside from border issues discussed previously, the thrust of transportation policy has to be to increase efficiency and lower costs of transportation, particularly between countries and to countries outside the region. Multimodal transport systems need to be developed linking rail and road in particular and also integrating with air transport in specific instances. The rail network, which was expanded significantly in the 1990s remains underutilized. Linkages with countries outside the region have to be strengthened for utilization rates to increase. Regional cooperation and integration are slowly emerging as a priority for the five countries and their neighbors (UNDP, 2005). International trade in Central Asia has risen in this (21st) century after falling dramatically following the breakup of the Soviet Union in the early 1990s. But poorly managed borders still hamper trade. In a survey commissioned for the UNDP report the great majority of Central Asians noted problems in crossing borders, and many reported harassment and corruption. Improved border management, with faster transit, fewer hassles and bribes, would increase trade and reduce costs.

The UNDP Report (UNDP, 2005) estimates that upgrading Central Asia's extensive transportation network would save the region some $ 300 million a year in transport costs. Only 25% of roads in Kyrgyzstan and 20% of roads in Tajikistan are in good condition and as a result high transport costs and long transport delays reduce any competitive export advantage. The report estimates that transport costs and times from Central Asia to Moscow, Istanbul and Western Europe, as well as within the region, could be reduced by half with better transport and transit management. A 50% reduction in trade costs, including reduced tariff rates, and especially transit and transport costs, according to the report, would increase real GDP in 2015 by about 20% in Kazakhstan and by 55% in Kyrgyzstan, the report says. As a result, real consumption would be 13% higher in Kazakhstan and 66% higher in Kyrgyzstan.

Unofficial trade, also known as the "shuttle trade", accounts for even more activity. Unrecorded imports of consumer goods, gasoline and diesel fuel were estimated to be equal to 17% of Kyrgyzstan's recorded imports in 2003, while unrecorded exports of the small-scale sewing industry and re-exports of consumer goods amounted to 20% of recorded exports according to the report. Much of the shuttle trade is carried out by women. "In short, shuttle trade is an essential part of trade in Central Asia and has important implications for the reduction of poverty", says the report.

For improved trade, transport and transit, the report makes the following recommendations:

- Prioritize multilateral trade liberalization within the WTO framework.
- Simplify and coordinate customs and border crossing rules and procedures and improve governance at customs, police and border protection services.
- Closely cooperate to create "borders with a human face" to facilitate cross-border movements of people, goods and vehicles.
- Facilitate, rather than interfere with shuttle trade.
- Develop regional transport corridors and improve conditions for competitive transport services.
- Simplify and improve regional transit systems for road transport — upgrade regional air transport to provide access to each other's principal airports.
- Enhance competition in rail and air transport — in particular, by restructuring state-owned railways and airlines.

Furthermore, internal volatility in several countries has held back the continued development of smooth transit and efficient transportation to surrounding regions. Smooth and efficient transportation flow will resume only when these bottlenecks are removed.

In SAARC, transport infrastructure development has been hampered by the continuing differences between Pakistan and India.

Border crossings between the two countries are difficult and trade between them is much lower than expected by trade analysis such as the gravity model. Border crossings between Nepal and India are not as difficult as well as between India and Bangladesh and there are rail links as well. However, a comprehensive trade facilitation agreement has not been reached although SAFTA has emphasized the importance of such a strategy.

For the Asian region as a whole the Intergovernmental Agreement on the Asian Highway Network was adopted in 2003. The Agreement identifies internal linkages with other countries and contains road designs and standards and provides a roadmap for connecting 32 countries and 140,000 km of roads. This agreement has been joined by the Trans-Asian Railway network which was updated in 2003 to include new linkages and now comprises 80,000 km of track in 25 countries. These networks will be improved and extended over the next decades. The Railway network will get improved rolling stock, upgraded safety, and better tracks to accommodate faster trains. The Highway network will be upgraded and connected with more feeder roads. Both systems will link to port facilities where appropriate. If these plans for roads and rail are implemented in a timely manner, costs of transport will be reduced even further giving Asian exporters and consumers an added competitive advantage and lower prices.

As noted in the previous section, developments in transportation and port upgrading in the Asian region has been remarkable. To put the rapid growth of the region's container capacity in perspective, consider the top 20 container terminals in the world and their throughput between 2002 and 2004 as shown in Table 2.11.

The top six ports are all in Asia and two of the top five are in China. All of the NIEs and two Southeast Asian economies have at least one port in the top 20. Laem Chabang in Thailand handles more traffic than Tokyo and 13 of these top twenty are in Asia.

Port traffic in China is projected to reach 7.5 billion tons by 2010 and China is planning to expand its port facilities in Fujian and Xiamen. Taiwan has plans to expand its port facilities as well and the rest of the region is also expected to maintain steady growth in port facilities. Containers are getting larger and port facilities are expanding

Table 2.11 Top 20 Container Terminals and their Throughput, 2002–2004 (in million TEUs)

Rank	Port	2002	2003	2004
1	Hong Kong (China)	19.14	20.82	21.93
2	Singapore	16.94	18.41	20.60
3	Shanghai, China	8.81	11.37	14.57
4	Shenzen, China	7.61	10.70	13.65
5	Busan, Korea	9.45	10.37	11.43
6	Kaoshiung, China	8.49	8.81	9.71
7	Rotterdam	6.52	7.10	8.30
8	Los Angeles	6.11	6.61	7.32
9	Hamburg	5.37	6.14	7.03
10	Dubai	4.19	5.15	6.43
11	Antwerp	4.78	5.44	6.06
12	Long Beach	4.52	4.66	5.78
13	Port Klang, Malaysia	4.50	4.80	5.24
14	Qingdao, China	3.41	4.24	5.14
15	New York	3.75	4.04	4.40
16	Tanjung Pelepas, Malaysia	2.67	3.50	4.02
17	Ningbo, China	0.00	2.77	4.00
18	Tianjin, China	0.00	3.01	3.81
19	Laem Chabang, Thailand	2.66	3.18	3.62
20	Tokyo	2.71	3.28	3.58

Note: TEU is a measure of number of containers handled.
Source: US Bureau of Transportation Statistics, 2005; http://www.aapa-ports.org/industryinfo/statistics.htm.

to accommodate this by deepening berths and adopting better technology to handle them. Expectations of growing trade with China seem to be the major driving force behind these future expansion plans.

Several years ago Yusuf and Evenett (2002) wondered whether East Asia would be able to keep pace with new innovations in ICT and transportation as well as industrial innovations following the Asian financial crisis. They noted that

> Supply chains are relatively inefficient compared with those in Europe. On average, distribution and logistics account for 12 % of the free on board price of steel in East Asia as against 4 % in Europe.... In East Asia freight costs

> account for 8 % of total value of input .... compared with 4 % for the industrial countries (p. 63).

Although the figures in Table 2.11 are based on volume and not value, we wonder whether Asia is still at such a competitive disadvantage. As noted in the technology section of this chapter, the richer countries in the region are also keeping pace with new technological developments in communications and business applications of new technology.

Environment

Leaving aside global issues such as global warming and the exploitation of ocean resources (which are discussed in Chapters 3, 6 and 7) and focusing instead on regional problems of water use and air pollution, there are several unresolved items on the agenda. With the exception of the Bramaputra, water use agreements have already been signed for many of the major rivers flowing through the region. It is time to further develop the Bramaputra basin by bringing together India, China and Bangladesh to discuss outstanding issues, including China's plans for hydroelectric generation and diversion of flow from the river. This meeting could be sponsored by a neutral international body or group of agencies such as the ADB, the World Bank or the UN.

Air pollution from China to neighboring countries is the most pressing environmental problem that can be addressed by regional cooperation. Acid rain from other locations in the region does not present a serious problem as of the present moment. India produces a substantial amount of emissions from power plants. However, its coal produces less acid rain and there have not been reports of mercury poisoning in neighboring countries. Meetings should be held between China and the affected countries to find new approaches to dealing with the problem including possible subsidies from Korea, Japan and the United States to install proper equipment to reduce the pollution and lobbying efforts to convince China to install more such equipment.

Bibliography

Abella, M. and K. Lonnroth (1995). *Orderly International Migration of Workers and Incentives to Stay — Options for Migration to the Arab World: Experience of Returning Migrants.* Japan: United Nations University Press.

Agrawala, S. P. and J. S. Sehra (2002). Construction of Broadband Network Environments in Asia. http://asiaitssummit.nic.in [Retrieved 25 October 2006].

Anwar, M. (1994). *Indonesia in ASEAN: Foreign Policy and Regionalism.* New York: St. Martin's Press.

ASEAN (2006). ASEAN Plan of Action in Transport and Communications (1994–1996). http://www.aseansec.org/7373.htm [Retrieved 11 October 2006].

ASEAN Secretariat (2006). http://www.aseansec.org/64.htm [Retrieved 11 October 2006].

Asian Development Bank (ADB) (2002). *Asian Development Outlook.* Manila, Philippines: ADB.

Asian Development Bank (ADB) (2006a). *Asian Development Outlook.* Manila, Philippines: ADB.

Asian Development Bank (ADB) (2006b). *Workers' Remittance Flows in Southeast Asia.* Manila: ADB.

Bhattacharya, S. K. and N. B. Biswa (2006). Prospects and Challenges of Regional Cooperation and Integration in Trade, Investment and Finance in Asia: An Empirical Analysis on BIMSTEC Countries and Japan. CESifo Working Paper Series No. 1725. Asian Development Bank, CESifo (Center for Economic Studies and Ifo Institute for Economic Research).

Burgess, R. and V. Haksar (2005). Migration and Foreign Remittances in the Philippines. IMF Working Paper No. 111. Washington, DC: International Monetary Fund.

Dasgupta, A. and N. M. Maskay (2003). Financial Policy Cooperation in SAARC — A First Step Toward Greater Monetary Integration in South Asia. *South Asian Economic Journal*, 4(1), 133–143.

Dobson, W. and S. Y. Chia (eds.) (1997). *Multinationals and East Asian Integration.* Ottawa: International Development Research Centre.

Dowling, J. M. and R. Valenzuela (2004). *Economic Development in Asia.* Singapore: Thomson Learning.

Dowling, J. M. and G. Wignaraja (2006). Central Asia After Fifteen Years of Transition: Growth, Regional Cooperation, and Policy Choices. Working Paper Series on Regional Economic Integration No. 3, Office of Regional Economic Integration, Asian Development Bank.

Economist Intelligence Unit (2005). The 2005 e-Readiness Rankings: A White Paper from the Economist Intelligence Unit. London: Economist Intelligence Unit.

Eichengreen, B. (2001). Hanging Together? On Monetary and Financial Cooperation in Asia. Mimeo, University of California, Berkeley.

Fabella, V. and S. Madhur (2003). Bond Market Development in East Asia: Issues and Challenges. EDR Working Paper No. 35. Manila: Asian Development Bank.

Fay, M. and T. Yepes (2003). *Investing in Infrastructure, What is Needed from 2000–2010.* Washington, DC: World Bank.

Frankel, J. and S. J. Wei (1998). Regionalization of World Trade and Currencies: Economics and Politics. In J. Frankel (ed.), *The Regionalization of the World Economy.* Chicago: University of Chicago Press.

Goldstein, M. (1998). *The Asian Financial Crisis: Causes, Cures, and Systematic Implications.* Washington, DC: Institute for International Economics. http://www.iie.com/publications/bookstore/publication.cfm? pub_id-22 [Retrieved 1 March 2003].

International Monetary Fund (IMF) (2005). IMF Direction of Trade Statistics (DOTS). CD-ROM, April 2005.

International Telecommunication Union (ITU) (2004). http://www.itu.int/ITU-D/ict/statistics/ [Retrieved 2 January 2006].

Kawai, M. (2005). East Asian Economic Regionalism: Progress and Challenges. *Journal of Asian Economics,* 16(1), 29–55.

Khadria, B. (2006). Uncharted Contours of a Changing Paradigm: Skilled Migration and Brain Drain in India. *Harvard International Review.* http://hir.harvard.edu/articles/print.php?article=1445 [Retrieved 1 November 2006].

Legot, P., D. Arner and L. Qiao (2004). Asia's Bond Markets: Reforms to Promote Activity and Lessen Financial Contagion. HKIEBS Working Paper 1090, University of Hong Kong, Hong Kong.

Livelihood Connect (2006). Support for Sustainable Livelihood. http://www.livelihoods.org [Retrieved 1 November 2006].

Maddison, A. (2003). *The World Economy: Historical Statistics.* Paris: OECD.

McDonald, P. and R. Kippen (2000). The Implications of Below Replacement Fertility for Labour Supply and International Migration, 2000–2050. In Proceedings Population Association of America 2000 Annual Meeting, Los Angeles, USA.

Morgan Stanley (2006). http://www.morganstanley.com/GEFdata/digests/20060804-fri.html#anchor.

Office of the United States Trade Representative (USTR) (2000). National Trade Estimates Report on Foreign Trade Barriers. http://www.ustr.gov/reports/nte/2000/contents.html [Retrieved 1 November 2006].

Organisation for Economic Co-operation and Development (OECD) (2003). *Recent Trends in International Migration in Asia and Central and Eastern Europe.* Paris: OECD.

Park, Y. S. (2001). Development of Asian Bond Markets. In Y. H. Kim and Y. Wang (eds.), *Regional Financial Arrangements in East Asia.* Seoul: Korea Institute for International Economic Policy.

Pomfret, R. (1996). ASEAN: Always at the Crossroads? *Journal of Asia Pacific Economy,* 1(3), 365–390.

Rana, P. (2002). Monetary and Financial Cooperation in East Asia: The Chiang Mei Initiative and Beyond. PECC Finance Forum Conference, Honolulu.

Rana, P. (2005). Economic Integration in East Asia — Trends, Prospects and a Possible Roadmap. Paper presented at third High-Level Conference on Building a New Asia: Toward an Asian Economic Community, Taiyuan, China, 15–16 September.

Rana, P. B. and J. M. Dowling (2006). South Asia — Rising to the Challenge of Globalization. Mimeo, Asian Development Bank.

Robinson, S. and K. Thierfelder (1999). Trade Liberalization and Regional Integration: The Search for Large Numbers. International Food Policy Research Institute, Trade and Macroeconomics Division, Discussion Paper 34.

Saith, A. (1999). Migration Processes and Policies: Some Asian Perspectives. *Asian and Pacific Migration Journal*, 8(3), 285–311.

Sakakibara, E. and S. Yamakawa (2003). Regional Integration in East Asia: Challenges and Opportunities- Part I: History and Institutions. Reprinted as World Bank Policy Research Working Paper No. 3078. Available at SSRN: http://ssrn.com/abstract=636440; Part II: Trade, Finance, and Integration reprinted as World Bank Policy Research Working Paper No. 3079. Available at SSRN: http://ssrn.com/abstract=636441.

Schiff, M. and A. Winters (2002). Regional Cooperation and the Role of International Organizations and Regional Integration. World Bank Policy Research Working Paper No. 2872, Washington, DC: World Bank.

Soesastro, H. (2001). Towards an East Asian Regional Trading Arrangement. In Simon S. C. Tay, Jesus Estanislao and Hadi Soesastro (eds.), *Reinventing ASEAN* (pp. 226–242). Singapore: Institute of Southeast Asian Studies.

Soesastro, H. (2003). An ASEAN Economic Community and ASEAN + 3: How Do They Fit Together? Asia Pacific Economic Paper 338, Asia Pacific School of Economics and Government, Australian National University.

United Nations Development Program (UNDP) (2005). Bringing Down Barriers: Regional Cooperation for Human Development and Human Security. *Human Development Report on Central Asia*. Geneva: United Nations.

United Nations Economic and Social Commission for Asia and the Pacific (UNESCAP) (1987). *Trans-country: Power Exchange and Development.*

United Nations Economic and Social Commission for Asia and the Pacific (UNESCAP) (2002). Regional Financial Cooperation in East Asia: The Chiang Mai Initiative and Beyond. *ESCAP Bulletin 2002/2003 Asia-Pacific Economies: Sustainable Growth Amidst Uncertainties.*

United Nations Economic and Social Commission for Asia and the Pacific (UNESCAP), International Telecommunication Union (ITU) and Asia Pacific Telecommunity (APT) (2003). Expert Group Meeting on Regional Roadmap Towards Information Society in Asia and the Pacific and Fifth Regional Interagency Working Group on ICT, 13–15 August 2003, Bangkok. http://www.unescap.org/oes/state/st030813.pdf [Retrieved 1 January 2006].

United Nations Economic and Social Commission for Asia and the Pacific (UNESCAP) (2004). *Meeting the Challenges in an Era of Globalization by Strengthening Regional Development Cooperation*. New York: United Nations.

United Nations Economic and Social Commission for Asia and the Pacific (UNESCAP) (2005). *Statistical Abstract of Transport*. New York: United Nations.

United Nations Economic and Social Commission for Asia and the Pacific (UNESCAP) (2006). *Financing Regional Development in Infrastructure Development Including That Related to Disaster Management*. New York: United Nations.

US Bureau of Transportation Statistics and http://www.aapa-ports.org/industryinfo/statistics.htm.

Waddington, C. (2003). International Migration Policies in Asia. Paper presented at Regional Conference on Migration, Development and Pro-Poor Policy Choices in Asia, Department for International Development, UK.

Wickramesekera, P. (2002). Asian Labor Migration: Issues and Challenges in an Era of Globalization. International Migration Papers No. 57. Geneva: International Labour Office.

World Bank (2005). Migrant Labor Remittances in the South Asia Region. Finance and Private Sector Development Unit/South Asia Region, Report No. 31577, February.

Worldpress.org (2006). Oil- Central Asia Pipelines. http://www.worldpress.org/specials/pp/pipelines.htm [Retrieved 2 October 2006].

Yang, Y. K. (2006). The Taiwanese Notebook Computer Production Network in China: Implications for Upgrading of the Chinese Electronics Industry. Paul Merage School of Business, University of California.

Yusuf, S. and S. J. Evenett (2002). *Can East Asia Compete? Innovation for Global Markets*. Washington, DC: World Bank.

Chapter 3

The United States, Europe and Developing Asia

The wheel of fortune turns round incessantly, and who can say to himself, "I shall to-day be uppermost."

— Confucius (551–479 BC)

Introduction

Developing Asia is fast becoming a major force in the global economy. The region has nearly 70% of the world's population and more than half of the increase in world population between 2005 and 2025 will come from Asia. Around 25% of global output is produced in developing Asia. It contributes an even larger share of world trade at 39% and this share has more than doubled in the last three decades (IMF, 2006). As the region has grown in size and importance it has begun to play a larger role in global affairs. Many countries in Asia belong to the World Trade Organization and their voices are more forceful in trade negotiations.

The region's relationships with industrialized countries have developed as its economic stature has evolved. Many developing Asian economies can now be classified as belonging to the group of industrial countries. All four of the East Asian NIES (Hong Kong, Korea, Taiwan and Singapore) fall into this category and Malaysia and Thailand have fast approached the lower threshold of per capita income that qualifies countries to be designated as industrial. Poland has the lowest per capita of EU members at $13,300 and Malaysia is just below at $12,100, while Thailand is at $8300 (CIA, 2006; see also Chapter 1) though the World Bank still classifies Malaysia as an upper middle income country and Thailand as a lower middle income country. Relationships between these higher income countries in developing Asia and Europe and North America will have a different orientation as a result of their graduation to industrial country status. Trade and investment issues are the focus of most discussions between higher income developing Asian economies and industrial countries.

For the other countries in the region that fall into the low income or lower middle income categories and which continue to receive development assistance, the relationship is different. The agenda for discussion revolves more around development and policies for the developing Asian economies.

China and India are the exceptions. Because of their size and importance to the global community and to the rest of the region they have a wider array of issues to resolve with Europe and the United States — trade issues with China, and brain drain and intellectual property with India and China.

Economic factors are of primary importance in discussing relationships between industrial countries and developing Asia. However, it is also important to recognize the role of political and cultural influences in the current international environment where terrorism and security have become critically important. The cultural mix within countries has an impact on their interactions with other economies and regions. Environmental considerations are also important in considering relationships between these three groups of countries.

We explore economic development issues in the next section, followed by political and cultural factors and the environment. The way forward looks at each of these issues in turn.

Economic Development Issues

International imbalances

The United States has been running a current account deficit for the last decade or more and the deficit has been on an alarming upward trend since the war on terror began in 2001. To finance this deficit, the US borrows in international markets by issuing government bonds. The Asian economies have been accumulating large foreign exchange reserves and current account surpluses since the end of the Asian financial crisis. These reserves are being used, in part, to finance the US current account deficit. The details of the build up of the US current account deficit, holdings of US assets by foreign central banks and increases in current account surpluses and build up of foreign exchange reserves in Asia are displayed in Tables 3.1 and 3.2.

Observers have said time and again that these imbalances are not sustainable in the long run, particularly given the rapid increase in these imbalances in the past few years.

However, these imbalances can persist as long as the Asian economies are willing to fund the US deficit. There are several alternative to the status quo which would result in a lower level of US current account deficits along with a lower level of Asian surpluses. Aside from temporary import restraints on Chinese products which have been suggested by some, there are three longer term solutions possible.

Table 3.1 Global Current Account Balances of East Asian Economies and the United States (in US billion dollars)

Country	2002	2003	2004	2005	2006
China	35.4	45.9	70.0	76.5	81.3
Hong Kong	12.6	16.2	15.9	16.3	17.0
Indonesia	7.8	7.3	7.3	6.3	2.7
Japan	112.6	136.2	271.8	157.2	173.1
Korea	5.4	12.1	26.8	26.1	23.1
Malaysia	8.0	13.4	15.7	17.4	16.9
Philippines	4.4	3.3	3.9	2.4	2.0
Singapore	15.7	27.0	27.9	27.2	28.7
Thailand	7.0	8.0	7.3	3.5	2.8
Taiwan	25.6	29.3	19.0	22.6	21.9
Others	2.7	2.0	2.1	2.1	2.1
Total	237.3	300.3	367.4	357.4	371.1
Share of US current account deficit	50%	57%	55%	49%	49%
United States	–473.9	–530.7	–665.9	–724.5	–749.8

Note: Others include Brunei, Cambodia and Vietnam.
Source: IMF (2006); Bergsten (2005).

Table 3.2 Foreign Exchange Reserves in US billion dollars

Country/Region	2001	2004	2006	Share of change between 2001 and 2004
Japan	388	824	850	31
China	212	610	988	28
Taiwan	122	242	265	8
Korea	102	198	230	7
Singapore	75	112	135	3
Hong Kong	111	124	131	1
India	45	125	160	6
Malaysia	30	65	79	3
Thailand	32	49	61	1
Philippines	13	13	19	9
Indonesia	27	35	38	1
Latin America	128	189		4
Industrial non Asia	330	358		2
All others	70	171		7
Total	1685	3105		

Source: IMF (2006); Bergsten (2005); Economist Intelligence Unit (2006).

The first alternative is a withdrawal of funds by Asian creditors from the US bond market. This would result in an increase in US interest rates sufficient to attract alternative funds to finance the deficit from other surplus countries but would cause a recession in the US and the world economy. Lower import demand from Asian economies would follow, along with slower growth and smaller current account surpluses.

The second alternative is a rearrangement in exchange rates similar to the realignment of the Plaza Accord that took place in 1985 when the dollar/yen exchange rate was driven down. Yen appreciation resulted in slower export growth, a reduction in the Japanese current account surplus, and slower growth in the Japanese economy.

The third alternative is a reduction in the US current account deficit that could be accomplished by a combination of reduced taxes on exports to China, slower growth to reduce imports, a cut in the government fiscal deficit and other measures that would reduce the saving and investment imbalances in the US by raising the saving rate. (The low US saving rate is due in large part to the negative saving rate of the government.) This would almost certainly lead to slower growth in the US economy through the multiplier effect on consumption and investment which would be offset in the medium term by a reduction in the external imbalance between exports and imports. The risk in this alternative is a hard landing and recession, both in the US and globally.

Options one and three run the risk of triggering a recession in the US and the global economy. This is not a prospect that endears either alternative to policy makers. This leaves alternative two, a path that has been followed vigorously by the US government with little success so far. Before the modest yuan appreciation in 2004 the Chinese currency had depreciated by 10 between 2001 and 2004 by following the weakening US dollar and thereby further strengthening its competitive position. Economists including C. Fred Bergsten of the Institute of International Economics in Washington, DC along with many other economists feel that a further currency strengthening of 20% to 25% is needed to bring about an appropriate adjustment. This would avoid further erosion in the US current account deficit and also protectionist reaction in the US and possibly Europe. Looking back at history, a concerted effort to talk "up" the renminbi reminds one of the Plaza Accord agreement which resulted in an eventual 50% appreciation of the Japanese yen. We all know the resulting impact on the Japanese economy.

Are the Chinese willing to swallow the same medicine? More appropriately, should they be forced to? Chinese officials are concerned that a significant currency revaluation would hurt exports and contribute to further job losses and economic instability. After all Japan was a mature economy that had been able to grow rapidly for several decades with an exchange rate fixed at 360 yen per dollar. China is still a poor economy with 20% of the population (close to the population of Western Europe) in poverty. Nevertheless, the United States has become increasingly insistent that China revalue its currency in light of continuing high bilateral trade deficits with China of around $200 billion per year which have alarmed a wide range of business interests, labor unions and politicians. A bill now in the US Congress (September 2006) would raise tariffs on Chinese goods of nearly 30%. Although the bill is not expected to pass the House of Representatives it would probably be vetoed by President Bush. Nevertheless, a measure of this type has not had much support until recently and reflects growing dissatisfaction with exchange rate negotiations between China and the United States. President Bush is sending Treasury Secretary Henry Paulson to China as part of a team to engage in a strategic dialogue on this and other economic matters that are crucial to better US/Chinese relations.

In summary, options one and three may drive the world economy into recession while option two could do significant harm to the Chinese economy. Maintaining the status quo risks a severe crisis later on and imposes large opportunity costs of keeping excessive foreign exchange reserves.

Is there any other solution, or combination of solutions, that is preferred? One approach would be to do a bit of everything: (i) The US could impose export taxes on many goods currently exported to China and also prohibit exports of some strategic materials. These regulations could be reworked to facilitate more US exports to China. The repeal of the Jackson–Vanik amendment that requires annual renewal of MFN for China should be part of this initiative; (ii) The US financial hemorrhage began when the Bush administration came into office and began lowering taxes without cutting spending. Can the Congress grab the spending reins and work to restore the budget surplus the current administration inherited in 1990? Growth momentum may be slowed somewhat, so these budget adjustments have to be made carefully and the Federal Reserve has to be ready to provide sufficient monetary stimulus to achieve a soft landing. (iii) Some further appreciation of the reminbi is possible without having a large negative impact on Chinese exports. Furthermore, a needed shift from export and investment led industrial growth to consumption and services led growth can be facilitated by such an exchange rate adjustment. For more discussion on this point see Chapter 7 on India and China.

Reduction in the US current account and saving/investment imbalances would further reduce trade tensions with other surplus countries in Asia, including Japan, Taiwan and Korea thus creating more room for them to develop an agenda for using their reserves in a more constructive fashion after providing sufficient cover for risk.

Trade issues

There are several areas of negotiation in the WTO Doha negotiations that involve the Asian economies, industrial countries in Europe and the United States. These issues include tariffs and subsidies for agriculture products; liberalization of services; issues of intellectual property and trade facilitation; and global and regional environmental issues. These discussions have been put on hold since negotiations were suspended in June 2006. The main sticking points have been agricultural subsidies by several rich countries including the US, France, other countries in Europe, and Asia, and the unwillingness of developing countries to lower tariffs and grant more access for trade in manufactured goods and services.

Most of the agricultural subsidies in industrial countries go to richer farmers and these subsidies are generally thought to be inefficient and highly distorting. The OECD estimates that public support for farmers in OECD countries costs a family of four nearly $1000 a year in higher prices and taxes, and from $1 in price support, only $0.25 winds up in the farmer's pocket. In the US the biggest and richest 25% of farmers received 90% of all support provided. In Europe it is 70% of all support provided. In the United States over 50% of farm subsidies are concentrated in only a few of the over 400 congressional districts. Tens of thousands of small farmers get no benefits from farm subsidies. Nearly all OECD countries apply tariffs of over 100% on some agricultural products. The OECD estimates that cutting all agricultural tariffs and subsidies by 50% would result in an increase of world income of $26 billion (OECD, 2006; *Economist*, 2006).

Aside from agriculture, OECD has made some estimates of loss to the global community if WTO members fail to reach a trade agreement and of the gains that could be made if agreement is reached. If no agreement is reached the global economy will suffer although the poorest will probably suffer the most. Full tariff liberalization for industrial and agricultural goods would bring increased international trade and economic activity worth an estimated $100 billion. The benefits from liberalizing trade in services are estimated to be nearly $500 billion. Trade facilitation could save another $100 billion by removing procedural barriers. OECD estimates that developing countries would gain more than developed countries by as much as two-thirds of the total gain (see Gurría, 2006 and OECD, 2006). Estimates from resolving environmental issues including global warming are over a longer time horizon and more difficult to measure. They could be substantial if Kyoto protocol or other global solutions are not implemented. The loss in income generated by a wide ranging trade agreement of $700 billion is a little less than 2% of global GDP of $44 trillion measured in exchange rate terms and a little less than the GDP of India. It is a significant sum.

Failure of the Doha round would further weaken the prospects for multilateral trade negotiations and strengthen the hand of proponents of regional trade agreements and bilateral preferential trade agreements (PTAs). These issues were discussed at more length in Chapter 2. It is likely that more PTAs will be signed and the possibilities of greater inefficiencies arising as more rules of origin and product standards are developed in conjunction with these agreements. There are ongoing meetings to explore ways to salvage a global trade treaty involving a group of 21 developing nations called the G-20 and also the Cairns Group of agricultural exporters which includes Australia, Canada, New Zealand and 16 developing countries. Further discussion on salvaging Doha is presented in the outlook section of this chapter.

Immigration

Migration has become a headline item in recent years, both because of the migrant flow from Eastern Europe into Western Europe and from Mexico to the United States and because of the fight against terrorism that began following the bombing of the World Trade Center in New York City in 2001. General issues of labor migration within Asia were discussed in Chapter 2. In this section we discuss issues of migration between Asian countries and Europe and the United States. Permanent or long-term migration of skilled and professional workers is the most pressing issue facing the United States and European countries. OECD estimates that the foreign labor force increased dramatically in the last couple of decades (see Table 3.3). Some of these workers in Europe are from other European countries; others are from Asia and other developing regions. The United States has had a deficit of domestically trained doctors and nurses and other health care professionals for many years. Immigrants from Asia have helped to make up the shortfall and the numbers are growing. This migration is beneficial for the migrants and there is no shortage of potential migrants from Asia.

Most of the migrants to the United States and Europe from Asia have either a professional or tertiary degree. There are very few without

Table 3.3 Migration and Migration Rates to the United States and Europe 1990 from Asia as Percent of Workforce

Country	Migration rates (in %) to the United States		Migration rates (in %) to OECD		Migration to US as percentage of total OECD	Migration to US as thousands
	secondary	tertiary	secondary	tertiary		
Bangladesh	0.0	0.6	0.2	2.5	25.9	12.4
China	0.1	1.4	0.2	3.0	51.5	404.6
India	0.1	1.1	0.3	2.6	44.1	304.0
Indonesia	0.1	1.4	0.1	1.5	90.5	32.2
Korea	1.2	5.9	3.3	16.2	36.0	377.9
Malaysia	0.2	5.0	1.2	26.1	18.1	15.3
Pakistan	0.2	2.4	0.5	7.2	35.2	52.7
Philippines	4.5	6.8	6.2	9.5	71.6	728.4
Sri Lanka	0.1	3.8	0.5	27.3	14.1	8.7
Taiwan	0.8	8.8	0.8	8.8	100	153.0
Thailand	1.5	1.2	1,7	1.5	87.6	53.1

Note: Migration rates are rounded and midpoint estimates of the range in the original table.

Source: Carrington and Detragiache (1999).

such qualifications, primarily because of immigration requirements that screen out the unqualified. In the United States, the bulk of the immigrant population are in the health professions, business, information technology and engineering. The bulk of total migration from Asia, including irregular migration, is to other Asian countries or to the Middle East. Only 7% of contract workers from the Philippines went to North America (Wickramasekara, 2002a). Brain drain or brain gain is the major migration issue facing industrial countries as they negotiate with their Asian counterparts. Other immigration issues such as child labor, working conditions, abuse, extortion, and labor contracting agencies were discussed in Chapter 2.

Richer receiving countries like the US that attract skilled and professional workers have been willing to let markets work to pick and choose which potential migrants are qualified after certain requirements are fulfilled. A similar system is used in Europe although migrants from Eastern Europe may be subject to fewer restrictions. Many skilled and professional migrants started out as students and remained after their studies to work. Others were trained in their home country and then underwent additional training or testing once they were admitted into the country. Sending countries benefit from out migration of professionals to the extent that they remit part of their incomes to the home countries. The sending country can also benefit in the intermediate to long run if the migrant eventually returns home to work, carrying with him the skill and contacts acquired while overseas. The receiving country is definitely a net gainer, since some if not all of the skill and training have been paid for elsewhere, either by the migrant or the government of the sending country. Often skilled migrants are discriminated against and paid at a lower rate than their domestic counterparts.

The size and time dynamics of skilled migration and the respective gains and losses to host and sending countries are not well understood nor is there a lot of data available for further study. Carrington and Detragiache (1999) have assembled estimates of migration rates by educational category. In all Asian countries migration rates to the United States are much higher for tertiary graduates than secondary school graduates. Immigration rates from those with only a primary education are negligible. In Taiwan, for example, migration rates for tertiary graduates in 1990 were in double digits for Korea, Malaysia and Sri Lanka in 1990 and just under 10% for the Philippines and Taiwan. Only in Korea and the Philippines were secondary graduates' migration rate above 2% (see Table 3.3).

The migration rates for tertiary graduates from Korea, Malaysia, Sri Lanka, and to a lesser extent, from Philippines and Taiwan are quite high, given the limited access to higher education in many Asian economies. The numbers of immigrants are also relatively large, given the small cohort of tertiary graduates in the total workforce.

How important is this brain drain for the sending countries' economies? In the growth models of Romer (1986), Lucas (1988), Mankiw *et al.* (1992), and others, human capital is critical to development. We know from the experience of Japan, Korea, and the other NIEs that human capita played a critical role in their development. The importance of human capital is also evident from the discussion in Chapter 4. So it is important not to minimize the damage from this loss in human talent.

So we ask the question, how can the loss in human capital in the sending country be reduced? Certainly remittances will help reduce the pain of the loss and efforts should be made to encourage and expedite the transfer of remittance as well as introducing ways of investing these funds for the well being of the individual recipients and the country as a whole. Some of these suggestions were discussed in the migration section of Chapter 2 as they relate to temporary migration. The same arguments apply for remittance income of professional and other tertiary graduates.

A more fruitful approach that goes to the root of the problem would be to encourage skilled and professional migrants to return home after a few years of work experience overseas, the so-called circular migration. Wickramasekara (2002b) has explored a number of ways in which host and sender countries can encourage return migration. Some of them are displayed in Table 3.4.

In most of the countries with high sending rates of tertiary graduates, many of these programs are already in place. There are lots of domestic educational and work opportunities for skilled and professional

Table 3.4 Ways to Encourage Return of Skilled and Professional Migrants

Sending countries	Receiving countries
Provide incentives to return home	Visa regime that encourages return home
Promote linkages with national abroad	Reduce student fees especially from major source countries
Promote short-term migration under GATS Mode 4	Encourage temporary migration rather than permanent migration
Provide more local opportunities by faster growth and an expanding labor market	Provide more training and education for local workforce
Give dual citizenship	Be aware of impact that skilled and professional migration has on sending countries
Create centers of excellence at home	
Develop information network of job opportunities in home country	
Develop domestic R&D and centers of excellence	
Develop incentives to attract FDI	

Source: Wickramasekara (2002b).

personnel in Taiwan, Korea and Malaysia, although less so in Sri Lanka and Philippines. Countries with low rates of skilled migration including India and Thailand have also invested heavily in tertiary education. There has been a recent boom in the IT industry in India and the rate of skilled migration to the United States may have also increased as a result. Similar migration of skilled Chinese IT professionals has also intensified. These outcomes and recent developments suggest that there are other factors at work encouraging high migration rates in these countries. In Malaysia there have been policies in place to encourage the educational participation of the ethnic Malay populations. These policies have introduced some incentives for ethnic Chinese to migrate. Perhaps the timing of these surveys in the late 1980s and early 1990s may have some bearing on the very high out migration in Malaysia at that time. It would be interesting to follow up the current migration rates of skilled and professionals.

Historically receiving countries have had little interest in the impact of skilled migration on the economies of sending countries. In the Philippines, for example, increased migration of nurses and doctors has created local health care shortages in Manila and the rest of the country (Tujan, 2002). Preventative health care in many of the poorer provinces has deteriorated and morbidity and mortality have increased in some areas. There are complex factors at work in this particular case and it is difficult to find a satisfactory solution. Remittance income from these health care professionals and other migrants is supplying needed foreign exchange and lifting living standards in the Philippines. At the same time the United States has not increased the number of domestically trained nurses and doctors to keep up with demand, preferring rather to attract health personnel from developing countries like the Philippines and India to such an extent that around 25% of all medical doctors practicing in the United States were trained elsewhere.

This brain drain in the health professions can be arrested if more resources were devoted to the training of US citizens. Other options are also available including reduced recruitment in vulnerable countries. The United Kingdom now has strict guidelines which state that developing countries should not be targeted for the recruitment of health care professionals. The Commonwealth Secretariat has adopted similar guidelines (Wickramasekara, 2002a).

In education, students from developing countries are often charged high fees to study in OECD countries. These fees are usually paid by sending country governments. Then many of the best of these foreign students are employed in the host country following graduation. This constitutes a double drain on the sending country and a free gain to the host country.

In a theoretical paper, Beine *et al.* (2003) concluded that only a small number of developing countries would benefit from an increase in the brain drain. These are, however, the largest countries in the sample including China, India, Indonesia and Pakistan from Asia, and Brazil from Latin America. The logic is that these very large countries can easily replace a small proportion of the skilled and professional workforce. There is also a bigger chance that migrants will eventually return home to start up their own companies or work or a foreign company that is investing in the home country. There are many examples of these returning migrants who have become rich in India and China in recent years. For smaller countries with more limited employment opportunities the brain drain harms the local economy more, particularly if growth is not robust and the migration creates shortages as in the case of the Philippines.

Genetically engineered crops

Genetically modified crops (GMC) can help raise yields, reduce the use of pesticides and herbicides, and increase the amounts of vitamins and minerals in foods. GMCs can also help to reduce disease. Cotton injected with a synthetic gene to protect against pests has raised yields and reduced the use of insecticides in China, Pakistan and the United States and the increased use of herbicide-tolerant soybeans has increased yields and reduced the use of herbicides (Fernandez-Cornejo *et al.*, 2000, summary). Rice rich in vitamin A, which helps to reduce the incidence of blindness, is being developed by IRRI in the Philippines.

In the United States, many GMCs have been developed, planted, and tested in the past decade. The impact of these crops on pesticide use, yields, and net returns have been studied as well as the impact on consumer's health and spending habits. A list of genetically modified crops being cultivated and sold in the United States has been prepared by the US Department of Agriculture. The list includes canola, corn, cotton, flax, papaya, potatoes, soybeans, squash, sugar beets, and tomatoes. Many of these GMCs are being widely used in the production of processed foods in the United States.

There is widespread mistrust of genetically modified crops in Europe (see for example Gaskell *et al.*, 1999) where they have been banned by the EU since 1998. There are three basic objections to the use of GMCs. The first is that genetically modified crops do not provide significant increases in yield or reduction in pesticide use. These allegations are being debated and further research is being carried out around the globe. Evidence does seem to suggest that farmers would not continue to use GMC unless they were increasing yields or otherwise reducing costs per unit output. The second argument is that there may be long-term negative impacts from using GMCs. Unlike traditional cross breeding that led to the green revolution in rice and wheat, GMCs are not developed from a natural process. So far there is no scientific evidence of these negative impacts. The third objection is that these new genetically modified seeds are patented and the big companies are benefiting from these developments, not the consumer. For example, genetically engineered rice is patented and seeds cannot be saved but rather bought each year. Furthermore, there are many patent holders involved and this could raise the price of genetically engineered rice and other products. This objection has not been systematically investigated. However, cultivation of GMCs has continued to increase, suggesting that consumers prefer them to traditional varieties. Since there is not yet any clear evidence that they are harmful to health, many observers suggest that the market should decide on the use of GMCs and GMCs are on the agenda for discussion in the latest round of WTO trade negotiations. However, the European Union has been pushing for strict controls over the importation of GM foods.

In Asia GMCs have been widely adopted in some countries but not in others. Genetically modified cotton is being grown in China and Indonesia with mixed results. Other GMCs are also being tried out. Generally richer countries in Asia are more reluctant to support GMCs. Korea and Japan are more skeptical and less willing to promote and encourage them than China and countries in South and Southeast Asia. Farmers are also more conservative and entrenched in these richer countries as they are in Europe (FAO, 2004).

Aside from the health, yield and environmental benefits of GMCs in Asia there are financial considerations. As the use of genetically engineered crops has spread throughout India a disturbing trend toward higher debt has emerged as farmers have taken loans from moneylenders to buy genetically modified seeds. While bringing hope of higher yields, higher indebtedness has also brought with it higher risks of default and bankruptcy. *The New York Times* (Sengupta, 19 September 2006) reports that one shrewd moneylender compels farmers to sell their cotton to him (at lower than market prices) in exchange for access to credit at high rates of interest and most farmers have no other credit alternatives. Keeping this in mind it is important for developing countries to negotiate lower seed prices in exchange for market access for genetically modified seeds.

Lines have been drawn between opponents and supporters of GMCs. In Thailand there have been protests against genetically engineered rice and also against the FAO, which has promoted its use. Many environmental organizations including "Greenpeace" and "Save the Earth" are opposed to the use of GMCs. Nevertheless the use of GMCs is increasing in developing Asia and the use of GMCs has the support of the United States and the FAO. Further research and greater cooperation between industrial countries and developing Asian economies would be fruitful in exploring the scientific, health, and economic impacts of GMCs.

Health

There are many ways in which industrialized countries can cooperate with Asian governments in providing health services. In this section we concentrate on the provision of antiretroviral therapy (ART). ART consists of the use of at least three antiretroviral (ARV) drugs to maximally suppress the HIV virus and stop the progression of HIV disease. These drugs are produced and marketed by western drug firms and this is where cooperation between the United States, Europe and Asia comes into play.

Huge reductions in death rates and suffering are possible with the use of a potent ARV regimen and ART in developed countries. As a result, AIDS-related mortality in Europe and the United States has dropped by more than 70%. Effective use of ART caused a dramatic drop in mortality and in the number of opportunistic infections. As a result HIV has become, for many people, a chronic infection managed on an outpatient basis (Chirac, 2002, Chapter 14).

About 38.6 million people are now living with HIV, of whom roughly 37 million live in low- and middle-income countries. WHO estimates that at least 6.5 million of these people are in need of ART. While WHO is providing countries with ongoing guidance, tools and

Table 3.5 Rates of ART by Developing Regions (%)

Subsahara Africa	17
Latin America and Caribbean	68
East, South and Southeast Asia	16
Europe and Central Asia	13
North Africa and Middle East	6

Source: WHO (2006).

support in delivering and scaling up ART within a public health approach (WHO, 2006), only 1.65 million people had access to ART in low- and middle-income countries, a fraction of those with HIV/AIDS (see Table 3.5).

UNITAD is the international drug purchase facility established by Brazil, France, Chile, Norway and the United Kingdom. It is a funding mechanism designed to accelerate the access to drugs and diagnostics for AIDS, Malaria and Tuberculosis. A trust fund was established in 2006 to support its efforts.

Global efforts to address HIV/AIDS have advanced in recent years. Greater international political commitment has been accompanied by increased financial resources through the Global Fund to Fight AIDS, TB and Malaria, the United States President's Emergency Plan For AIDS Relief (PEPFAR), continued funding through World Bank loan and grant instruments, increasing investments by bilateral donors and contributions from private foundations. For example, since 2001, the Global Fund has attracted US $4.7 billion in financing through 2008. In its first two rounds of grant-making, it has committed US $1.5 billion in funding to support 154 programs in 93 countries worldwide (see website http://www.theglobalfund.org/en/about/how/).

On the recommendations of the Global Task Team on Improving AIDS Coordination among Multilateral Institutions and International Donors, the international community is making more concerted efforts to improve coordination and cooperation among the United Nations partners, donors and governments so that the best use is made of these resources in countries (WHO, 2003).

How effective are these programs? The estimated number of HIV/AIDS cases has dropped in some countries and increased in others. These estimates are only rough approximations of the number of cases compiled by WHO and national authorities. Comparisons of 1999 and 2005 figures suggest that some progress is being made in several countries including Cambodia, Philippines and Thailand. In China, WHO suggests the epidemic is getting worse although data from 1999 suggest it may be moderating. There has been less progress in other countries, particularly India, where the estimated number of cases has increased by over 40% in 6 years and in Indonesia by 70% since 1999 (see Table 3.6).

Dealing with AIDs requires a multifaceted approach, including prevention, treatment and care. Greater focus on implementation is required as well as the strengthening of human and institutional resources. Bringing AIDS under control will require tackling with greater resolve the underlying factors that fuel the AIDS epidemic

Table 3.6 Asian HIV/Aids Cases (in thousands) — 1999 and 2005

Country	1999	2005
Bangladesh	4	12
Cambodia	220	130
China	1000	840
India	4000	5700
Indonesia	100	170
LaoPDR	1.4	3.7
Malaysia	49	69
Nepal	34	75
Pakistan	74	85
Philippines	28	12
Republic of Korea	3.8	13
Singapore	4	5.5
Thailand	775	580
Vietnam	100	260

Note: 2005 estimates are midpoints of an estimated range of cases of HIV and AIDS.
Source: ADB (various issues); WHO website.

including societal inequality and injustice and the lack of access that accompanies these inequities and injustices.

Despite some progress, the vast majority of those suffering from HIV/AIDS still do not have access to it. Where ART is widely available, the incidence of opportunistic infections and associated morbidity and mortality are greatly reduced. Even where ARV treatment is available, treatment of other infections such as pneumonia and tuberculosis is also in short supply and more attention needs to be paid to joint planning and collaboration between TB and HIV programs and integration of TB and HIV services.

Countries in Latin America already have extensive ARV programs and high rates of treatment (Table 3.6). Other countries planning to extend their programs are currently in the process of negotiations, under the umbrella of the Accelerating Access Initiative (AAI) coordinated by UNAIDS and WHO in partnership with pharmaceutical companies (Chirac, 2002). The AAI is a cooperative endeavor of UNAIDS, the WHO, UNICEF, the UN Population Fund, the World Bank, and seven research-based pharmaceutical companies (Abbott Laboratories, Boehringer Ingelheim, Bristol-Myers Squibb, GlaxoSmithKline, Gilead Sciences, Merck & Co., Inc., and F. Hoffmann–La Roche).

Pharmaceutical companies in the AAI program work with governments and international organizations to find ways to broaden access to ARTs while ensuring rational, affordable, safe, and effective use of drugs for HIV/AIDS-related illnesses. The drug companies involved have offered to improve access and availability to a range of medicines by providing more affordable prices in developing countries.

These efforts are bearing some fruit. More than 80 countries plan to implement HIV treatment programs and wish to collaborate with the AAI. Of these countries, 49 already have national plans in place and have reached agreement on prices with the individual companies concerned. Treatment rates have increased in the past few years including an eightfold increase in Africa between 2000 and 2005 (information on AAI taken from International Federation of Pharmaceutical Manufacturers (IFPMA) website at http://www.ifpma.org/index.aspx).

Yet a lot more must be done to raise the rest of the world to the Latin American standard of compliance. Latin America has been able to achieve high rates of treatment in a short period of time. Asia should be able to match this rate of success, even with higher rates of infection. We will explore alternatives in the way forward section.

Income convergence

There have been many studies of income convergence, within countries and between countries. The theoretical arguments for convergence go back to the work of Robert Solow (1956), even though technology in his model is exogenous. With endogenous technological change, countries or regions with low levels of capital endowment per capita will grow faster than countries or regions with high levels of capital endowment because of diminishing returns to capital. Simple tests of this proposition have been developed and tested by Barro and others. Evidence suggests that in countries or regions that are "similar enough" in terms of initial conditions there is a tendency for convergence to occur. Convergence, albeit slowly, is observed in the states of America, the provinces of Germany, and the prefectures of Japan (Barro and Sala-i-Martin, 1995).

The members of the EU exhibit a tendency toward income convergence as does an extended group of countries composed of the OECD countries and the Asian NIEs plus China. For the rest of the globe there is divergence resulting in a gap between the richer and poorer countries that continues to grow.

We have a particular interest in a subgroup of these poor countries, namely the group of South Asian nations. What has kept these countries from growing rapidly and what can be done to accelerate their rates of growth? Much has been written about the causes of this slow growth and we do not intend to retrace these developments. Many observers have focused on the limitations of government policy and the regulatory regime and some of these issues are discussed in other chapters of this book. In this section we highlight the contribution of human capital and the role of agglomerations of workers in concentrated work environments. Typically we think of cities as the focus of work environments. Recently, special economic zones such as the Silicon Valley in California and special zones in India, China and other Asian countries have been developed that focus industrial activities in a region that may contain some small cities, but not necessarily qualifying as an urban agglomeration. These special zones, many of which focus on exports, were discussed further in Chapter 2.

In the growth models of Romer (1986), Mankiw *et al.* (1992), Lucas (1988), and others, these agglomerations of human capital and the externalities of human capital formation play a critical role in growth. Of course the usual contributions of labor and capital are also important. It is likely that these educational externalities have been slow to develop in South Asia, either because of an unfavorable regulatory environment that retarded the growth of innovation and opportunities for the build up of human capital skills or because of fundamental deficiencies in the educational systems in these countries. In the chapter on human capital formation we note the extreme differences in government spending per capita on primary, secondary and tertiary education in South Asia compared with East Asia. East Asia is more egalitarian. Conversely, South Asia spends a lot on a relatively small cadre of students trained at the tertiary level. This educational elitism has been a contributing factor to slower growth in South Asia until recently.

To redress these historical inequities South Asia must devote more resources to primary, secondary and technical education. In India particular emphasis should be put on the four or five provinces where most of the poor live and particular poverty pockets identified in the other countries in the region. What can industrial countries do to assist in this effort? Development assistance is needed to provide demonstration schools that will show governments how to improve the learning, provide technological support by way of computers, internet, and general telecommunications upgrading, assistance in upgrading classrooms, and building new classrooms. Emphasis in aid needs to be shifted away from physical infrastructure projects toward education and health. Northern European and Nordic countries have demonstrated that such an emphasis does have a beneficial impact on growth as discussed in Chapter 4.

Political Issues

North America, Europe and Asia are the most powerful economic and political regions in the global community. Their ability to forge a political consensus that brings peace and prosperity to the global community is paramount. Without their leadership and with the support of the United Nations, the political landscape could easily deteriorate into a series of squabbles between regional foes. There are many international bilateral diplomatic challenges ongoing in Asia. These include the continuing dispute between India and Pakistan over Kashmir, the development of nuclear weapons by North Korea and Iran, and the China confrontation with Taiwan. There are also economic issues of trade involving all three regions that resulted in the breakdown of the Doha round of trade negotiations.

In addition there are subtle issues about the growing economic and political power and influence of China and India. Leaving the Doha round of WTO trade negotiations aside, this section will focus primarily on (i) the several bilateral disputes that have arisen within Asia and the role that can be played by outside observers in the region and in Europe and North America; (ii) the balance of political and economic power as China and India become more influential on the global stage.

China and the West

China has emerged as a growing force on the global development stage. Its share of world income and trade has grown spectacularly. It now competes for raw materials with industrialized countries throughout the world. It has become a member of the WTO, joined with ASEAN, Japan and Korea to form an extended regional trade group (ASEAN + 3) and has entered into a number of bilateral trade agreements with other countries within and outside the Asian region (see Chapter 2). It has joined WTO and seems to be complying with the requirements for membership. China has developed a more engaging diplomatic style in the past decade. It signed a code of conduct with ASEAN in 2002 that should help to manage offshore territorial disputes over the Spratley, Senkaku and Paracel Islands. It has reached agreements with Russia, India, and the states of Central Asia that suggest its borders are secure and unlikely to be subject to further dispute and are more secure than ever. China has also initiated talks with ASEAN to develop greater mutual security ties. China has also begun to play a more significant role in UN activities. Although it still refused to censure Iran for plans to build nuclear weapons, it has increased its participation in peace keeping operations, for example in East Timor, and for weapons inspections in Iraq in 2002. It has also ratified arms controls agreements and signed the nuclear test ban treaty in 1996 (Medeiros and Fravel, 2003).

This global outreach has been associated with the maturation of the Chinese diplomatic corp and additional confidence in the role of multilateral institutions in maintaining regional and global stability. In Central Asia, for example, China led the establishment of the Shanghai Cooperation Organization to settle territorial disputes and to address the problem of terrorism and regional trade. It is true that China has a mercantilist interest in central Asia, which has vast reserves of oil and natural gas. Yet it seems to be willing to pave the way for trade by following a political ambit.

China has also developed a program of development assistance for poorer countries in the Asian region (Periez, 2006). China has been making loans to Cambodia, Laos, Myanmar and the Philippines for infrastructure investment in roads, ports and bridges at attractive terms with no strings attached. Although these economies still borrow heavily from development institutions like the Asian Development Bank (ADB) and the World Bank, loans from China are taken to supplement funds from international donors. These initiatives echo the suggestions made in Chapter 2 that surplus countries should undertake more lending programs for less fortunate countries in the region.

In summary, China has begun to show a more moderate face in recent diplomatic maneuvers. It is beginning to realize that its economic fortunes can be further enhanced by pursuing regional cooperation and diplomatic rapprochement rather than a bellicose confrontational stance. Taiwan issues are still there and China remains an authoritarian state run by a cadre of party members and government technocrats even though it is far less centralized and less dependent on individual personality. Decisions and policy options are often discussed after input from scholars and policy analysts, many of whom have extensive experience in the West. Issues like internet access and the continual incursion of western political, social and economic ideas have created some backlash. Yet, the entire political, social and economic system seems to be more flexible and less doctrinaire. There is also a realization that the United States is still the predominant world hegemon and that it would be folly to challenge that global leadership. This is not to say that China has weakened its resolve. As one observer said, "Today China is certainly smarter and more sophisticated — but not necessarily kinder or gentler" (Medeiros and Fravel, 2003, p. 7). As China matures economically, politically and diplomatically, it is incumbent upon the United States and Europe to ensure that China's emergence as a world leader is consistent with global stability and security.

India and Pakistan

Animosity between India and Pakistan runs deep. They have fought three times since separation in 1947. The success of SAARC is threatened by these differences, and trade between the two countries is a fraction of what would be expected by two large countries sharing a long common border. India has helped promote an alternative to SAARC that excludes Pakistan. It is called the Bay of Bengal Initiative for Multisectoral Technical and Economic Cooperation (BIMSTEC). BIMSTEC adds Thailand and Myanmar to the members of SAARC while dropping Pakistan. BIMSTEC and SAARC were discussed in further detail in Chapter 2. Outside support for each of the protagonists has not helped them resolve their differences. China is supporting Pakistan, and Russia and India are long time allies. The United States has tilted toward India in recent months by supporting its nuclear program while withholding support for the Pakistan program.

The issue of Kashmir has been a bone of contention for many years and has contributed to the poor relations between the two countries.

India controls two-thirds of the territory in this disputed region and has been reluctant to entertain any changes in the status quo which divides the territory by a tacitly agreed line of control which acts as a border. There have been periodic attacks in the India-controlled section of Kashmir by Pakistan-based militants that have contributed to tension and which have served to destabilize any efforts for a peaceful settlement. The events following the attack on the World Trade Center in New York City on 11 September 2001 have complicated the political landscape. Partly in response to these changes, Pakistan's President Musharraf offered a series of diplomatic initiatives in early 2004 which were designed to normalize Kashmir and which were backed by his promise to reduce infiltration of Pakistani supported militants across the Line of Control (the informal border that separates Kashmiri Pakistan from Kashmiri India). These initiatives included a plan to demilitarize all or parts of Kashmir, and dividing Kashmir into seven geographical regions — two now under Pakistani control and five under Indian control — with joint governance or governed according to a United Nations mandate. Indian reaction to these proposals has been muted, despite the fall in border incidents following Musharraf's pledge to reduce support for the militants (Lavoy, 2006 and Behera, 2002).

The role of the United States, Europe, and other Asian countries has been limited aside from diplomatic and military support of the two countries by their allies. The United States, which has provided military and humanitarian aid to both countries, moved closer to Pakistan as it has battled Islamic terrorists in Afghanistan and along the Pakistani border. Several terrorists have been captured and killed by Pakistan military forces. In a recent visit to both countries President Bush did not explicitly discuss Kashmir, arguing that it was a bilateral issue that had to be resolved by India and Pakistan themselves. The chances for further progress to resolve the Kashmir issue are discussed in the section on future prospects.

China and Taiwan

Economic relations between the economies of Taiwan and China have improved over the years, particularly as China has opened up to the rest of the world and its industrial sector has grown rapidly. Taiwan and China both belong to the WTO, and trade and investments have increased as a result. Although official trade and investments by Taiwan in China is banned, third-country trade through Hong Kong has increased rapidly and there is also smuggling of an undetermined amount. Estimates of total trade range from 50 billion to 80 billion, although official statistics record trade through Hong Kong at less than half that amount. Investments have been estimated at over $100 billion in the past decade. Whatever the political climate, economic ties have continued to grow, sometimes sporadically, as part of the greater China area comprising Hong Kong, Taiwan and China.

Politically, China continues to hold the position that Taiwan is part of China, while Taiwan maintains its independence. China has been able to isolate Taiwan diplomatically. The ADB is the only international organization to which Taiwan belongs outside the World Trading Organization where it has membership under a different name. It only has a free trade agreement with Panama and limited diplomatic coverage worldwide. Nevertheless, Taiwan is a viable democracy and has the support of the United States as an independent entity.

This uneasy relationship has persisted for several decades, despite periodic threats and counter threats and military preparedness in both countries. The United States is in an unenviable position of backing both countries. This has created a policy that Nancy Tucker calls "strategic ambiguity" where the United States convinces Beijing that it will defend Taiwan if China attacks while convincing Taiwan that the US will not defend it if it provokes a Chinese attack (Tucker, 2005). Periodically Taiwanese politicians have moved to solidify their independence from China by holding a plebiscite and drafting a new constitution. China is adamant about leaving the door open for reconciliation.

There seems to be little to be gained by changing the status quo at this time. Both countries are thriving, living standards are rising, and trade is increasing. Growth in the Taiwan economy is slowing as it matures and its labor force ages and population growth diminishes. It is already a mature economy and growth of 4% seems to be acceptable. Aside from hubris, China also has nothing to gain by rocking the boat. Its economy is growing and it is becoming more active in international organizations and in the global community. It has begun to replace a blunt and sometimes belligerent foreign policy with a smoother and more sophisticated approach. A confrontation with Taiwan would betray that image unless it was provoked by overt policy changes that encourage independence (Medeiros and Fravel, 2003). Statements such as the assertion by Taiwan's President Chen Shui-Bian in a meeting at the end of December 2003 that Taiwan could make a unilateral decision to change the status quo found a quick rebuke by President Bush. The United States does not recognize Taiwan as an "independent entity". For the time being, the one China policy of the mainland and the self-determination of Taiwan continue to exist side by side — A possible diplomatic cul de sac, yet one which has not inhibited growth in trade and also one which has contributed to increased standards of living in both countries. As the expression goes, "if it ain't broken, don't fix it".

Environmental Issues

We have discussed environmental issues for Asia both in a separate chapter on the environment (Chapter 6) and also in the chapter dealing with China and India (Chapter 7). In this section we explore environmental issues from a more global perspective through the prism of Asia and the industrial countries. The focus is on the interrelationships between these three regions and the environmental issues they face, both individually and collectively. To introduce the subject, consider two measures of environmental load. The first is energy use per capita, which translates all energy use in a country into a common measure of kilograms of energy equivalent use per capita. These figures have been assembled by the World Bank and are displayed in Table 3.7 for selected countries and regions.

The second is the ecological footprint, which is displayed in Table 3.8 and measures how much land and water area a human population requires to produce the resources it consumes and to dispose or absorb its wastes. By considering the relationship between consumption and waste absorption we get an estimate of the sustainability

Table 3.7 Energy Use in Terms of Kilograms of Oil Equivalent Use per Capita for Selected Regions and Economies — 1990 and 2002

Country or region	1990	2002	Annual growth (%)	Per capita income in 2002 PPP
East Asia and Pacific	721	904	1.9	
South Asia	393	468	1.5	
High incomes	4859	5395	1.0	
Europe	3583	3895	0.8	
USA	7722	7943	0.4	41,800
India	430	513	1.6	3300
China	775	960	1.7	6800
Singapore	4384	6078	1.7	28,100
Korea	2161	4272	5.7	20,400
Japan	3610	4058	1.1	
Germany	4485	4198	–0.3	
UK	3686	3824	0.4	
Low income	463	493	0.6	
Middle income	1372	1338	–0.2	

Source: World Bank (2006), and Tables 1.1 and 1.2 from Chapter 1.

of the ecosystem. The ecological footprint is a more general concept than energy use. Both energy use and the ecological footprint are related to levels of income. Richer countries use far more resources than poor countries. However, these two measures also tell something different about resource use. Together they give a good indication of the relative resource use and waste production. For energy use we have estimates more than a decade apart. For the ecological footprint we have a point estimate. For individual countries there are some time series estimates of the ecological footprint that we will also refer to in this discussion.

The information derived from these tables and other historical evidence regarding the global environment and ecological impact can be summarized in a few major points.

(i) The global environmental footprint is larger than the regenerative power of the planet. This is evident from the large number of

Table 3.8 Ecological Footprint, Population and Biocapacity, 2005, for Selected Countries

2002 data	Population (millions)	Total ecological footprint (global ha/ person)	Biocapacity (global ha/ person)	Ecological reserve/ deficit (global ha/ person)
World	**6225.0**	**2.2**	**1.8**	**–0.4**
Australia	19.5	7.0	11.3	4.4
Bangladesh	143.8	0.5	0.3	–0.2
Cambodia	13.8	0.5	0.7	0.2
Canada	31.3	7.5	15.1	7.6
China	1302.3	1.6	0.8	–0.8
Germany	82.4	4.4	1.8	–2.6
India	1049.5	0.7	0.4	–0.4
Indonesia	217.1	1.0	1.0	–0.1
Japan	127.5	4.3	0.8	–3.5
Korea Republic	47.4	4.3	0.6	–3.8
Malaysia	24.0	2.4	3.3	0.9
Nepal	24.6	0.6	0.5	–0.2
Pakistan	149.9	0.6	0.3	–0.3
Philippines	78.6	1.0	0.6	–0.4
Russia	144.1	4.4	7.0	2.6
Sri Lanka	18.9	0.8	0.4	–0.5
Switzerland	7.2	4.7	1.6	–3.1
Thailand	62.2	1.4	1.0	–0.4
United Kingdom	59.3	5.6	1.6	–4.0
United States of America	291.0	9.7	4.7	–4.9
Vietnam	80.3	0.8	0.8	0.0

Source: European Environment Agency in coordination with Global Footprint Network, The National Ecological Footprint, and Biocapacity Accounts, 2005 edition (http://www.eea.europa.eu/highlights/Ann1132753060).

negative numbers in the final column of Table 3.8 for many developed countries and also developing countries including China and India. In aggregate, humanity's ecological footprint is over 20% larger than what the planet can regenerate. This imbalance is maintained by using up the world's ecological resources.

In addition to the growing depletion of non-renewable resources such as minerals, ores and petroleum, renewable resources are also being depleted. These include fisheries, global warming, species extinction, deforestation, and the depletion of water resources.

Some countries with vast land mass and few people per square kilometer like Canada, Russia and Australia still have a large biocapacity and are still in a strong positive balance. Most other industrial countries and some developing countries are in a negative balance.

(ii) The responsibility of reducing environmental damage is a global task that requires cooperation of everyone on the planet, rich and poor alike.

(iii) Rich countries use many more resources per capita than poor countries, both in terms of energy used and in terms of depletion of other natural resources. Some rich countries such as Korea are continuing to use more energy. Other rich countries such as Japan are managing to conserve energy by applying a variety of control measures, pricing policies and technology.

(iv) Energy use in rapidly growing countries in Asia is increasing faster than in other regions.

Figures 3.1 and 3.2 show the nature of increasing demand on the global environment relative to its carrying capacity. Over time global biocapacity per person has been falling as population has been increasing. However, demand has continued to increase although more slowly now than in the past. Carbon dioxide is responsible for the bulk of the increase in demand. The other components are falling as reflected in Figure 3.3. CO_2 emission has been growing and is now by far the largest component of the global environmental footprint.

Figures 3.4 and 3.5 present comparisons of China's and India's footprints. The rapid increase in CO_2 emissions in China is due to its increased pace of industrialization than India. While other factors in the two countries are similar, Korea's footprint is even more dramatic than China's. Korea's industrial expansion has been going on for a longer time with little effort to control environmental damage.

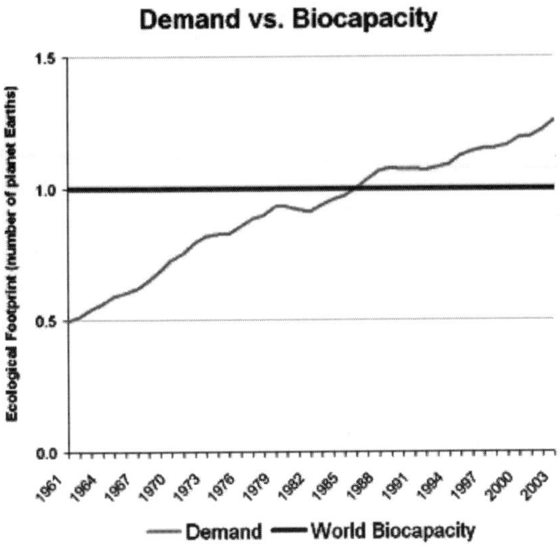

Figure 3.1 Global Demand and Biocapacity

Source: Global Footprint Network website (http://www.footprintnetwork.org/gfn_sub.php?content=global_footprint).

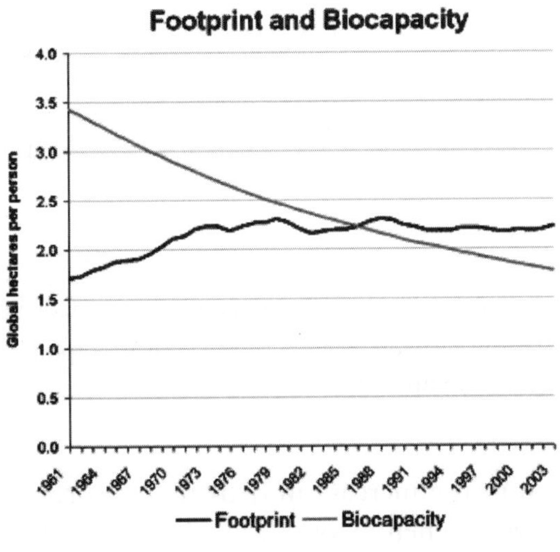

Figure 3.2 Global Footprint and Biocapacity

Source: Global Footprint Network website (http://www.footprintnetwork.org/gfn_sub.php?content=global_footprint).

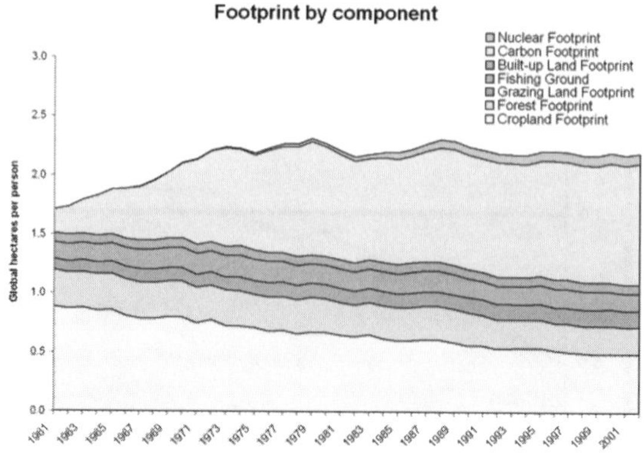

Figure 3.3 Global Footprint by Component

Source: Global Footprint Network website (http://www.footprintnetwork.org/gfn_sub.php?content=global_footprint).

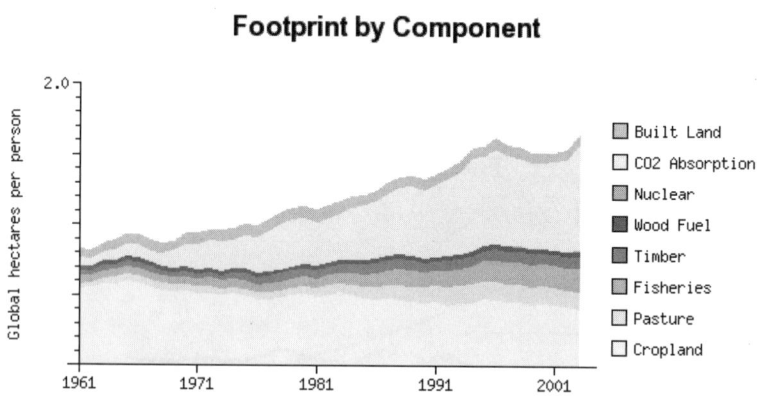

Figure 3.4 China's Global Footprint by Component

Source: Global Footprint Network website (http://www.footprintnetwork.org/gfn_sub.php?content=global_footprint).

The only hiccup in the increase in demand in Korea was during the Asian financial crisis, when income growth was negative in 1998.

The global footprint for all the regions of the world compared with population is displayed in Table 3.9. The United States' footprint has

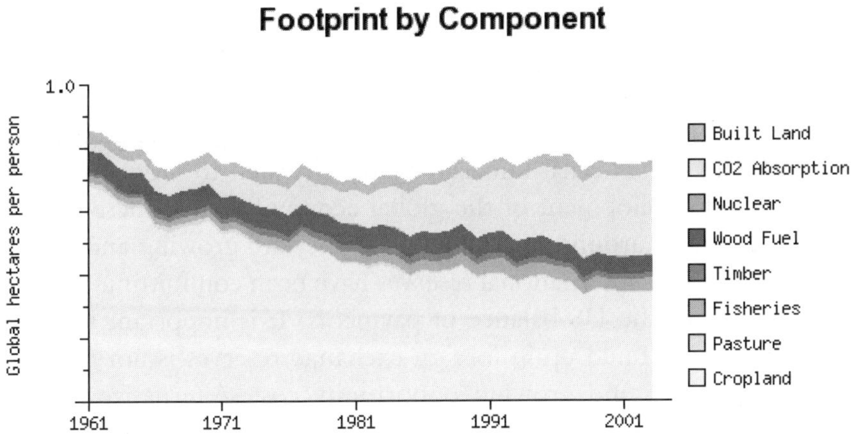

Figure 3.5 India's Global Footprint by Component

Source: Global Footprint Network website (http://www.footprintnetwork.org/gfn_sub.php?content=global_footprint).

Table 3.9 World's Global Footprint

Region	Population in millions	Global hectares per person
North America	319	9.1
Europe	307	5.0
Middle East and Central Asia	185	2.5
Latin America	496	2.2
Asia Pacific	3230	1.5
Africa	614	1.4

Source: Global Footprint Network website (http://www.footprintnetwork.org/gfn_sub.php?content=global_footprint).

been relatively constant for the past few years. Although still higher than any other region, its share of global environmental demand has not been increasing. The biggest risk to global ecological stability is the continued increase in demand in Asia, simply because the region's population is so large.

Issues and Prospects

International imbalances

In the first section of this chapter several alternatives were suggested to deal with the external imbalances that threaten to unhinge stable growth and development of the global economy. While these imbalances have been around for several years they are growing and countries with surplus international reserves have been committing then to the support of the US balance of payments. It is important to note that these large build up in foreign exchange reserves is not coming without sizeable and growing opportunity cost. Alternative investments in developing Asian countries might be more profitable. Furthermore these reserves could be put to better use by funding the growing infrastructure needs in the region, particularly in the poorer countries that do not have the resources to make these large needed investments. We discussed possible solutions which would recycle these funds in Chapter 2 dealing with regional cooperation.

Three alternatives to deal with these imbalances were suggested — a withdrawal of funds by Asian creditors from the US bond market, a rearrangement in the dollar remimbi exchange rates, and a reduction in the US current account deficit. It was argued that currency realignment would entail less risk for the global economy but more risk for China. The challenge is to convince China that currency appreciation is in their best interest in the long run. However, the Chinese are unlikely to offer more than token currency revaluation without some demonstration that other means of adjustment are also being undertaken — measures such as reducing the US current account deficit by raising taxes and cutting spending. Without such efforts it is unlikely that the Chinese will be willing to bear the full brunt of the requirement adjustment.

A longer term historical view of the current global financial arrangements has been suggested by Michael Dooley and his associates at Deutsche Bank (Dooley *et al.*, 2003a,b) and Deutsche Bank (2004) and the popular account in the *Economist* (2006). The argument put forward is that the current global financial system has many similarities to the Bretton Woods system of international monetary arrangements that established fixed exchange rates among the major currencies after World War II. This system involved a center country and a group of countries on the periphery. In the 1950s the center country was the United States and the periphery countries were the countries in Europe and Japan that have been ravaged by war and were recovering. The periphery countries chose a strategy that undervalued currencies, put in place capital controls, and stressed the accumulation of reserves. The United States served as a financial intermediary and helped to establish their still fragile banking and financial systems. The United States also made available long-term loans to the periphery countries through direct foreign investments and aided through the Marshall Plan. This system existed for a time. As the periphery countries became further established, the countries in it graduated to the center. Eventually the fixed rate strategy was abandoned in favor of a flexible rate system.

The current counterpart of the Bretton Woods system (sometimes called Bretton Woods II) puts the Asian economies at the periphery with OECD countries at the center. Exchange rates are flexible among center countries but relatively fixed between the United States, the main center country, and the periphery. Asian countries have built up reserves, financed the US current account deficit, and raised exports to the center. Growth has accelerated in the periphery countries and remained strong for the last two decades. Rather than taking an aggressive strategy to invest surpluses on their own, particularly in light of somewhat weak financial systems, the Asian economics have been content to rely on FDI from the center to build up an internationally competitive capital stock. Europe followed a similar strategy for a time, after World War II, eventually moving toward freer markets and relaxation of financial controls. This led to the eventual collapse of the fixed rate system in the early 1970s since there was no periphery to replace Europe and Japan at the time. Nevertheless, these Bretton Woods arrangements lasted over 20 years.

The current global system is complicated by the fact that the United States has been joined by Europe and Japan at the center. Aside from that, the relationship between the center and the periphery is the same and it is similar to the original Bretton Woods system.

Developing Asia is at the periphery and OECD is at the center. Asian countries keep their currencies undervalued against the dollar and accumulate current account surplus and foreign exchange reserves. China, Taiwan, Hong Kong, Singapore and Malaysia manage their dollar exchange rates by rigid exchange controls. Others such as Japan and Korea employ a controlled float.

As Dooley *et al.* suggest, "the dynamics of the international monetary system, reserve accumulation, net capital flows and exchange rate movements are driven by the development of these periphery countries" (2003a, p. 2).

Exchange rate changes within the center will have an impact on the dynamics of this system. If the Euro appreciates against the dollar, then Europe will lose international competitiveness to Asia and it will tend to grow slower. The same holds for Japan.

In addition to the center and the periphery there are other countries in the global system including Latin America, Australia and Russia as well as Eastern Europe. To the extent that these countries float their currencies against the US dollar or move with the Euro, their exports could be displaced by Asian exports. If they elect to fix exchange rates against the US dollar, they could join the Asian periphery.

This system as described may continue for the foreseeable future although there are still a number of problems to be resolved. The argument that China and other Asian countries can rely on FDI to build their capital stock is also becoming less true. FDI now accounts for less than 5% of fixed investment in China. Reform in the Chinese financial system (and in other Asian economies as well) is needed to achieve better allocation of investment resources and higher interest rates that slow down investment by making local authorities more sensitive to the cost of capital.

Pressure on Asian economies to revalue will continue and trade barriers may be erected by the center, threatening a return to a period of isolationism and lower global growth. This will be offset to some extent by a continuation of offshore investments in Asian economies by Japan and other OECD countries. One needs to only look at the Asian automobile industry to see this trend.

Other issues are also relevant. Should the Asian surplus countries begin to undertake a modern Marshall Plan of aid to poorer Asian countries as suggested in Chapter 2? What impact would this have on the global system? Can the current system accommodate a modest shift away from the Bretton Woods II model to accommodate the growing power of the oil exporters as a result of higher oil prices? These emerging oil economies now have surpluses of over $400 billion and these funds could easily be invested in other currencies other than the US dollar. They are not part of the periphery as described here, and could be a destabilizing force.

International trade

Following the breakdown in the Doha round of trade negotiations it is difficult to say what will come next on the international trade scene. We can be pretty sure that the thrust toward more unilateral agreements will intensify further. Agreements between trade groups such as ASEAN, ASEAN + 3, SAARC and BIMSTEC with each other and with individual countries will continue. Furthermore, the number of agreements between South and East Asian countries is likely to accelerate. We are already seeing the fruits of automobile trade between India and Thailand following the recent signing of an FTA between the two countries. A number of sectors were identified for possible cooperation within BIMSTEC at a recent meeting in Dhaka, including automobiles and parts, pharmaceuticals, gems and jewelry, textiles and agricultural products. Trade facilitation was also identified in customs, banking, the internet, intellectual property and labor mobility (see BIMSTEC website for details).

Eventually the WTO should be able to engage in a new round of international trade negotiations. When that time comes it is important for the developing Asian economies to be prepared to negotiate on issues that are of vital interest to the region, including agriculture, services and intellectual property. Presenting a united front in these negotiations will be crucial if the interests of Asian countries are to be well articulated and developed if Asia is to get the most out of these negotiations.

Immigration

Globalization has intensified the flow of resources between countries. This holds for all factors of production including skilled and professional labor. Letting market forces do the job of allocating these scarce resources is not recommended. There are externalities that require interventions to regulate the flow of human resources. In the case of sending countries, skilled migration can create shortages in key occupations that have secondary impacts on the labor market and economic development in the sending country. The health professions are a good example of these externalities. There are also labor market issues in recipient countries including impact of skilled migrants on local labor markets, supply of new graduates to fill existing positions and the transfer of resources implied when well-trained overseas workers are employed.

Resolution of these issues requires some government interventions. Encouragement of circular migration is one suggested solution that should be explored further by both sending and recipient countries. In some areas such as science, information technology, and engineering this is already happening in Taiwan and India and also beginning in China. This cycle has been documented by Saxenian (2006). Saxenian argues that after working in the United States for a time, many scientists engineers and information technology experts return to their own countries to start new companies using the knowledge and expertise they have acquired abroad. She details a number of individual cases where this has been fruitful for both the individual and the sending country. In the IT field, Bangalore, Taiwan and Shanghai are now becoming new silicon valleys in the East. The exchange of ideas and technology is facilitated further by the internet itself and cheap real time communications via the internet such as Windows Messenger or Skype. As a result returnees remain well connected with old friends and colleagues and are easily able to keep up with the latest developments in their fields on a global level.

Opportunities are beginning to develop in other fields and need further support from governments to develop strong information and communication channels. In medicine, there are already detailed communications between different hospitals, surgeons and researchers which can allow returning physicians to remain in contact and also to share their expertise with others domestically and internationally. Similar developments are also evolving in business and other professional areas.

Circular migration can also be encouraged by the formation of research and development centers within Asian countries. Government seed money is needed and a focused plan of action in particular areas has to be formulated. Some examples are research on biotechnology in Singapore and chip development in Korea. The challenge will be for countries at somewhat lower levels of development like the Philippines and Sri Lanka to be able to encourage skilled and professional migrants to return home. Sometimes the opportunity to help at home would be enough incentive for mature professionals who are already thinking of retirement to return to their home country. In cases of younger professionals it may require a monetary sweetener and an opportunity to do original work that will help the country.

Health

The focus on HIV/Aids is important because of the assistance that drug companies in the United States and Europe can give to developing countries. WHO (2005) and other experts have documented the close relationship between illegal drugs, unsafe needles and unsafe sex in spreading AIDS in Asia and elsewhere. Programs to control the access to drugs, providing clean needles and better education can help to slow the spread of AIDS at the national level. Cooperation with drug authorities from the United States and Europe can also control or reduce the flow of illegal drugs into the rest of Asia from the known drug producing centers in China, Myanmar, and other parts of the "golden triangle". These are worthy endeavors and should be adequately funded to assist in reducing the spread of AIDS.

A strategy of increasing the supply and reducing the cost of ARV drugs to the Asian countries where the AIDS epidemic seems to be spreading more rapidly will achieve the greatest benefit while minimizing cost, since there are economies of scale to supplying large amounts of drugs to a limited number of markets. The WHO database suggests that seven countries have the most cases — Cambodia, China, India, Indonesia, Pakistan, Thailand and Vietnam. In addition, regions with the highest incidence of HIV/AIDs should be identified and targeted for distribution. These include the Ho Chi Minh City in Vietnam, Uttar Pradesh, Bihar and Tamil Nadu in India, Karachi and Northwest Frontier in Pakistan, Southwest China and Bangkok, and Jakarta.

Additional resources by western governments and charitable foundations in addition to the drug companies can be made available for the WHO effort and those of individual governments. PEPFAR, or the longer name, the "President's Emergency Plan" allocates US funds to reduce the spread of AIDS. It has an African focus although it includes India. The Gates Foundation has programs to treat AIDS victims. Groups like Act Up are seeking other routes, persuading big employers of third-world workers — Coca-Cola, Unilever, Anglo-American's mining subsidiaries, and the like — to pay for drugs for their infected workers. The most aggressive charity on this front is Doctors Without Borders, which won the 1999 Nobel Peace Prize for its work in war-torn areas. It has 1500 patients in eight countries receiving ARVs at a cost of $500,000, a little over $300 per patient per year — still a stiff price for poor countries with limited health care budgets (see McNeil, 2002 for more charitable ideas).

Then there are the pharmaceutical companies that originally balked at charging lower prices for drugs in poor countries. Is there more to do besides the highly publicized Accelerating Access Initiative (AAI)? Can smaller drug companies join the effort? Would drug companies with products still under patent be willing to let other companies prepare and sell these drugs in the developing countries as generic products? These are ideas that should be explored in order to increase the rate of ARV use in Asia to much higher levels. Combined with reductions in infections, these initiatives can reverse the level of human misery and suffering that this deadly virus has brought with it.

Genetically modified crops

Though there have been many breakthroughs, GMCs are still in their infancy. So far most of the developments have been carried out on the crops grown in the United States and Canada. Only 10% of field trials have been carried out in developing countries. There is a host of possible GMCs that can be developed for different soil and climatic conditions. Unlike the green revolution where research was carried out by government-sponsored organizations whose mission was to disseminate information about the new varieties, current GMCs are being developed by a small group of private companies that are selling the seeds and monopolizing the technology. Profit should not be the only motive for developing new GMCs, and regulations that permit new technology to be transferred to developing countries for a nominal cost are needed. The United States has not put its weight behind the research effort to develop new technology and Europe is also not interested. How will new research be funded? The richer countries of the Asian region do not seem willing to take the lead. Can China and India, and the rest of South Asia and Southeast Asia take the lead? This is an interesting question.

Currently less than $50 million is spent annually on biotechnology research in the developing world compared to about $200 million for government-sponsored biotechnology research in the United States. This is dwarfed by spending by the private sector which runs over $2 billion. Despite such a small budget, there have been breakthroughs in rice biotechnology in developing countries, where GM crops are raising yields. Developments in Mexico have raised resistance to aluminum toxicity and raised crop yields. Further developments in cereal yields could benefit the poor in developing countries around the world. Developing crops that are resistant to drought and soil toxicity, and varieties that need fewer applications of pesticides are needed in poor countries. The price signals need to be rethought to provide the kinds of research that will be beneficial to poorer countries, not the rich. The trend toward stagnating research on agriculture in poor countries and regions (Africa and South Asia in particular) and booming research in developed countries has to be reversed. To do this the involvement of the private sector is required. These developments cannot depend on international public sector agencies like CGIAR or national efforts in large countries like Brazil, China, India and Mexico. One option would be to offer incentives to private sector firms for the development of new genetically engineered crops that will be beneficial to developing agriculture. The incentives can be from a consortium of developing countries and the international community, and should have specific targets for small farmers and poor areas. The agenda is stated well by the ethics group at Nuffield College, Oxford (Nuffield Council on Bioethics, 1999):

> Developing countries and multilateral agencies need to devote much large shares of effort to agricultural research that not only develops new GM crops but derives criteria for such development from the stated requirements of small farmers in staples-based systems. These should take into account farmers from rain fed and intercropped areas and the needs of the poor countries for employment-intensive production and processing of cheap food stables (p. 68).

Due care will need to be taken that these GM crops are field-tested and given health and safety clearance in much the same way as the GM crops developed in the United States are tested. The focus of research will be much different — more in the spirit of the first green revolution of the 1960s with research and field trials being undertaken in poorer countries rather than in industrial countries.

India and Pakistan

Pakistan is in a delicate position of having to balance the Islamic factions (including terrorists elements) that seek to further disrupt Kashmir and Pakistan generally, against the government's commitment to remain an ally of the United States and to fight terrorism. In retrospect, political initiatives by President Musharraf in Kashmir in 2004 seem to have been made in good faith as terrorist attacks were reduced. Further progress in resolving the Kashmir dispute will depend primarily on the two governments and the Indian government's reaction to these Pakistani initiatives. Although diplomatic efforts by the United States and other outside powers may help to move the peace process forward, there is no clear case for high level diplomacy unless it involves bringing the two sides together to discuss how Kashmir issues can be resolved. Efforts to increase trade and generally reduce tensions may also help to set the scene for more fruitful diplomatic moves. Stronger growth and improving living standards in both countries as they adopt more liberalized trade and industrial policies may contribute to a thaw in diplomatic relations.

China and Taiwan

For the immediate future, maintenance of the diplomatic status quo seems the best course to follow. At the same time, greater business, economic as well as social and cultural ties should be encouraged. Chapter 1 projects income per capita in Asia to the year 2020 or 2025. There are strong tendencies toward convergence in incomes within the Asian region. Certainly the southern coastal provinces of China will, by that time, benefit from rapid growth in per capita income that should raise living standards to a level that is comparable to the standards in Taiwan. Perhaps by that time as well, political development may make it more feasible for an agreement to set aside their differences and to integrate Taiwan with the mainland of China can be reached. In the meantime pressure from outside the region, including the United States and its European allies, should be exerted to persuade both sides that there is much merit in continuing the current stable, yet uneasy balance that has existed for so long (Swaine, 2004).

This is not to say that continued efforts should not be made to quietly soften the two alternative positions regarding Taiwan governance. In addition, cultural and social exchanges should be encouraged as well as reducing barriers for foreign direct investment, tourist and business travel, and other exchanges that foster cooperation and trust while defusing political tensions.

China and the West

High level dialogue between senior officials from the United States, the European Union, and China would be a beneficial step toward resolving a variety of issues including currency valuation, international trade, computer software, intellectual property, agricultural policy, and military spending. A team from the United States headed by the secretary of the treasury Henry Paulson is currently undertaking such talks, with particular focus on currency issues. These talks could be extended to cover other issues and the dialogue could be widened to include the European Union. Singling out China for such dialogue seems to be appropriate at the present time, since it is playing an increasingly important role in international trade and strategic relations within the Asian community. In the future other high level dialogue between the European Union, United States, and Asian economies may provide a useful platform for discussing a variety of issues that would be difficult to address at the global level or at the individual level of a free trade agreement. These might include various intellectual properties and migration issues with India and environmental issues with ASEAN and Korea.

Environment

In Chapter 6 a series of proposals is discussed to reduce global warming that involved Asia and industrial countries. In this section we explore possible ways in which industrial countries in Europe and North America can assist developing Asia in implementing new technologies to reduce global warming and clean up the environment. In Chapter 6 we noted six policies that could help to bring carbon emissions down over the next decades to 2050 that have been suggested by Socolow and Pacala (2004, 2006). Here we look at an extended list of proposals that includes several other possible courses of action to save energy and reduce carbon emissions. Each technology would be phased in and would prevent the release of 25 billion tons of carbon over the next 50 years. The technologies are arranged according to five categories: end-user efficiency and conservation, power generation, carbon capture and storage, alternative energy sources, and agriculture and forestry. Each category has several different options and these are displayed in Table 3.10.

Table 3.10 Fifteen Ways to Save 25 billion tons of Carbon Emissions Over the Next 50 Years

End-user efficiency and conservation	Power generation	Carbon capture and storage (CCS)	Alternative energy sources	Agriculture and forestry
Double average fuel economy of global automobile fleet	Raise efficiency at 1600 large coal-fired plants by 40% to 60%	Install CCS at 800 large coal-fired plants	Drive all cars on ethanol using one-sixth of world's cropland	Expand conservation tillage to all cropland
Cut miles driven in half	Replace 1400 large coal-fired plants with gas-fired plants	Install CCS at coal-fired plants that produce hydrogen	Increase wind power 80-fold to make hydrogen for cars	Stop all deforestation and continue replanting
Cut electricity use by 25%		Install CCS at coal to syngas plants	Increase solar power 700-fold	
			Increase wind power 40-fold to displace coal	
			Double nuclear capacity to displace coal	

Source: Socolow and Pacala (2006).

The basic thrust of the Socolow and Pacala argument is to replace some coal with alternative fuels and to sequester carbon from the remaining coal-fired plants. Oil is to be replaced by ethanol or mixed with ethanol in all vehicles in a phased manner.

This is an ambitious plan and involves the cooperation of the global community and a commitment of significant resources to achieve these objectives. It requires long range planning and commitment of resources for an extended period of time. In the meantime, developing countries like China and India are clamoring for more cars and more energy-intensive technologies. As noted in Chapter 6, car sales in China are expected to reach 6 million in 2006, an increase of over 100% in the past three years. Western automobile manufacturers are committed to China as the next growth market. Volkswagen, General Motors and Honda have the most momentum. The largest Chinese automaker First Auto Works has partnerships with Toyota and Volkswagen. Other companies, including Nissan, with investments of about $2 billion and Hyundai with investment of $1 billion are also trying to break into the Chinese market (Lienert, 2003). As the number of vehicles increases so will the level of emissions.

Foreign automobile makers can make a major contribution to reducing these emissions by working with the Chinese automakers and government authorities in a number of ways: car makers can reduce emissions by increasing fuel economy; turning to alternatives including ethanol and other low carbon fuels; installing catalytic converters and other means of reducing CO_2 emissions; urging authorities to raise gasoline taxes; encouraging fewer trips by car; helping to publicize the importance of environmental protection; and developing alternative transportation modalities. Government policies can support these efforts by adopting legislation that rewards those that reduce emissions and penalizes those that do not.

These efforts need not be confined to China. Along with the rest of Asia, including Japan, vehicle production reached nearly 26 million vehicles in 2005, more than Europe (21 million) and more than the United States (19 million) according to the Organization of Automobile Manufacturers (OICA) (Table 3.11). Japan and Australia

Table 3.11 Worldwide Vehicle Production in 2005 and Change from 2004 (in million units)

Country or Region	2005	Percent Change from 2004
Europe	20.8	0
USA	12	0
Rest of North and South America	6.9	5.7
Asia	25.8	6
China	5.7	9
India	1.6	8
Japan	10.8	3
Thailand	1.1	21
South Korea	3.7	7

Note: No other Asian country produced more than 900,000 vehicles in 2005.
Source: OICA (2006).

already have strong environmental regulations on emissions. There are fewer regulations in India, Thailand and other Asian countries and yet demand is rising faster in Asia than the rest of the world. By cooperating to reduce emission and with the support of stronger government regulations, car makers can make a substantial contribution to reducing global CO_2 emissions.

In the past few decades, industrialized countries have been able to slow the growth in ecological damage and carbon emissions even though they have a large ecological footprint and high energy use per capita. The US share of global emissions was 39% in 1952. It fell to 23% by 2002 and continues to fall. New technology, better pricing, markets for pollution using market permits, and other developments have helped the United States and European countries to slow the pace of environmental degradation. The US footprint is still much bigger than other industrialized countries, in part because automobile efficiency and gasoline and car taxes are all low by comparison with Europe and Japan. The United States is at the forefront of many of the new technologies that can break the cycle of attachment to fossil fuels. Yet, fuel economy in the United States lags behind all other industrial countries and China by as much as 70% — 24 miles per gallon (mpg) in the United States, 34 mpg in China, 44 mpg in Europe, and 47 mpg in Japan according to Gore (2006, p. 272).

Partnerships between developing Asia, the United States and Europe can be beneficial for the global economy and generate profit for US and European entrepreneurs. In China, the US Environmental Protection Agency (EPA) has undertaken a number of projects to assist in developing technologies to deal with air pollution and carbon emissions resulting from the use of coal. A Memorandum of Understanding between the EPA and China's State Environmental Planning Agency (SEPA) was signed in 2003 (see US Embassy, 2006). This cooperation includes reduction in methane emissions and recovery, acid rain reduction, development of wind power, and waste water treatment among others. Additional cooperation to reduce coal emissions, explore ways to sequester carbon emissions, and other applications of technology are possible between the United States and China, and also between Europe and the developing countries of Asia.

Alternatives to the private automobiles have to be developed if vehicular pollution is to be reduced. These include various forms of mass transit, including trains and buses, where Europe has been a leader for decades. By marrying western technology with Asian needs, it may be possible for these newly emerging economies to avoid many of the problems that industrialized countries have had to face over the last 50 years. This is the time to act. China is already producing several million cars a year and India is not far behind. In 2005 China produced nearly 5 million vehicles making it the third leading producer of automobiles after the United States and Japan (see Chapter 6 for more discussion).

Another approach is to use market mechanisms to control pollution. Carbon purchase contracts are now being widely used in industrial countries as a market tool to reduce the total level of greenhouse emissions. Overall pollution ceilings and standards are set by the individual country governments and pollution permits are issued up to that amount. In Chapter 6, we mention another initiative by the World Bank whereby industrial countries can buy pollution rights from developing countries. This Global Environment Facility can encourage the flow of resources to developing countries to reduce emissions. Generally tradable pollution permits or pollution contracts have a number of different applications in Europe and the United States.

These and other market-based solutions along with mechanisms for determining the size, location, and timing of emissions being developed by companies that specialize in these trading permits can be shared with developing countries in Asia. Reducing the carbon footprint is becoming more popular in Europe and is slowly spreading to the United States. In Europe, carbon emissions are regulated by the Greenhouse Gas Emission Trading Scheme, a mandatory system that not only imposes caps on emissions but also allows trades with those who are emitting lower than allowable maximum emissions. This trading scheme is part of the way the EU hopes to meet its obligations under the Kyoto Protocol. For example, paying for air pollution created by air travel by business executives is one way that businesses are paying for their greenhouse emissions. Other service industries are becoming involved (Phillips, 2006) in this type of scheme.

In the industrial sector, more than 9400 industrial plants in the European Union are given CO_2 emission quotas which allow plants to emit CO_2 up to a permissible level determined by the quota. The aim of emissions trading systems is to generate incentives for companies to reduce their emissions and impose costs on those that are less efficient. Emission permits are traded in a market. Plants that emit less than their quota can sell the surplus. The system is still being developed and permits granted may not yet be constraining the level of emissions. For example, in 2005, EU industrial plants pumped out 1.785 billion tons of CO_2 greenhouse gases into the atmosphere while national authorities had given them allowances for 1.829 billion tons according to data supplied by the European Commission (see http://www.eubusiness.com/Environ/060515095822.j3umicjn).

The advantages of economic incentives like tradable permits over other systems of controlling emissions like emission standards is that over time they provide a continual incentive to reduce emissions. This helps to promote new technology, and achieves greater flexibility in reducing emissions. In the United States and Europe, tradable permits for nitrous oxides and sulfur oxide emissions have been quite effective in reducing emissions (Harrington and Morgenstern, 2004). Private companies like Natsource Europe are helping Europe's power companies, and other heavy industries buy and sell carbon credits.

The expertise accumulated by the European Union and the United States can be shared with Asian countries to help them meet their Kyoto Protocol obligations as well as build up expertise in monitoring and managing the production of greenhouse gases.

Moving beyond the Kyoto Protocol requires compliance by the United States and Australia, and also the large developing countries such as Brazil, India and China that have up to now been exempt. It also involves more stringent standards to reduce carbon emissions and a concerted effort on several fronts to marry new innovations and technology with market-based policies and selective control legislation to guarantee a minimal level of compliance. A menu of possible paths to achieving lower global emissions has been presented in Table 3.9 above and also in Tables 6.6 and 6.7 in Chapter 6. What items to select from this menu, how to introduce policy and price changes that will support these choices and appropriate technologies to be used, will depend upon initial conditions, climate, industrial development, state of technology, and implementation capacity. For the individual country the challenge will be to select an appropriate agenda that is doable and practical and which can be implemented with available resources. For the global system and the three regions of Europe, United States and Asia, the challenge is to develop an ongoing dialogue and mutual support network that transfers appropriate technology and expertise along with policy modalities that are appropriate and effective for Asia.

Bibliography

Asian Development Bank (ADB) (various years). *Key Indicators.* Manila, Philippines.

Barro, R. J. and X. Sala-i-Martin (1995). *Economic Growth.* Singapore: McGraw-Hill.

Bay of Bengal Initiative for Multi-Sectoral Technical and Economic Cooperation (BIMSTEC) (2006). http://www.mfa.go.th/bimstec/bimstecweb/html/sectors.html [Retrieved 11 November 2006].

Behera, N. C. (2002). Kashmir: Redefining the US role. Brookings Institution Policy Brief No. 110.

Beine, M., F. Docquier and H. Rarroport (2003). Brain Drain and LDCs Growth: Winners and Losers. Institute for the Study of Labor, Discussion Paper No. 819.

Bergsten, C. F. (2005). The Trans-Pacific Imbalances: A Disaster in the Making? Speech at 16th general meeting of PEEC, Seoul, S. Korea.

Carrington, W. J. and E. Detragiache (1999). How Extensive is the Brain Drain? *Finance and Development*, 36(2). http://www.imf.org/external/pubs/ft/fandd/1999/06/carringt.htm [Retrieved 9 May 2006].

Central Intelligence Agency (CIA) (2006). *World Fact Book 2006*. Washington, DC: CIA.

Chirac, P. (2002). Increasing the Access to Antiretroviral Drugs to Moderate the Impact of AIDS: An Exploration of Alternative Options. In G. A. Cornia (ed.), *AIDS, Public Policy and Child Well-Being*. UNICEF-IRC.

Deutsche Bank (2004). The Revived Bretton Woods System: Alive and Well, Global Markets Research.

Dooley, M., D. Folkerts-Landau and P. Garber (2003a). An Essay on the Revived Bretton Woods System. NBER Working Paper No. 9971.

Dooley, M., D. Folkerts-Landau and P. Garber (2003b). Dollars and Deficits: Where Do We Go from Here? Deutsche Bank Global Markets Research.

Economist (2006). Uncle Sam's Teat, 9 September.

Economist (2006). A Survey of the World Economy, 16 September.

Economist Intelligence Unit (2006). EIU Country Data database. http://www.eiu.com [Retrieved 20 December 2006].

EU Business Week (2006). Permits Surplus Raises Questions About EU Emissions. http://www.eubusiness.com/Environ/060515095822.j3umicjn [Retrieved 25 October 2006].

Fernandez-Cornejo, J., W. D. McBride, C. Klotz-Ingram, S. Jans and N. Brooks (2000). Genetically Engineered Crops for Pest Management in U.S. Agriculture. Agricultural Economics Report No. AER786, May.

Food and Agriculture Organization (FAO) (2004). *State of Food and Agriculture 2004*. Rome: Food and Agriculture Organization of the United Nations.

Gaskell, G., M. W. Bauer, J. Durant and N. C. Allum (1999). Worlds Apart? The Reception of Genetically Modified Foods in the Europe and the US. *Science*, 285 (July), 384–387.

Global Footprint Network (2006). http://www.footprintnetwork.org/ [Retrieved 11 November 2006].

Global Fund Website (2006). http://www.theglobalfund.org/en/about/how/ [Retrieved 26 October 2006].

Gore, A. (2006). *An Inconvenient Truth*. Emmaus, NJ: Rodale Press.

Gurría, A. (2006). Doha: The Low Hanging Fruit. http://www.oecd.org/document/4/0,2340,en_2649_201185_37295108_1_1_1_1,00.html [Retrieved 6 November 2006].

Harrington, W. and R. D. Morgenstern (2004). Economic Incentives versus Command and Control: What's the Best Approach for Solving Environmental Problems? *Resources* (Fall/Winter), 13–17.

International Federation of Pharmaceutical Manufacturers (IFPMA) (2006). http://www.ifpma.org/index.aspx [Retrieved 11 November 2006].

International Monetary Fund (IMF) (2006). *World Economic Outlook*. Washington, DC: IMF.

International Organization of Motor Vehicle Manufacturers (OICA) (2006). http://www.oica.net/correspondents survey [Retrieved 21 June 2006].

Lavoy, P. R. (2006). Pakistan's Kashmir Policy After the Bush Visit to South Asia. *Strategic Insights*, V(4).

Lienert, D. (2003). The Rising Chinese Car Market. *Forbes Magazine*. http://www.forbes.com/2003/12/15/cx_dl_1215feat.html [Retrieved 11 November 2006].

Lucas, R. (1988). On the Mechanics of Economic Development. *Journal of Monetary Economics*, 22, 3–42.

Mankiw, N., G. D. Romer and D. N. Weil (1992). A Contribution to the Empirics of Economic Growth. *Quarterly Journal of Economics*, 107(2), 407–437.

McNeil, Jr, D. G. (2002). Overseas/AIDS; One by One, Charities Attack the AIDS Juggernaut. *New York Times*, 18 November.

Medeiros, E. S. and M. Taylor Fravel (2003). China's New Diplomacy. *Foreign Affairs*, 82(6), 22–35.

Nuffield Council on Bioethics (1999). *Genetically Modified Crops: The Ethical and Social Issues*. London: Nuffield Council on Bioethics.

Organisation for Economic Co-operation and Development (OECD) (2006). The Doha Development Round of Trade Negotiations: Understanding the Issues. http://www.oecd.org/document/45/0,2340,en_2649_201185_35738477_1_1_1_1,00.html [Retrieved 11 October 2006].

Periez, J. (2006). China Competes with West in Aid to its Neighbors. *Internatonal Herald Tribune*, 18 September.

Phillips, B. (2006). Paying the Freight for Polluting the Air: Europe Takes the Lead. *New York Times*, 18 September.

Romer, P. M. (1986). Increasing Returns and Long-Run Growth. *Journal of Political Economy*, 94(5), 1002–1037.

Saxenian, A. L. (2006). *The New Argonauts: Regional Advantage in a Global Economy*. Cambridge: Harvard University Press.

Socolow, R. H. and S. W. Pacala (2004). Stabilization Wedges: Solving the Climate Problem for the Next 50 Years with Current Technologies. *Science*, 305(5686), 968–972.

Socolow, R. H. and S. W. Pacala (2006). A Plan to Keep Carbon in Check. *Scientific American*, 295(3), 50–57.

Solow, R. M. (1956). A Contribution to the Theory of Economic Growth. *Quarterly Journal of Economics*, 70(1), 65–94.

Somini, S. (2006). On India's Despairing Farms, a Plague of Suicides. *New York Times*.

Swaine, M. D. (2004). Trouble in Taiwan. *Foreign Affairs*, 83(2), 39–49.

Tucker, N. B. (ed.) (2005). *Dangerous Strait: The U.S.-Taiwan-China Crisis*. New York: Columbia University Press.

Tujan, Jr, A. (2002). Health Professional Migration and Its Impact on the Philippines. *Asian Pacific Research Network Journal*, 6 March. http://www.aprnet.org/index.php?a=list&t=journals [Retrieved 6 April 2006].

US Embassy (2006). Japan Balancing Growth, Environmental Protection: A Challenge for China. http://toyo.usembassy.gov/e/p/tp-20060203-03.html [Retrieved 14 October 2006].

Wickramasekara, P. (2002a). Policy Responses to Skilled Migration: Retention, Return and Circulation. International Migration Programme, Geneva: ILO.

Wickramasekara, P. (2002b). Asian Labor Migration: Issues and Challenges in an Era of Globalization. International Migration Papers 57. Geneva: ILO.

World Bank (2006). *World Development Indicators Online*. Washington, DC: World Bank.

World Health Organization (WHO) (2003). Towards Universal Access by 2010: How WHO is Strengthening Health Services to Fight HIV/AIDS.

World Health Organization (WHO) (2005). AIDS Update. http://www.who.int [Retrieved 23 May 2006].

World Health Organization (WHO) (2006). Antiretroviral Therapy. http://www.who.int/hiv/treatment/en/ [Retrieved 25 October 2006].

Chapter 4
Poverty and Income Distribution

In our world of 6 billion people, one billion own 80% of global GDP while another billion struggle to survive on less than a dollar day. This is a world out of balance.

— James D. Wolfensohn (former President of the World Bank), 2003

Introduction

Raising incomes and reducing the incidence of poverty is probably the most important goal of economic development in the low income economies of Asia. This chapter discusses poverty and the associated issue of income distribution from several different perspectives. We begin by explaining the different concepts of poverty measurement and then go on to investigate alternative theories of poverty. This is followed by an outline of the vulnerable groups and general reasons why people fall into poverty. We then turn to the experience of poverty in Asia and policies that have been adopted for reducing poverty. After discussing income distribution issues we turn our attention to the future including millennium development goals and macroeconomic and sector policies that can be adopted to achieve poverty reduction objectives.

In order to better understand the concept of poverty and poverty reduction we have to adopt a common definition and standard for measurement of poverty. In its broadest sense, poverty is the inability to access resources to be able to enjoy a minimal standard of living. To make this definition more concrete and measurable we have to define what a minimal standard of living means. Furthermore, we have to be able to relate this definition to different cultures and income groups.

The concept of a poverty line or poverty threshold has been adopted to convey the inability to achieve a minimal standard of living to maintain health and well being. Below this poverty threshold families are unable to provide sufficient food, clothing and shelter to sustain acceptable levels of health and general well being. Poverty lines have been used by international agencies such as the World Bank, the Asian Development Bank (ADB), and the United Nations to track progress in reducing poverty in developing economies around the globe. National governments have also developed poverty thresholds to accomplish similar objectives. In what follows in this chapter, we will also use the poverty line methodology as our primary index of poverty. The head count ratio (HCR) is the proportion of poor people or families relative to total population and is the primary measure of the extent of poverty

$$\mathrm{HCR} = q/n, \tag{4.1}$$

where q is the number of people (or families) in poverty and n is the number of people or families in the society.

This is not to say that there are other ways to measure poverty. The depth of poverty, or how far a family is below the poverty threshold, is also important. If most people are very close to the poverty line, it would take a minimal level of resource transfers to raise them above the poverty threshold. On the other hand, if many families are close to starvation, the level of resources required to reduce poverty would be substantially higher. Simply looking at the HCR, differences between two societies would not be discernable. The issue of whether we are considering individuals or families is also important. Conventional approaches treat family units rather than individuals in the family. By lifting family income above the poverty threshold it is assumed that everyone in the family is treated equally and benefits equally from an increase in income or consumption for the family as a whole. This is not necessarily true if there is systematic discrimination within the family, e.g., male members get more or children get less or the male breadwinner gets a bigger share. While these issues are important in themselves we will focus our attention to trends in head count ratios within countries and comparisons among countries in the Asian region.

Although it focuses on the entire range of incomes received in the society, the concept of income inequality is also related to poverty. To supplement notions of a poverty threshold some scholars have focused on the share of income earned by the lowest 20% of the income distribution or the Gini coefficient as a measure of relative deprivation in a society. As the share of the lowest 20 falls and the Gini coefficient rises, the distribution of income becomes more inequitable. The rich earn a larger share of income at the expense of the poor. We will investigate the historical relationship between income distribution and poverty in Asia as part of our analysis of trends in poverty.

Poverty Theory

Economic theory pays little attention to poverty and income distribution as part of the standard build up of factor payment equilibrium in the theory of the firm and the household. Labor is paid the value of its marginal product and income distribution and the level of living standards are a byproduct of this process. Those with high skill and lots of capital to work with will get paid more than those with few skills and minimal capital endowment. Using this marginal productivity theory and considering the pattern of economic growth and structural change, economists have developed some ad hoc ways to describe the pattern and incidence of poverty.

Vicious cycle and trickle down

According to the vicious cycle poverty begets poverty. Each generation of poor is unable to mobilize enough physical or human capital to raise their marginal productivity and living standards enough to pass on any additional capital to future generations. Educational attainment is low and continues to be inadequate to enable future generations to earn a living wage. The poor lack health care and this contributed to a low standard of human capital accumulation. The poor have minimal physical assets and are unable to provide their children with any physical capital either. Each generation remains poor and very few poor families are able to break out of this cycle. Land redistribution or the redistribution of other physical assets does not take place to a significant extent even with the aid of industrial countries. Education provides the primary mechanism for breaking the poverty cycle. By building human capital it is possible to break the cycle.

Trickle-down theories of development suggest that the process of economic growth can be a powerful enough force to lift living standards of the poor by raising employment, extending education to the poor, and by providing more physical capital for everyone to work with as labor productivity is generally raised throughout the society. It is not direct programs targeting the poor that reduces poverty. Rather, it is the powerful process of rapid economic growth that raises income for all, including the poor through the trickle-down mechanism.

Relating income distribution with growth and poverty

The Kuznets curve describes the relationship between income distribution and income over time. The title comes from the theory of Nobel Price winner Simon Kuznets who proposed in 1955 that in a two-sector growth model labor moved from the agricultural sector, where income inequality was low, to the industrial sector, where income inequality was high. As a result there would be deterioration in the overall income distribution in the industrialization phase of economic development. Later on, as the service sector became more important and economic progress became more widespread, Kuznets argued that this pattern would tend to reverse. The result of these shifting factors over time is that income distribution or income inequality is related to income by an "inverted U" shaped curve. Income inequality increases, peaking at a point where income is in a middle range and then decreases as income increases to higher levels.

Using time series data for individual countries, there is little evidence of a Kuznets curve for countries that have a large number of time series observations, except England, where there was an upsurge in inequality during the industrial revolution. Most other countries, both developed and developing, do not have sufficient data to permit a robust test of the hypothesis. Analysis of cross-section data for a wide array of countries at different levels of income suggests that there could be a Kuznets curve (Lindert and Williamson, 1985). However, some observers suggest that this is because income inequality is high in Latin America and the income of these countries are in the range where the curve peaks. When Latin American observations are removed, the Kuznets curve disappears.

The conflicting results of different empirical works on the Kuznets curve suggest that the matter is not completely settled although the bulk of evidence leads to the conclusion that its impact on income distribution and poverty is minimal. Therefore the tendency for income distribution to deteriorate during the crucial stages of the growth cycle as countries move from agriculturally based production systems to industrial economies with larger urban populations is small.

Role of economic growth

Many development economists put overwhelming stress on the importance of economic growth as the critical factor in reducing poverty. For some it is enough to stimulate economic growth and then let poverty take care of itself, much as the classical economists did when they neglected the distribution of income. Some of the chief proponents of growth as the primary driver of poverty reduction work for the World Bank and other international organizations (Dollar and Kraay, 2001; Ravallion and Chen, 2004). They are joined by Bourguignon and Morrison (2002). Financial columnists such as the highly respected Martin Wolfe also hold this view: "There's a proven way to rid the world of poverty" (Wolfe, 2004). Wolfe also argues that the world distribution of income has also improved dramatically in the past few years because of the rise of India and China. In any event there is strong econometric evidence that growth and poverty reduction are closely related at the aggregate level when economic growth and poverty are regressed on each other both over time within individual countries and in cross-section analysis across a panel over countries.

Role of microeconomics and labor market policies

While not denying the importance of growth and general macroeconomic health in reducing poverty, market factors and government policies at the microeconomic level are also important in the view of many development economists. Taken broadly, these issues and policies relate to human resource development and labor markets as well as product markets and financial markets.

Financial markets

The poor have limited access to formal financial markets. Because they have low and fluctuating income they are poor credit risks. Furthermore, since they would borrow very little, compared with richer customers, the administrative costs of lending to the poor are high. As a result, banks and others financial intermediaries are not willing to lend to the poor. The poor have to turn to moneylenders, landlords, friends or pawnshops to access credit. These credit sources are expensive and could be unreliable. When efforts have been made to develop a credit relationship with the poor, many of these stereotype views prove to be erroneous. The poor are usually good credit risks and administrative costs are not substantially higher than for other borrowers. Credit can be made available to the poor at reasonable rates (there is the experience of Grameen Bank and its look alikes) allowing the poorer strata's society to even out income streams and prevent households from falling into poverty.

Human resource development

Where health and educational facilities are widely available and where children are required to attend school, human development indicators are high relative to levels of per capita income, and income is rather evenly distributed. China and Vietnam are two countries in Asia that exemplify these characteristics. When economic growth began to accelerate in the 1980s and 1990s as market-friendly macroeconomic policies were put into place, the rate of poverty fell dramatically.

The interaction of favorable education and health indicators interacted with market-oriented product and labor markets to yield a favorable result both in terms of economic growth and poverty reduction. While income distribution did deteriorate in both China and Vietnam, the overall effect has been very favorable to poverty reduction. In the earlier era of the 1960s and 1970s, an equitable distribution of income and focus on universal education in the miracle economies of East Asia (Taiwan, Korea, Singapore and Hong Kong) also helped to pave the way for more rapid growth once the macropolicy environment became favorable (Quibria, 2002; Dowling and Valenzuela, 2004).

Labor markets

The early theories of development beginning with Arthur Lewis, stressed the importance of having a pool of unskilled labor willing to work for subsistence wages. The movement of this "reserve army" of the underemployed and unemployed to the cities enabled newly forming industrial establishments to produce a wide array of labor-intensive products that were not only consumed locally but also exported to the rest of the world. These theories were proven to have powerful explanatory power as first Japan and then the newly industrializing economies of East Asia were able to industrialize, grow rapidly, and lift standards of living quickly. More recently, China has used the same model to raise the living standards and develop its eastern provinces into modern industrial centers. A key component of the developments in all these countries has been the mobility of labor and the lack of government control of wages. It is true that "sweat shops" that exploit labor, provide minimal benefits, employ child labor, and skimp on worker safety exist and even flourish. These are certainly negatives that have to be redressed. Nevertheless, it is useful to contrast and compare the experience of these countries with India, where a much more restrictive labor market has been at least partially responsible for the inability to provide jobs for new entrants into the labor force and the continued high level of unemployment and poverty. The position for maintaining a free and competitive labor market is certainly controversial. However, from the point of view of poverty reduction, free labor markets do seem to work better than restrictive labor policies that maintain high wages and benefits for organized workers at the expense of creating employment for the rest of the workforce. When combined with health and education policies, restrictive labor policies also discriminate against the poor, resulting in an intractable poverty problem.

Trade policies and poverty

In the era of globalization, trade policies have tended to focus on the various rounds of trade negotiations under the World Trade Organization (WTO) as well as bilateral trade agreements between countries in the region and elsewhere. How do these negotiations and trade agreements interact with domestic policy to affect the level and incidence of poverty? There is no simple answer to this question. The impact on poverty will vary depending on the circumstances of individual countries as well as the time frame of analysis. Short run and long run impacts may be much different. Nevertheless we can infer a lot from the interaction between international trade and poverty by looking at a few fundamental economic relationships and policy guidelines.

Initially we have to analyze the way markets work and how international prices and quantities traded change as a result of trade liberalization. For example, if trade liberalization for agricultural goods increases the demand for these products in international markets, what will be the outcome for the poor? It will depend upon many factors, including the competitiveness of markets, the elasticity of demand and supply for goods, and factors of production as well as the degree to which the poor have access to these markets. If the poor are able to gain greater market access and can effectively sell more than they did before liberalization, their incomes and consumption will rise and so will incomes of those in their communities who supply the goods they purchase (see Timmer (1997) and Mellor and Gavian (1999) for analysis of these so-called second-round effects). The impact on the rural poor is likely to be large, particularly if the elasticity of labor supply is not infinite. Restrictions on market access and monopoly elements that restrict the impact on the poor will have milder or even negative impacts on their well being.

To understand further the complexity of the poverty/trade relationship we can look at one of the cornerstones of trade theory, the Samuelson–Stopler theorem. This theorem says that an increase in the price of labor-intensive goods raises real labor income and reduces real returns to capita. Since people in poverty are usually working in labor-intensive industries, this theorem provides a critical link with poverty. However, the theorem depends on a number of assumptions that may not be true in practical applications. These include the fact that the model applies in a basic two good and two factor market world and that there is mobility of labor between regions and markets. Furthermore, the Samuelson–Stopler theorem assumes goods being traded are homogeneous, there is substitutability between domestic and foreign goods, and that goods and factor markets are perfectly competitive. While the relaxation of the assumptions of the Samuelson–Stopler theorem does not necessarily vitiate its conclusion that greater international trade will reduce poverty, we have to recognize that trade's impact on poverty could be muted (Winters, 1999) and the pattern, strength, and timing of the linkage between poverty and trade will vary.

Another aspect of the trade/poverty nexus relates to the impact of trade reforms on taxes and spending. Lower levels of social spending resulting from reduced revenue from trade taxes could have a negative impact on the poor. However, this does not necessarily have to be the case. Governments have other revenue sources. If they have a commitment to poverty reduction they will be able to offset the loss of trade taxes and maintain social spending at desirable levels.

The poor have few assets and are therefore much more vulnerable to external shocks that have negative impacts on their income and employment. To the extent that more open and trade-oriented economies experience shocks, poverty will be adversely affected. In the Asian crisis of 1997 many countries experienced sharp declines in output and employment. Countries that had trading regimes that were less open, including China and India, were less affected. Furthermore, there may be transition shocks as a result of movement to a more open trading regime. Analysis of the Asian financial crisis suggests that, for the most part, economic disruptions were temporary and the level of poverty began to fall again a year or so later.

To conclude this section, it is my belief that we have to understand that if policy makers keep the poor in mind when undertaking trade reform, the impact of their reforms will be more powerful in providing a mechanism for reducing poverty. Establishing market access for the poor, understanding the effect of monopoly power as well the size of demand and supply elasticities can provide a better understanding of the trade/poverty connection. Sequencing major trade liberalization is probably desirable in order to reduce the shocks to the poor as well as to gain further knowledge of the channels of transmission and appropriate policies to follow. Monitoring and insuring the proper functioning of markets is particularly important. In addition, the government should be willing to supply infrastructure that enables the poor to access markets and provide credits that enables (for example) farmers to hire additional laborers, buy fertilizers, or expand capacity in light of greater demand.

Vulnerable Groups

There are vulnerable groups that have significantly higher rates of poverty than others. To reduce poverty significantly these groups will have to be targeted.

Women

Women in Asia are discriminated against in many different ways and as a result are among the poorest groups in society, even though they constitute around half of the total population[1]. Both subtle and overt discrimination against women is manifest in a number of ways, including lower wages for similar work (between 30% and 50% on average)

[1]Amartya Sen (1992) identified a particular characteristic of the gender distribution in South Asia which he described as the missing women phenomena. Elsewhere in the world women outnumber men by a ratio of around 106 to 100 while in south Asia the ratio is 94 to 100. Where are all the missing women, estimated to be around 75 million? While we do not know the specific answer, researchers speculate that it is due to a combination of infanticide and neglect.

and lower levels of education and skill which results in women being relegated to low skill and low wage occupations. Women also have higher rate of infant mortality, poorer health care, less education on average and shorter life expectancy. They also consume fewer calories than male counterparts within the family. As a result, more than 70% of the poor in Asia are women and rates of poverty for women are significantly higher than for men in most Asian economies. In Cambodia, women constitute 53% of the workforce, yet earn 30% –40% less than men. Gender inequality is also reflected in educational attainment with male school enrolment being 50% higher at age 15 and 300% higher at 18 (see Mollett's (2002) study commissioned by BOND[2]).

Because of various social conventions and pressures, the extent of discrimination against women appears to be higher in South Asia than in other regions and these prejudices reflect a higher incidence of poverty for women in this region than in other parts of Asia.

Gender inequality is perpetuated by the lending programs of international donors. Bond, a network of NGOs in the United Kingdom finds that only a minute fraction of EC aid finances gender equality programs which is inadequate for meeting internationally agreed targets for women's rights.

Landless and those lacking property rights

In rural Asia, land ownership is the major determinant of prosperity. Marginal farmers, hired labor, and tenants comprise the bulk of the rural poor in Asia. According to IFAD (2002), 75% of landless households are in the lowest expenditure quintile in South Asia. In Bangladesh, 70% of the poor and 80% of the seriously poor are landless, and poverty falls dramatically as the size of owned land increases (Table 4.1). In Korea, nearly 90% of the poor are landless. In India, households cultivating no land or with virtually no land had poverty incidence of over 40% compared with a 15% poverty incidence for families cultivating more than 4 ha. In the Philippines, a substantial

[2]BOND is a network of over 260 development NGOs in the United Kingdom.

Table 4.1 Poverty and Land Ownership in Bangladesh — Late 1980s

Landholder class by acres of land owned	Percent of population	Headcount index of poverty (in %)
Landless (0–0.04)	13.5	61.4
Near Landless (0.05–0.49)	31.5	53.9
Marginal (0.50–1.49)	19.2	43.4
Small (1.50–2.49)	11.3	34.2
Medium (2.5–7.49)	18.8	26.6
Large (7.50 and above)	5.3	10.1
Rural Bangladesh	100.0	47.5

Source: IFAD (2002, p. 30, Table 2.2).

majority of the poor owner-cultivators have small farms located in rain-fed areas, with poverty limiting their capacity to improve productivity and limited access to credit constraining them from adopting modern technology.

Lack of land access is compounded by lack of access to irrigation. In Bangladesh, nearly 80% of farmers without irrigation are poor compared with only 51% of farmers with access to irrigation. Similar results hold for other countries in South Asia and Southeast Asia (Table 4.1)

In a broader sense, lack of property rights inhibit the ability of residents to take full advantage of the opportunities that full ownership provides. Tenants without ownership in northern Thailand are fearful of land expropriation if they leave the premises and pass up opportunities for seasonal work in other regions of the country. Without secure land ownership, farmers are unwilling to undertake land improvement investment or buy new seed varieties or investigate new farming techniques and application of fertilizers and pesticides. And lenders are reluctant to extend credit to farmers who do not have secure tenure (see next section for further discussion of credit). Furthermore without tenure, minorities are often displaced without proper compensation. In India, the Katkan people have been displaced, and in Malaysia there has been extensive logging of tribal lands. In Laos, the Nam Theum 2 dam project has displaced many, and in China the Three Gorges Project has created significant social disruption.

Bypassed Regions, Ethnic Minorities and Lower Classes

Bypassed regions are those parts of a country or region that are geographically isolated from the rest of society. They can be remote mountains or jungles without good transportation or communication, or backward regions where soil fertility is low and there is limited use of irrigation. These regions are often settled by ethnic minorities although this need not always be the case. For example, northeastern Thailand has long been a backward region with low income per capita and a high incidence of poverty. Generally, backward regions have been settled by ethnic minorities who have been driven there by lowland settlers. They are poor because of discrimination and also social exclusion. Tribal people have been forced to remote and mountainous regions of the Philippines, Thailand and Vietnam, and also in India and parts of China. Poverty levels are high among these bypassed people.

Poverty is also considerably higher in lower castes in India, particularly the lowest caste called Dalits (untouchables) also known as scheduled castes. Health indicators are weak, illiteracy rates high, and 92% of Dalits are laborers. Furthermore, life expectancy is lower and infant mortality is higher than for other classes in society. Along with scheduled tribes[3], the scheduled classes are widely subject to social exclusion despite efforts of the government to promote a classless society. Together, scheduled castes at 16% and scheduled tribes at 8% comprise about 25% of the population of India. Scheduled tribe and scheduled class poverty rates are over 50% compared with a national average of less than 40%. (For further details on poverty in India, see World Bank (1998)).

In Vietnam, tribal rates of poverty are also considerably higher than the national average (see IFAD (2005) report for further details).

[3]Scheduled tribes include communities which did not accept the caste system and preferred to reside deep in the jungles, forests and mountains of India, away from the main population. The Scheduled Tribes are also called Adivasi, meaning aboriginals.

Human development indicators such as average level of education, levels of literacy, life expectancy, and infant mortality are also lower for indigenous people/ethnic minorities and scheduled tribes. In China, per capita in southwestern provinces dominated by ethnic minorities' per capita income is only 30% of the national average.

Children and the Elderly

Young children and the elderly are poorer than other groups because they do not have income and must depend upon others for their survival. For a variety of reasons families tend to be larger in poor communities resulting in more children who are poor than in well-to-do families. Because there are generally no social safety nets in Asia, particularly in the poorer countries, the aging poor become poorer as they lose their ability to work and earn income. The need to support several generations within family-oriented communities puts further pressure on the resources of poor villages and their constituent families. Low family income and high incidence of poverty also result in higher infant mortality and morbidity in poorer regions (Tables 4.2–4.4). Government resources for education and health are also focused on urban areas where politicians and the urban elites control government allocation of resources. Rural residents and the poor in urban areas do not receive enough resources to raise living standards so that children and the elderly can be properly cared for.

Artisanal Fishermen

About 90% of the world's 30 million fishermen work in Asia and roughly 80% of them are small-scale artisanal fishermen. Without

Table 4.2 Mortality of Children Across Regions per 1,000 Live Births (1998)

	South Asia	Sub-Saharan Africa	East Asia and Pacific
Infant mortality	69.8	91.7	43.4
Under-five mortality	101.6	166.3	71.2

Source: Navaneetham and Jose (2005, p. 9, Table 2).

Table 4.3 Malnutrition of Children (Under Five) Across Regions (%)

Anthropometrical indicator	South Asia	Sub-Saharan Africa	East Asia and Pacific
Moderate to severe wasting	15.5	8.7	7.7
Severe stunting	25.7	13.9	12.6
Moderate to severe stunting	44.8	32.8	32.9
Severe underweight	17.5	7.4	5.5
Moderate to severe underweight	46.5	25.7	26.3

Source: Navaneetham and Jose (2005, p. 10, Table 3).

Table 4.4 Other Related Indicators of Child Nutrition (%)

Other related indicators	South Asia	Sub-Sahara Africa	East Asia and Pacific
Incidence of diarrhea	14.9	22.3	9.3
Acute respiratory infection	17.3	16.4	12.2
Immunization against measles	54.7	60.4	75.6
Immunization against DPT3	54.3	57.1	70.3
Low birth weight	33.3	13.3	12.4

Source: Navaneetham and Jose (2005, p. 11, Table 4).

better management of the oceans, there will be not enough fish for food (FAO, 1998; Indo-Pacific Fishery Commission, 1994).

Population growth, open access to the sea, and the belief of unlimited fishing resources in the sea have doubled the number of fishermen since 1970 according to the FAO. On the other hand, fishery resources are limited and are depleting fast in most coastal areas in Asia. Coastal fishermen using traditional fishing methods are among the poorest groups in South and Southeast Asia. They suffer from increasing population pressure and declining productivity as fisheries have become over-fished and these traditional fishermen face competition from large fishing boats and environmental advocates who want to preserve coastal environments. Their range is also restricted to a narrow band of coastal waters since their boats are not big enough to range further out to sea. These fishermen are particularly vulnerable in Bangladesh, the Philippines and Thailand where storms and other natural disasters can create further disruptions to their livelihood.

Rural and Urban Poverty

The incidence of poverty in rural areas is significantly higher than in urban areas. This is one reason why there has been out-migration from the countryside to the cities of Asia. This pattern of migration has accelerated in the past few decades. Population in urban areas has increased by 3.75% per annum in the past 25 years and is expected to double between 2000 and 2030 from 665 million to 1.2 billion even as there is an absolute decline in rural populations (World Bank, 2003).

Nevertheless, rural poverty is the focus of most policy in the poorest countries in Asia since the majority of the poor still reside in rural settings. This pattern is particularly prevalent in South Asia and in the Mekong countries (Table 4.5). The incidence of poverty is not always consistent with other sources. For example, overall poverty rates in

Table 4.5 Rural and Urban Poverty in Asia — 1990 and 2000

Country	Period	Rural poverty (first year)	Rural poverty (second year)	Urban poverty (first year)	Urban poverty (second year)
Bangladesh	1992, 2000	46.0	37.4	23.3	19.1
Cambodia	1994, 1997	43.1	40.1	24.8	21.1
Indonesia	1990, 1999	14.3	20.2	16.8	15.1
India	1988, 1999	39.1	27.1	38.2	23.6
Malaysia	1989, 1999	21.8	13.2	7.5	3.8
Pakistan	1991, 2003	25.2	38.6	26.6	22.4
Philippines	1991, 2000	55.1	48.8	35.6	18.6
Korea	1988, 1996	7.2	9.4	8.9	6.1
Sri Lanka	1991, 1996	22.0	27.0	15.0	15.0
Vietnam	1993, 1998	66.0	45.0	25.0	9.0

Source: UNESCAP (2004).

Table 4.6 Rural and Urban Poverty in Asia — 1990s

Country	2001 Gross national income per capita PPP dollars	Survey year	Rural poverty using national poverty lines	Urban poverty using national poverty lines	National poverty using national poverty lines
Nepal	1360	1996	44	23	42
Bangladesh	1600	2000	53	37	50
Pakistan	1860	1999	36	24	33
Vietnam	2070	1998	37	7	29
India	2820	2000	30	25	29
Sri Lanka	3260	1996	27	15	25
China	3950	1998	5	2	5
Philippines	4070	1997	51	22	37
Thailand	6230	1996	16	10	13

Source: World Bank (2006).

India reported in Tables 4.6–4.8 are higher than those reported in Table 4.5 for either rural or urban poverty rates. Furthermore, the rural poverty rate for Pakistan for 1991 is suspiciously high. However, the general pattern of higher poverty in rural areas tends to be confirmed by these tables and other information compiled for individual countries by Dowling and Valenzuela (2004) and the World Bank (2006).

Circle of Poverty

The reasons why people are poor and unable to escape from the poverty trap were discussed in the previous section. They have little human or physical capital and they are often subject to discrimination, marginalization and exploitation. Furthermore, they are unable to accumulate enough assets, either physical capital or human capital in the form of better education and better health to escape from poverty. As a result, the poverty cycle is perpetuated from generation to generation. In addition, there are other extenuating circumstances that help perpetuate the circle of poverty.

Table 4.7 Poverty Indicators for Asian Economies — Percentage of Population Below One Dollar a Day Poverty Line in Selected Countries (1975–2002)

Country	1975	1985	1990	1993	1996	1998	2000	2002
Bangladesh				35.9	29.1		36.0	
India			46.6	51.1	46.2	44.2	34.7	
Nepal					37.7			
Pakistan			47.8	33.9	31.1			
Sri Lanka			3.8		6.6			
Cambodia			48.3		36.7	37.7	44.6	44.2
Indonesia	64.3	32.2	20.6	14.8	7.8	12.0	9.9	7.2
Lao PDR			53.0	48.8	41.3	36.6	35.1	30.0
Malaysia	17.5	10.8	2.0	1.0	<0.5	<0.5	<0.5	<0.5
Philippines	35.7	32.4	19.1	18.4	14.8	13.7	13.5	11.5
Thailand	8.1	10.0	12.5	6.0	2.2	3.3	5.2	2.4
Vietnam			50.8	39.9	23.6	16.4	15.2	13.6
China			31.5	29.0	16.4	16.1	15.4	12.7

Source: World Bank (2002); UNESCAP (2004); Quibria (2002).
Note: Korea, Hong Kong and Singapore had negligible levels of poverty for all years using the one dollar a day cut off line.

Table 4.8 Poverty Indicators for Asian Economies — 2 Dollars a Day Poverty Line in Selected Countries

Country	1990	1995	2000	2002
Bangladesh	86.4	77.8	82.8	
India	88.2	87.9	79.9	
Nepal	88.0	82.5		
Pakistan	87.9	84.7		
Sri Lanka	40.6	45.4		
Cambodia	83.7	76.9		
Indonesia	71.1	50.5	59.5	53.5
Lao PDR	89.6	83.1	77.4	76.6
Malaysia	18.5	14.0	9.8	9.6
Philippines	53.5	46.5	47.1	44.7
Thailand	47.0	28.2	35.6	27.7
Vietnam	87.0	69.4	63.5	58.2
China	69.9	51.6	44.8	37.3

Source: UNESCAP (2004).

Limited access to modern technology

Poorer farmers adopt new technology at a slower rate than the non-poor. In Bangladesh and the Philippines, farmers plant a smaller proportion of their holdings in modern high-yielding varieties and are slow to implement new methods of crop rotation and/or introduce new crops (ADB, 1995). This slow rate of adaptation of new technologies is the result of a combination of low literacy, lack of contact with farm extension centers, limited communication including a poor farm to market road network, poor telecommunications, and other impediments that slow the flow of new information and expertise. Furthermore, there is a lack of credit to enable farmers to adopt new varieties, undertake farm improvements, and buy new seeds, fertilizers and pesticides.

Family size

Poverty is generally correlated directly with family size and inversely with the number of income earners in the family. In Bangladesh, poor families have slightly smaller families than the non-poor but more children under the age 10, fewer males above 16, and a higher child/woman ratio. In Nepal, the poor have larger families and more children than the non-poor. These same generalizations hold for Thailand and the Philippines as well. Furthermore, the dependency ratio — the proportion of non-earners in the family to the total number of household members — is higher in poor relative to non-poor families. Part of the reason for this, perhaps counterintuitive result, is that the poor have limited access to birth control information and are unable to afford them even if they were available and had knowledge. Furthermore, the poor see children as sources of labor and helping to augment family income. Spending on education is low and children begin working as soon as they reach puberty or even before. As a result the circle of poverty is perpetuated.

Access to markets

Whether they work in rural or urban areas, the poor are often cut off from the formal market economy. We saw in the previous section that the poor are often concentrated in backward geographic locations. In China, progress from the reforms of the 1980s were limited in rural areas where land quality was poor and irrigation limited or where rural communities were not well integrated into wider labor and product markets. The southwestern provinces dominated by ethnic minorities are one example. In Pakistan the more arid areas of Baluchistan are more poverty-prone while the northeastern provinces of Thailand have significantly higher incidences of poverty than the rest of the nation. In urban areas the poor most often work in the informal sector where they work long hours and receive no benefits (ADB, 1995). Recent immigrants from the countryside are often young and illiterate. They are competing with others for available jobs, being part of the rapid inflow of new immigrants to the cities. As a result, wages are very low, unemployment rates high, and working conditions often unsafe. Child labor is also widespread.

Access to credit

Credit is likely to be more available to the poor living in relatively wealthy communities with well-developed capital markets where commercial transactions and credit markets are commonplace and a variety of financial instruments are available and widely used. In this setting, credit may be available from the landlord, moneylenders, pawnshops and other informal lenders. It will be priced at higher rates than credit extended in the formal sector to qualified borrowers because of higher risk due to lack of an established credit history and higher administrative costs because of the size of the loans. In more remote areas, where there are fewer lenders, both formal and informal, credit from the private sector may not be forthcoming. In these circumstances, it may be useful to consider the provision of subsidized credit through government lending programs. However, the provision of such credit, perhaps at subsidized rates, should be carefully considered and used only in circumstances where the poor are prevented from exploiting available investment opportunities because of credit constraints. These programs should not be used to subsidize consumption. It would be better to enable the poor to cope with risks of income volatility by reducing this volatility rather than a credit subsidy to maintain consumption.

Education and health

Land redistribution and capital accumulation by the poor is limited. Greater accumulation of human capital is, therefore, the primary way for the poor to raise their living standards and the living standards of the next generation born into poverty. Several studies by the ADB in two studies of poverty (ADB, 1992, 1994) and summarized in a summary volume titled *Escaping the Poverty Trap* (ADB, 1995) demonstrate several features of human capital and the ramifications for the poor.

Firstly, the impact of schooling duration on both post-school wage levels and labor productivity is considerable. Even returns from primary schooling are high for both boys and girls. Furthermore, the gender wage differential tends to fall with more schooling. Another aspect of education and health care for poor girls is that the opportunity cost of teenage girls attending school is high because they often have to take care of sick younger siblings. An indirect benefit of providing better health facilities for young children and infants would be more schooling, higher wages, and increased productivity for their older sisters.

Secondly, increases in schooling quality and duration provide additional benefits in terms of positive impact on wages and labor productivity. These studies also found that productivity gains may be larger from concentrating scarce schooling resources on fewer high-quality schools than spreading them thin across many low-quality schools.

Thirdly, there are synergies between health and educational expenditures. Better health and nutrition lead to higher productivity and wages and also to better performance in school. When these are combined with longer schooling duration and better quality schools the productivity gains and impact on wages and poverty reduction are even more powerful. Rather than substitute for each other, better schools and better health provide synergy and increasing returns to investments in terms of the beneficial impact on poverty, increases in wages and labor productivity. There are also highly beneficial results from increasing women's education in terms of family health for this and the next generation, lower infant mortality, and a reduction in population growth. In Chapter 6 we coin the term super-externality to describe the various benefits to the environment from reforestation. This term applies equally to the education of girls since there are so many secondary benefits to society from women's education in addition to the direct benefits to individual communities and families.

Poverty in Asia

Many of the general principles, observations and conclusions discussed apply to Asia and many examples from the Asian experience were used to illustrate various aspects of poverty. In this chapter we explore the historical experience of poverty in Asia over the last few decades as well as the key relationship between poverty reduction and income growth.

Poverty and income

There are several studies that look into the relationship between income growth and poverty reduction both on a global scale and for the Asian region (UNESCAP, 2004; Dollar and Kraay, 2001a). The general conclusions of these papers is that income growth and poverty are inversely related. How strong the relationship is and the importance of other intervening variables such as trade openness is subject to further investigation. We focus on the findings for Asian economies since the 1980s whenever possible and for more recent years where historical data are lacking.

A summary of poverty indicators compiled by the World Bank (2002), UNESCAP (2002), and Quibria (2002) are displayed in Table 4.7. There are inconsistencies between the three sources, and even within the World Bank. Nevertheless a relatively consistent pattern of change in poverty over time is evident, even though the levels of poverty in a particular year might vary. A rather modest one dollar per day benchmark for poverty is used with the result that several countries have virtually eliminated poverty by 2002. Between 1990 and 2002 poverty has been reduced by half in China, Indonesia, Malaysia, Thailand and Vietnam, and virtually eliminated in Korea. Sri Lanka shows low levels throughout the sample period although there is not much reliable data after 1996. This means that these countries have already achieved the Millennium Development Goal of reducing poverty by half between 1990 and 2015 although we could question whether a one dollar is a realistic poverty threshold. Using a two dollar per day threshold poverty proportions are considerably higher, as reflected in Table 4.8.

Progress in reducing poverty has been less satisfactory in Cambodia, Lao PDR and the Philippines among Southeast Asian economies as well as in the remainder of South Asia outside Sri Lanka. In India, poverty has fallen rather slowly and still remains above 40% of the population even using a one dollar a day threshold. Furthermore, much of the poverty is persistent. According to sample data compiled by Gaiha and Deolalikar (1993), 88% of sample households had been poor at least some time in the last 9 years (between 1975 and 1984) and 61% had been poor at least half the time. Furthermore, since most of the lower 3 or 4 deciles of the income distribution are poor and the probability of transition out of these lower deciles is small, most families in these income categories will remain in poverty most of their lives. Poverty is highest in four provinces (Bihar, Orissa, West Bengal, and Assam) which are also the provinces with the lowest rates of economic growth (Srinivasan, 2004). In Pakistan, the last reliable figures are nearly a decade old. However, they do reflect a gradual decline in poverty since the early 1980s. In Bangladesh there has been virtually no change in the incidence of poverty since the early 1980s and perhaps even a small increase.

Many countries have compiled data based on their own poverty lines and these figures are compiled in Table 4.9. The pattern is similar to that compiled by the World Bank with some variations. In China, and to a lesser extent India, national figures show a much

Table 4.9 Poverty Indicators for Asian Economies — National Poverty Lines

Country	1990/91	1995	2000	2002
Bangladesh	47.8	36.0	33.7	
India	38.9	36.0	26.1	
Nepal	41.4	42.0		
Pakistan	26.1	28.7	32.1	31.8
Sri Lanka	20.0	25.0		
Cambodia		39.0		
Indonesia	15.1	11.3	18.2	
Malaysia	17.1	9.6	8.1	
Philippines	45.3	40.6	34.0	
Thailand	27.2	16.3	14.2	9.8
Vietnam	58.2	37.4	32.0	28.9
China	9.4	7.1	3.4	3.0

Source: UNESCAP (2004).
Notes: There may be some discrepancies with Table 4.6, which is from the World Bank. Bangladesh is the most noticeable.

lower incidence and a decreasing trend in poverty while in Sri Lanka, the Philippines, Thailand and Vietnam, national figures are much higher than the dollar a day index. In Indonesia, national poverty estimates do not decrease as rapidly as the dollar a day figures. There is a jump in both series in 1999, reflecting the impact of the Asian financial crisis. National figures are higher in Malaysia, particularly in the latter part of the 1990s. National estimates show poverty increasing in Pakistan. No estimates are available for the dollar a day index. There are discrepancies between World Bank estimates displayed in Table 4.6 and Table 4.9, which relies on figures from the United Nations. World Bank estimates are higher for Bangladesh. In Nepal, the most recent databanks show only one observation in the mid-1990s.

The incidence of rural poverty is substantially higher than urban poverty in most countries, although India, Indonesia, Korea and Pakistan are exceptions. Measurement differences may account for some of this, since urban poverty rates are typically lower than rural rates. Where urban rates are lower, the result is all the more onerous since in the poorest countries there is a higher proportion of the population living in rural areas than in urban areas.

Transition Economies in Central Asia

The transitional economies in Central Asia had a different poverty experience from the rest of Asia. In the Soviet Era up to the early 1990s before the breakdown of the Soviet Union, recorded poverty levels were low. However, in the first half of the 1990s there was a general increase in poverty as a result of the transition shock that resulted in large-scale loss of employment, income, and purchasing power in all Central Asian Republics (CARs). Underlying factors included the breakdown of Council for Mutual Economic Assistance production and distribution network, the collapse of the social security system, higher inflation caused by disruption in production of key goods and loss of control over the money supply, civil strife in several countries, and out-migration of skilled workers, primarily those of Russian origin. Since the mid-1990s national poverty line estimates suggest a decline in poverty in all CARs where data are available (Table 4.10). Nevertheless, the pace of decline has been slow and poverty remains quite high at about 40% of the total population in the CARs as a group[4]. Lower inflation rates, stronger economic growth, creation of new external trade linkages, restoration of social and political stability, and the upturn in agricultural

---

[4]On the broader issue of achieving the Millennium Development Goals (MDGs), the CARs have made progress but there is an unfinished agenda. To quote an authoritative study: "...the countries of greatest concern are Tajikistan, Uzbekistan, Kazakhstan, and Armenia. The first two have high and increasing rates of malnutrition. Education standards have also slipped and the virtual collapse of the social sector in some countries has resulted in a general deterioration of health indicators" (UNESCAP–UNDP–ADB, 2005, p. 3). The study calls for a reshaping of national and local institutions involved in service delivery to achieve the MDGs. Key issues include making services available, improving quality, reducing barriers to accessing services, and broadening the range of providers.

Table 4.10 Poverty and Income Distribution

	National poverty incidence (Population below poverty line, %)		GINI index	Share of bottom 20% of income recipients
	Earliest	Latest		
Oil exporters				
Azerbaijan	68.1 (1995)	49.6 (2001)	36.5 (2001)	7.5
Kazakhstan	34.6 (1996)	27.9 (2002)	31.3 (2001)	8.2
Turkmenistan	—	29.9 (1998)	40.8 (1998)	6.1
Non-oil exporters				
Kyrgyz Republic	51.0 (1997)	47.6 (2001)	29.0 (2001)	9.1
Tajikistan	—	56.6 (2003)	34.7 (1998)	8.0
Uzbekistan	—	27.5 (2000)	26.8 (2000)	9.2
CARs		39.9	33.2	8.0
Others				
Thailand	13.1 (1992)	9.8 (2002)	43.2 (2000)	6.1
Malaysia	—	7.5 (1999)	49.2 (1997)	4.7
Indonesia	15.7 (1996)	18.2 (2002)	34.3 (2002)	8.4
Mongolia	36.3 (1995)	33.1 (1995)	44.0 (1998)	—

Source: ADB (2004); Alam *et al.* (2005).
Note: — = not available.

production and incomes have been contributing factors to the decline in poverty in the CARs.

The latest estimates of national poverty lines show the following:

- Poverty is generally higher in the non-oil exporters than the oil exporters.
- With poverty incidence in excess of 50%, Tajikistan is the poorest CAR.
- Azerbaijan (an oil exporter) and Kyrgyz Republic seem to have high poverty incidence in excess of 45%.
- In Kazakhstan, poverty levels are somewhat lower at around 28%, having come down from over 40% in the late 1990s.
- Uzbekistan and Turkmenistan (an oil exporter) also appear to have low poverty levels.

Rural poverty seems to be a significant problem in some CARs as the sector has absorbed many urban unemployed. Income disparities and poverty are more pronounced in some sub-regions (e.g., the Ferghana Valley and border regions of Kazakhstan, the Kyrgyz Republic and Tajikistan).

Income Elasticity of Poverty Reduction

Because income growth is such a powerful force in reducing poverty we investigate this relationship in a bit more detail. Dollar and Kraay (2001a) look at the share of the lowest 20% of the population in the income distribution compared with income over time for a wide panel of developing countries over the past few decades. They find a very tight fit (*R* squared of 0.88 for 137 countries) in a regression of the log of per capita income in the lowest quintile as a function of the log of per capita income and also for a regression in first differences.

The results strongly suggest that within countries incomes of the poor on average rise equiproportionately with average income. These results assume that the lowest quintile of the income distribution is coincident with the level of poverty in the society or at least closely correlated with it. However, our investigation of the poverty proportion above suggests that there are wide variations in the proportion of the population below the poverty line in Asia although this does not necessarily negate the Dollar and Kraay results. Hasan and Quibria (2002) provide a better test of the relationship between income growth and poverty reduction by a direct regression of the log of income on the log of poverty for a wide range of developing countries. They are also able to isolate the response of poverty to growth for different regions using fixed effects. They find a much stronger response of poverty reduction from income growth in East Asia than in South Asia or the other developing regions of Latin America or Africa. 1% increase in per capita incomes in East Asia is associated with 1.6% reduction in the level of absolute poverty. In South Asia the response is only half as strong, 0.8% reduction in poverty, while in Latin America it is 1.1% and in Africa 0.7%.

Why did East Asian countries have such a strong response while the other developing regions did not? Hasan and Quibria (2002) suggest that the answer lies in a combination of institutions and policies. We explore these factors individually in the next section of this chapter.

Policy Implementation to Reduce Poverty in Asia

A variety of macroeconomic and microeconomic policies were adopted in developing Asia over the last three decades. They have been outlined and discussed by many writers in connection with what has come to be known as the Asian economic "miracle", a term coined by the World Bank, which has stuck as a moniker describing the economies of East and Southeast Asia, including Korea, Taiwan, Hong Kong, Singapore, Indonesia, Thailand and Malaysia. Sometimes the Philippines is included. South Asia is normally excluded since it only began to grow in the 1990s. The World Bank coined the term in its 1993 volume, *The East Asian Miracle: Economic Growth and Public Policy.*

Dowling and Valenzuela (2004) suggest there were four primary factors responsible for growth and economic dynamism which were shared by all the miracle economies. They were:

- outward looking trade policies including emphasis on exports and foreign direct investment;
- macroeconomic policies including good governance and prudent policies to keep budget deficits in check and maintain a low inflation rate;
- strong educational emphasis combined with rapid labor force growth and increased labor productivity;
- labor market flexibility.

These policies had both direct and indirect effects on poverty reduction. Dowling and Valenzuela also stress several secondary factors including initial conditions such as a higher initial starting point for education and labor market skills, and a mixture of good industrial and agricultural policies that achieved a balance in growth as the economies industrialized.

To cap off the discussion of miracle economies' growth, Dowling and Valenzuela stress the importance of higher saving and investment rates and an increase in labor productivity, either as a result of technological progress through total factor productivity growth or as a result of growth in the number of hours worked and the build up of physical capital facilitated by high saving and investment rates.

How does all of these impact poverty? Certainly we know that higher growth leads directly to a reduction in poverty just by the calculations of the poverty elasticity such as those cited above by Hasan and Quibria (2002). There are also indirect impacts through the channels of international trade, flexible labor markets, better education and health, and good governance and lower levels of corruption. We turn briefly to these factors.

Openness to trade

Many economists have emphasized the importance of international trade in lifting incomes and reducing poverty. Using several measures of openness, Hasan and Quibria conclude that East Asia is the most open of the four regions and South Asia is the most closed (Table 4.11). These results, which show the very high openness of East Asia compared with other regions, are consistent with the work of Dollar and Kraay (2001a,b), Sachs and Warner (1995), and Srinivasan and Bhatwati (1999). While some writers have argued that openness can hurt the poor, the Dollar and Kraay, and Hasan and Quibria results suggest that this was not the case in East Asia. The Asian growth miracle of the 1970s, 1980s and

Table 4.11 Measures of Openness in Developing Regions

Region	Trade share (%)	Average tariff rate (%)	Sachs–Warner index of openness
East Asia	65.6	9.1	0.74
Latin America	55.9	10.6	0.21
South Asia	34.0	22.1	0.07
Sub-Sahara Africa	57.4	15.8	0.07

Source: Hasan and Quibria (2002).

early 1990s is also widely attributed to the development of outward orientated international trade (World Bank, 1993). The direct impact of openness on poverty reduction is more difficult to measure. Certainly the indirect impact of openness of growth and the subsequent reduction in poverty are well documented for Asian countries.

Flexible labor markets

Labor markets play a critical role in the transmission of economic growth and general increases in living standards of the poor. If labor markets are flexible then it is more likely that labor-intensive manufacturing industries will be able to get a foothold and provide a platform for further growth in income by raising exports. The details of this process in the Asian miracle economies have been discussed in a number of books and journal articles (World Bank, 1993). Labor unions were weak and labor mobility from the rural to urban areas was high. Worker safety is often neglected, unions are weak, child labor is commonplace, and benefits are minimal. Nevertheless, the jobs provided by footloose labor-intensive manufacturing establishments provided a steady flow of jobs for the poor and lower middle classes who had some education and minimal skills. Women were also drawn into the workforce in growing number, particularly in East and Southeast Asia. The degree of labor market regulation has been studied by the ILO, which has compiled a list of Labor Market Conventions and whether these conventions have been ratified and enforced. These conventions and also a Labor Market Rigidity Index are reported in Table 4.12 for the various regions. The extent of rigidity and

Table 4.12 Measures of Labor Rigidity in Developing Regions

Region	ILO conventions	Labor market rigidity index
East Asia	9.35	0.18
Latin America	37.69	0.32
South Asia	23.71	0.27
Sub-Sahara Africa	27.24	0.21

Source: Hasan and Quibria (2002, p. 6, Table 3).

number of conventions is much lower in East Asia than it is in the other regions. While human rights advocates might complain that violations of good labor practices were frequent, the fluidity of the East Asian labor market has been identified as an important factor in explaining why the income elasticity of poverty in East Asia has been so high and why poverty levels have fallen so rapidly.

Stable macroeconomic policies

Because they have limited resources, the poor are particularly vulnerable to fluctuations in income, employment and prices. Inflation, destabilizing macroeconomic policies, and slow economic growth all can have negative impacts on the incidence of poverty. We saw how the Asian financial crisis had detrimental effects on poverty indicators in the affected countries including Thailand, Indonesia and the Philippines. Using inflation as a proxy for macroeconomic stability, we observe that East Asia has had a much more stable performance than the other regions covered. Furthermore, many countries in the region have been able to build up substantial foreign reserves in recent years which afford additional protection against macroeconomic volatility. Budget deficits have also been generally modest aside from the period directly following the Asian financial crisis. All of these factors provide further evidence that the poor in Asia have not generally been adversely affected by volatile macroeconomic developments.

Education and health

Acquisition of human capital in the form of education, training, better nutrition and stronger health is the primary way that the poor can raise their standards of living. Land has become more fragmented and access to other forms of physical capital is limited. Unfortunately, data are not readily available which measure these aspects of the poor over time. What we do have are general measures of literacy, life expectancy and infant mortality. These are the raw materials used by the United Nations when they compile their yearly Human Development Report. Progress in these three areas is reported in Tables 4.13 and 4.14 for

Table 4.13 Literacy Rates for Adults, Females and the Young — 1980 and 2001/2003

Country	1980 adult literacy (those older than 14)	2001 adult literacy (those older than 14)	2003 adult female literacy (those older than 14)	1980 youth literacy (percentage of people 15–24)	2001 youth literacy (percentage of people 15–24)
Bangladesh	29	41	31	36	49
India	41	58	48	55	73
Nepal	22	43	35	33	62
Pakistan	28	44	35	37	58
Sri Lanka	85	92	89	93	97
Cambodia	55	69	64	67	80
Indonesia	60	87	83	89	98
Lao PDR	48	66	61	63	79
Malaysia	71	88	85	90	98
Philippines	88	95	93	95	99
Thailand	88	96	91	97	99
Vietnam	87	93	87	95	95
China	67	86	86	91	98
Korea	93	98		100	100

Source: UNESCAP (2004) and Save the Children (2006).

1980 and 2001/2003. Generally, when these measures improve we can infer that living standards for the poor would have also improved.

Literacy

The data in Tables 4.13 and 4.14 show that literacy rates have increased dramatically in the past 20 years or so, particularly in countries where literacy was weak in 1980, including China, India, Nepal, Pakistan, Indonesia, Lao PDR and Malaysia. The Malaysian and Chinese cases are particularly interesting since they both already had relatively high literacy rates in 1980 yet were able to increase them by 17 and 19 points respectively to reach close to 90% literacy in the period since 1980. It is also notable that youth literacy is substantially higher than adult literacy, reflecting the recent stress on universal

Table 4.14 Life Expectancy and Infant Mortality

Country	Life expectancy at birth in years (1980)	Life expectancy at birth in years (2001)	Infant mortality per 1,000 live births (1980)	Infant mortality per 1,000 live births (2001)	Infant mortality per 1,000 live births (2004)
Bangladesh	49	62	129	51	56
India	54	63	113	67	62
Nepal	48	59	133	66	59
Pakistan	55	63	105	84	80
Sri Lanka	68	73	35	17	12
Cambodia	40	54	110	97	97
Indonesia	55	66	79	33	30
Lao PDR	45	54	135	87	65
Malaysia	67	73	31	8	10
Philippines	61	70	55	29	26
Thailand	64	69	45	24	18
Vietnam	60	69	50	30	17
China	67	70	49	31	26
Korea	67	74	16	5	5

Source: UNESCAP (2004) and Save the Children (2006).

literacy and mandated education for the young. Aside from South Asia (Bangladesh, India, Nepal and Pakistan) where youth literacy is still relatively low, all the countries in the Asian region have achieved youth literacy of around 80% or more. Furthermore, although adult female literacy still lags overall literacy by around 10 points in South Asia, Sri Lanka excepted, the gap is much narrower in Southeast Asia and East Asia than in South Asia.

Life expectancy and infant mortality

Life expectancy was already high in several countries in 1980, including China and Korea in East Asia, Malaysia in South East Asia, and Sri Lanka in South Asia. By 2001 they were joined by Philippines, Thailand and Vietnam. Life expectancy still lags behind in South Asia

(Bangladesh, India, Nepal and Pakistan) where life expectancy is still around 60 and sometimes less. Infant mortality is closely related to life expectancy, reflecting the strength and coverage of health care programs. By 2004 infant mortality was below 20 per thousand (in industrial countries it was less than 10 and most less than 5) in Sri Lanka, China, Malaysia, Thailand and Vietnam, a significant improvement since 1980.

Human development and poverty

How do the changes in these human development indicators relate to changes in poverty over time? One approach would be to study this relationship by displaying the percentage drop in poverty for the major Asian economies against the percentage improvement in literacy or drop in infant mortality. The problem with such an approach is that literacy and infant mortality have a strongly nonlinear trend over time. At low levels of life expectancy large gains are possible through better child care, immunization, better nutrition and so on. At a more developed stage where health care has been well developed it is harder to make big jumps. The same is true of infant mortality. Below 10 deaths per 1,000 live births, further reductions are difficult. This is why the human development report chose to relate these variables to the best practice country. Then the performance of other countries is gauged relative to the best practice benchmark. Nevertheless we can draw some tentative inferences about the elasticity of poverty reduction with respect to gains in literacy and infant mortality from the analysis of these data.

First consider poverty reduction by examining the data in Tables 4.5–4.7. We can calculate roughly the average yearly drop in poverty for the individual countries and draw the following inferences. Vietnam and China have consistently strong reduction in poverty for the decade of the 1990s and on into the early years of this century. Indonesia also had good success in reducing poverty, although less so when a 2 dollar a day threshold is used. The Philippines had good success in reducing poverty using a one dollar a day threshold but less for the 2 dollar a day threshold. Thailand followed a pattern similar to the

Philippines although it was nearly twice as successful in lower poverty on an annual basis. Data for Korea and Malaysia also show strong poverty reduction.

Results in poverty reduction for countries in South Asia are less impressive. In Sri Lanka and Pakistan, poverty went up in some cases and improved very little in others. In India, progress in reducing poverty was slow — less than 1% per annum in the two dollar a day threshold results and 1.6% per annum using national poverty line figures. Only by using the one dollar a day World Bank calculations for poverty is there a strong result of nearly 3% reduction in poverty per annum for the decade of the 1990s. In Bangladesh, only the national poverty line results show strong poverty reduction. In the other two scenarios, poverty either goes up or falls very slightly.

In summary there has been minimal progress in reducing poverty in South Asia and stronger progress in Southeast and East Asia. How do these results fit with the education and life expectancy experiences? Life expectancy is lower and infant mortality is higher in South Asia than in either Southeast Asia or East Asia. Simple averages are displayed in Table 4.15.

Table 4.15 Average Regional Life Expectancy, Infant Mortality, and Adult Literacy in South Asia and Southeast Asia

Region	Average life expectancy at birth in 1980	Average life expectancy at birth in 2001	Average infant mortality in 1980 (out of 1,000 live births)	Average infant mortality in 2001 (out of 1,000 live births)	Adult literacy in 1980	Adult literacy in 2001
South Asia	53.2	61.7	120	67	30	46.5
Southeast Asia	61.4	69.4	52	25	80	91.8

Source: Tables 4.13 and 4.14.

Note: South Asia comprises Bangladesh, India, Nepal and Pakistan. Southeast Asia comprises Indonesia, Malaysia, Philippines, Thailand and Vietnam. Table values are simple averages.

Life expectancy and infant mortality lag around 20 years or more between the two regions. Infant mortality and life expectancy in South Asia in 2001 are not much different from the rates for Southeast Asia in 1980. The gap is substantially greater for education, which is key to why the rate of poverty reduction has not been more rapid in South Asia. Greater efforts have to be made to raise the level of literacy for South Asia to be able to translate growth into poverty reduction. However, the results for youth literacy are somewhat encouraging — just over 60% on average in South Asia, but still not as high as the average for Southeast Asia in 1980. This suggests that South Asia is still at least two decades behind Southeast Asia.

Good governance

In Chapter 5 we note that levels of corruption are substantially lower in East Asia than in South Asia and part of Southeast Asia where Indonesia and the Philippines also had poor governance records (see Table 4.16, excerpted in part from Table 5.1 of Chapter 5). We also noted in a previous section of this chapter that corruption and poor governance bear more heavily on the poor who are often taken advantage of by corrupt bureaucrats and law enforcement officers who extract bribes from the poor. The inability to develop a legal system that is honest and trustworthy has had long range impacts on the poor in countries with poor governance and this has also contributed to the continuation of poverty levels despite accelerating growth in recent years. Despite efforts to reduce corruption and strengthen good governance, the level of bribery has not changed substantially in the past few decades. To reduce poverty levels further, renewed efforts must be made to also reduce corrupt practices.

Income Distribution, Poverty and Economic Growth

Deterioration in income distribution can have a deleterious impact on poverty. There is some evidence that this has been the case in Asia, particularly in the last decade or so. There are various ways to measure this deterioration. Two widely used measures are the share of the

Table 4.16 Corruption Perception Index 2004 — Transparency International

Country	Corruption perception index ranking (out of 146 countries worldwide) — Low scores good, high scores bad
Singapore	5
Hong Kong	16
Taiwan	35
Malaysia	39
South Korea	47
Thailand	64
Sri Lanka	67
China	71
India	90
Nepal	90
Philippines	102
Vietnam	102
Pakistan	129
Indonesia	133
Bangladesh	145

Source: Transparency International (2004), pp. 8, 9.

lowest 20% of the income distribution and the Gini coefficient. The share of the lowest 20% is geared toward reflecting the impact of changes in the income distribution on the poor while the Gini coefficient is a more general measure of income distribution taking into account the full range of all incomes and their dispersion. In many cases the two measures will move in the same direction. For example, the Gini coefficients for the major Asian economies in the last decade are highly correlated with the share of the lowest 20% of income (Table 4.17).

The pattern of variation in income distribution has been studied by Quibria (2002) drawing on the database on Deininger and Squire (1996) for the 1950s to the early 1990s.

Piecing together this information with more recent information from the World Bank we can draw several conclusions. First there is no apparent relationship between the level of income distribution and the level of per capita income. Some poor countries have very favorable

Table 4.17 Income Distribution in Asia — 2004

Country	Share of lowest 20%	Gini coefficient
Bangladesh	9.1	31.8
India	8.8	32.5
Nepal		
Pakistan	9.3	30.6
Sri Lanka	8.3	33.2
Cambodia		
Indonesia	8.4	34.3
Lao PDR		
Malaysia	4.4	49.1
Philippines	5.4	46.1
Thailand	8.3	42.0
Vietnam	7.5	37.0
China	4.7	44.7
Korea	7.9	31.6

Source: World Bank (2006).

income distributions and share of the lowest 20% (Pakistan and Bangladesh) while some richer countries have poor income distributions (Malaysia). Second, income distribution has changed less in the past decade than it did in earlier years. Third, the overall variation in income distribution in the postwar era has not fluctuated much for most countries. In China and Thailand, in the 1980s and early 1990s, income distribution deteriorated and in Bangladesh and Sri Lanka it improved (Wu and Perloff, 2004).

In the case of the Central Asian Republics the focus on maintaining a favorable income distribution established during the Soviet era is still evident although the transition shock has had some unfavorable consequences. Overall income distribution figures suggest that income is reasonably equally distributed in the CARs with an average Gini coefficient of 33.2%. Furthermore, the share of the poorest 20% of the population at 8% is higher when compared with countries in Southeast Asia such as Thailand and Malaysia (Table 4.10).

Looking at growth in per capita income and changes in poverty there is no apparent relationship between changes in income distribution and poverty reduction. China had a deteriorating income distribution, yet achieved remarkable reductions in poverty. The same can be said for Thailand. On the other hand, poverty reduction in South Asia has been slow even though income distribution has been relatively constant or improving. This is not to say that an improvement in income distribution would not have helped China make ever further inroads on poverty. Also, deterioration in income distribution creates social tensions that can lead to a more volatile political and economic environment as social tensions can result from a growing distance between the absolute incomes of the rich and the poor. This deterioration in income distribution could occur if there were divergence in income over time among the regions of a country. Gajwani *et al.* (2006) investigate the pattern of regional income growth in India and China for much of the latter half of the 20th century. They find that the evolution of regional inequality is related to various economic policies. Because of the barriers to migration regional differences are significantly higher for China than India. Additionally, openness and decentralization have contributed to the rapid increase in coastal/inland disparities in the 1980s and 1990s. In India, as the determinants of regional comparative advantage have shifted from the quality of land to the level of human capital, the Indian states have become clustered into two clubs; the more educated and the less educated. The time pattern of dispersion of per capita income by province as well as growth convergence regressions can provide some further insights into changes in income distribution for other countries in the region. There are several studies including Garbaccio and La Croix (1998), La Croix *et al.* (1998), and Manasan and Chatterjee (2003) that conclude that fiscal policies such as revenue sharing are needed to overcome the growing tendency for fiscal decentralization that exacerbates regional inequalities.

These studies confirm the conclusions drawn from the analysis of the Gini coefficients described above. Dispersion in China and Thailand has been on an upward trend while there has been little change in other countries, including Malaysia and India. Other work by Dowling and Konchoro (2006) for Indonesia and the Philippines suggests that there has been some convergence in income within regions but divergence between the urban and rural regions.

To conclude this section, we reiterate that it is important to develop policies that will result not only in poverty reduction but also improvements in the income distribution. In cases where regional disparities have been growing, barriers to mobility of resources, whether labor or capital, need to be removed to allow market forces and comparative advantage work to smooth regional imbalances. Nevertheless, at low levels of income, reducing poverty is much more important than improving the distribution of income.

General goals, agenda, and prospects for reducing poverty

The United Nations in cooperation with the World Bank, bilateral lenders, and the regional development banks (ADB, Latin American and African Development Banks) have crafted a series of eight goals for reducing poverty by the year 2015. These Millennium Development Goals (MDGs) were formulated in response to the main development challenges for reducing poverty and raising global living standards. These MDGs were formalized in a declaration adopted by most of the world's nations during the United Nations summit held in 2000. There are eight main goals which are supported and measured by a further series of targets and indicators (see http://www.undp.org/mdg/basics.shtml for further details). The goals are to:

(i) eradicate extreme poverty and hunger;
(ii) achieve universal primary education;
(iii) promote gender equality and empower women;
(iv) reduce child mortality;
(v) improve maternal health;
(vi) combat HIV/AIDS, malaria, and other diseases;
(vii) ensure environmental sustainability;
(viii) develop a global partnership for development.

According to the medium term objective the goal is to halve the one dollar a day poverty level by 2015 from 2000 levels. Furthermore, a road map for implementing the millennium objectives has been formulated by the United Nations, and Poverty Reducing Strategy

Papers (PRSPs) are being formulated by countries as a requirement for receiving assistance from the World Bank.

As part of this exercise, several countries in the Asian region have already developed such strategies. They will be implemented in conjunction with their own domestically developed poverty programs. Looking at the historical record we have observed that East Asia and some countries in Southeast Asia were able to reduce poverty to low levels. The focus of future poverty reduction strategies and policies is therefore on South Asia, China, Indonesia and the Philippines, and the Mekong countries, particularly Lao PDR and Cambodia. We can also include the Central Asian Republics, where poverty increased dramatically following the break up of the former Soviet Union. In the next section, we will summarize the strategies that have already been formulated for these target countries, discuss other possible objectives for reducing poverty in the Asian region investment priorities and longer term objectives for reducing poverty.

Poverty Reduction Strategy Papers (PRSPs) and Poverty Reduction Objectives

PRSPs are designed to pay particular attention to a wide variety of issues related to poverty reduction within the broader framework of a national development planning agenda. Given that different countries have different agendas, the main characteristics of PRSPs are to be result-oriented, recognize the multidimensional nature of poverty, and coordinated participation of development partner, involving broad-based participation by civil society (for example, NGOs) and the private sector. Another important objective is to develop both medium- and long-term development goals. The PRSP is developed in consultation with the IMF and the World Bank, who also evaluate and endorse the plan as a condition for receiving concessional assistance. Most of the poorer countries in the Asian region that are receiving concessional assistance from the World Bank are developing or have already developed PRSPs. UNESCAP has provided a useful summary of many of the issues highlighted in these documents and how they dovetail with existing structural adjustment programs aimed at achieving macroeconomic stability and sustainability. As a result PRSPs are able to maintain focus on macroeconomic issues including price stability, low budget deficits and fiscal sustainability, appropriate exchange rate policies to maintain export competitiveness, and debt management aimed at sustainability and probity. Other PRSP policies are consistent with the policy agendas developed by the World Bank in its structural adjustment lending including fiscal reforms and trade reforms. In addition to general policies, the World Bank has emphasized building capacity of the poor to allow them to enter the mainstream of development. These include empowerment, opportunities for generating income, enhancing security, and increasing gender equality (see UNESCAP (2002) for more details on how these programs have been implemented in particular countries as part of the PSRPs).

Empowerment

In the context of poverty reduction, empowerment means getting the poor involved in the decision-making process. It also enables them to direct resources to their best use for reducing poverty and raising incomes. This could involve land reform and assuring secure land tenancy by legal means, providing basic education and skills to vulnerable groups, building capacity through increasing expertise with specialized training and support to enable them to be more productive and efficient. For farmers this would involve training, and extension service and provision of agricultural inputs to enable farmers to grow higher yields and more profitable crops. Whatever form it takes, rural reforms that benefit the poor have to involve them as stakeholders with a voice in decision-making and resource allocation using bottoms up approaches. Edicts from the central government have not always been effective in addressing poverty concerns, nor in allocating resources to targeted groups.

Increasing opportunities

To build capacity the poor must be provided greater scope for raising incomes, raising productivity, and connecting to urban centers in several different ways. By increasing the borrowing capacity of the poor they can even out income streams and obtain capital for investment. Microfinance is one way to provide this kind of financing not only to individual farmers but also to small-scale rural industries. Increasing rural/urban linkages by developing farm to market roads and better rural telecommunications coverage is another way to create opportunities. Providing education infrastructure and scholarships to students in remote areas also raises the level of human capital in poor communities. Vietnam has spearheaded the idea of creating rural industrial zone in poor areas to help create industrial employment in labor-intensive industries.

Enhancing security

Secure and reliable access to sustainable levels of nutrition, clothing and housing is important for the poor to enable them to live full lives with less stress and also to enable them to work and go to school on a regular basis free of illness. Programs to provide affordable housing and better health care for the poor have to be developed. Supplemental meals for school children and clothing subsidies should also be considered.

Gender equality

We mentioned earlier in the chapter the importance of women's enhanced knowledge of appropriate health and sanitation practices and its impact on infant mortality and slower population growth. These benefits hold, *a fortiori*, for the poor. Better access to education should be augmented by access to financial assets and agricultural educational programs. These will enable women to take more control of their lives and participate more fully in decision-making for their families and community. Legal reform to amend gender-biased laws, incorporating gender issues into school curricula, and promoting gender-specific agricultural extension programs are other ways to promote gender equality.

Other initiatives

AIDS, general health care and social safety nets are another priority area. The elderly, children, and those without earning capacity need to be guaranteed a minimum standard of living to help keep them out of poverty. Those displaced by natural disasters as well as refugees need immediate assistance. Targeted programs need to be well thought out and engineered to make sure that the poor actually receive the benefits. Many programs in the past such as low cost housing have sometimes been subverted by the non-poor. Provision of sites and services has often proved to be a better approach to satisfy housing needs of the poor. Food subsidies such as school lunches is a good way to provide nourishment to poor children and also guarantee that they attend school. Prevention programs for HIV/AIDS in countries such as Lao PDR, Cambodia and India are an important support for families affected by the disease. General preventative health care including improved sanitation and vaccination also needs to be strengthened.

Environmental protection and sustainable development

We touch on the relationship between poverty and the environment in Chapter 6. It is sufficient to note here that the work of the World Bank in implementing the PSRPs takes cognizance of the environmental damage that results from poor policy choices that fail to balance environmental concerns with livelihood programs for the poor. More secure land tenure arrangements can help, and better awareness of short-term damage that can result from shifting cultivation and indiscriminate burning of fields following harvests need to be stressed.

Other Programs and Policies for Attacking Poverty

PRSPs are only applicable for poor countries that are receiving concessional assistance from the World Bank. Furthermore, these policy programs are subject to a vetting process by the World Bank. They also are interrelated with the national poverty reduction strategies of developing countries. The interaction between the PRSPs and national plans and objectives constitutes the springboard for achieving the MDGs, particularly the objectives of reducing poverty by half between 2000 and 2015 as well as achieving universal primary education and boosting literacy. Looking at the levels of one dollar a day poverty displayed in Table 4.18 we see that the objective of reducing poverty by half will be difficult for the countries in South Asia as well as in Cambodia and perhaps Lao PDR given their experience in the past decade. Reducing poverty by half will require some rethinking of how to address poverty from the microeconomic perspective as well as promoting in economic growth to reduce poverty reduction through a trickle down approach.

Comparable projections for the Central Asian Republics are not available. However, using national poverty lines ADB has made forecasts to 2015 (Table 4.19).

These estimates suggest that poverty is expected to fall dramatically although perhaps not by half, between 2000 and 2015. Research by ADB (Dowling and Wignaraja, 2006) suggests that further reductions

Table 4.18 Poverty Levels for 1990, 2000 and 2015 — One Dollar a Day

Country	1990	2000	2015
Bangladesh	35.9	36.0	18.0
India	46.6	34.7	17.4
Nepal		37.7	18.7
Pakistan	47.8	31.0	15.5
Sri Lanka	3.8	6.6	3.3
Cambodia	48.3	44.6	22.3
Indonesia	20.6	9.9	4.9
Lao PDR	53.0	35.1	17.6
Philippines	19.1	13.5	6.8
Thailand	12.5	5.2	2.6
Vietnam	50.8	15.2	7.8
China	31.5	15.4	7.7

Source: Table 4.3.

Note: Data in the 1990 column is 1992 for Bangladesh. Data in the 2000 column is 1995 for Nepal, 1996 for Pakistan, and 1996 for Sri Lanka. Data in the 2015 column are half of 2000 poverty figures.

Table 4.19 Poverty Forecast to 2015 — Central Asia

Country	Year of national poverty estimate	Poverty incidence in base year[a]	Forecast poverty incidence 2015[b]
Oil exporters			
Azerbaijan	2001	49.6	29.3
Kazakhstan	2002	27.9	17.6
Turkmenistan	1998	29.9	21.9
Non-oil exporters			
Kyrgyz Republic	2001	47.6	20.7
Tajikistan	2003	56.6	23.7
Uzbekistan	2000	27.5	12.3
Total		39.9	20.9

[a]National Poverty Lines.
[b]Poverty incidence in 2015 based on projections using most recent poverty line estimates and staff estimates of elasticities of poverty reduction with respect to income.
Source: Dowling and Wignaraja (2006) for historical data; and staff estimates for forecasts.

are possible with policy adjustments including regional cooperation, industrial policy adjustments, and greater openness to trade. Much will also depend on energy prices and the ability of the CARs to translate higher oil revenue into overall economic development and poverty reduction.

Achieving universal primary literacy will also require renewed effort for South Asia and the Mekong countries of Cambodia and Loa PDR, as is evident from Table 4.13. With the exception of Sri Lanka, literacy levels for the adult population are below 50% for three of the five South Asian countries. Their youth literacy rates are somewhat higher but still far below the full literacy level that would be required for universal primary education by 2015. Since youth literacy rates are not available for years before 2001 it is difficult to judge whether the South Asian countries would be able to achieve universal primary literacy by 2015, even if that only meant universal literacy for the young. Many adults would remain illiterate unless a massive effort was made to increase adult literacy. For the other countries of the region adult literacy rates are above 90% except Lao PDR and Cambodia, where adult literacy is close to 70% and youth literacy is 80%. For these two Mekong countries the chances of reaching universal literacy by 2015 are, therefore, good and Vietnam is doing even better (Table 4.20).

One of the reasons why educational performance and literacy has lagged behind is the lack of funding for primary education in South Asia compared with East and Southeast Asia. Countries with low per capita incomes in South Asia and the Philippines spend the least amount of money on primary education. The cost of tertiary education is as much as 40 times that devoted to primary education in the case of India (Table 4.21). This compares to a ratio of less that four times for Korea and Thailand. Furthermore, the amount of spending per capita for elementary education in South Asia is considerably lower than it is in the rest of the Asian region, less that half the rate in Thailand, Malaysia and Korea.

Table 4.20 Literacy in 1980, 2001, and 2015

Country	2001 adult literacy (those older than 14)	1980 youth literacy (percentage of people 15–24)	2001 youth literacy (percentage of people 15–24)	Implied 2015 youth literacy rate if country is to have universal primary education
Bangladesh	41	36	49	100
India	58	55	73	100
Nepal	43	33	62	100
Pakistan	44	37	58	100
Sri Lanka	92	93	97	100
Cambodia	69	67	80	100
Indonesia	87	89	98	100
Lao PDR	66	63	79	100
Malaysia	88	90	98	100
Philippines	95	95	99	100
Thailand	96	97	99	100
Vietnam	93	95	95	100
China	86	91	98	100
Korea	98	100	100	100

Source: Table 4.5 and MDG projections.

Table 4.21 Operating Per Unit Costs of Education as a Percentage of Per Capita GNP During the Mid-1980s

Economy	Primary	Secondary	Tertiary
Bangladesh	6	30	285
China	7	23	199
India	6	17	231
Indonesia	13	23	91
Korea	17	23	71
Malaysia	14	21	190
Nepal	9	14	249
Philippines	6	9	50
Sri Lanka	6	9	83
Thailand	16	15	40
Average	10	19	149

Source: Dowling and Valenzuela (2004, Table 10.3).

Investment and Other Priorities for Achieving MDG Objectives

Writing in 2004, Bajpai *et al.* noted the problems of achieving MDG objectives by 2015, stressing the importance of India:

> South Asia as a whole seems likely to halve poverty levels by 2015, due in large part to India's progress on this MDG indicator. However, the situation is not as promising in other areas, such as reduction of high rates of infant, under-five and maternal mortality. There are wide inter-country, inter-state and intra-state differences in levels of progress towards the goals throughout South Asia. In a country as large and populous as India, tackling problems such as high maternal mortality rates at the state and even at the district levels could determine whether India as a whole achieves the MDGs. Over one billion of the roughly 1.37 billion residents of South Asia reside in India, which means that India's achievement of any of the goals brings South Asia as a whole a step closer towards regional goal attainment. For South Asia as a whole and for India in particular, public spending must be increased in the areas of health, education, rural infrastructure and agricultural research and development. The Indian government under-invests in all of these areas. Not only must public spending be increased, but spending must be accompanied

> by reforms. Decentralization would lead to greater control and oversight, conditions necessary for accountability (Bajpai *et al.*, 2004, p. 1).

We echo their sentiments 3 years later. Government investment priorities will have to be redirected to some extent if poverty reduction objectives are to be achieved. This will require more focused spending on the programs identified in PSRPs and other priorities set by the government to reduce poverty. More spending on primary education in rural areas, establishment of rural banking facilities catering to the poor, and social safety nets for targeted groups are a few of the key areas where additional spending is going to be required. Raising funds to implement these programs will require greater tax effort and tapping additional sources of revenue as well as a continuing commitment to reduce poverty. In particular, progressive taxation to take advantage of the greater earning capacity and higher taxable income of the wealthier classes will require a wider income tax base and better tax administration and collection. Closing loopholes, and reducing bribery and corruption within the bureaucracy will be key challenges if this redirection of resources is to be achieved. Legally, there are two important areas where laws need to be strengthened and enforced. First, establishment of secure land tenure, either by ownership or secure farming rights and tenancy arrangements is critical. Second labor market regulations have to be relaxed so that employers have greater freedom to hire and fire. Current regulations, particularly in India, serve as a strong impediment to rapid labor absorption into the workforce in the formal sector. As a result the beneficial results of trickle down from rapid overall economic growth are being muted. Imbalances in the incidence of regional poverty also need to be addressed. Heavily populated provinces in northeastern India, including Assam, Bihar, Madhya Pradesh and Orissa, and to a lesser extent Uttar Pradesh, have high incidences of poverty. In southwestern China, tribal minorities and others in designated poor counties, usually in remote highland regions, are in need of special assistance (the 592 group, see UNESCAP (2006), Chapter III). In Thailand, the provinces in the Northeast have high incidence of poverty as do parts of the Viscayas and Mindanao in the Philippines.

Longer term objectives

The commitment to reduce poverty does not end once a country has achieved the millennium development objective of reducing poverty by a half and achieving universal primary education and the other millennium objectives of greater gender equality and improving health indicators. Longer term objectives are needed to end poverty, not just to reduce it. This is not an impossible objective. In Europe and North America, widespread poverty is a memory. In Asia, Japan, Hong Kong, Singapore and Korea have essentially eliminated poverty, and Malaysia is moving quickly toward achieving that objective. To maintain emphasis, development plans must continue to focus on poverty reduction as a cornerstone of policy. Economic growth without equity and poverty reduction is not an acceptable objective anymore and trickle down is not enough. We now have the knowledge and ability to reduce and eventually eliminate poverty throughout the Asian region, not just in East Asia and in some parts of Southeast Asia. In the next section, we spell out the elements of such a long-term strategy.

An agenda for long-term poverty reduction

Agricultural development

Agricultural sector growth was a pivotal factor in the development experience of Asia during the last quarter of the 20th century. This strong growth was one of the distinguishing features of the Asian experience compared with the other developing regions of Africa and Latin America. The index of per capita food production increased by nearly 40% between 1975 and 1995 in Asia compared with almost no increase in Latin America and a decline of over 20% in Africa (Dowling and Valenzuela, 2004, Figure 4.2).

Strong growth in the rural sector, supported by vibrant agriculture is a key ingredient for further poverty reduction in Asia. Although the relationship is not strong in all countries, there is evidence that economies that have had weak agricultural performance in recent years such as Cambodia and Sri Lanka had less success in

Table 4.22 Agricultural Production and Poverty Reduction

Country	Average annual growth in agriculture (1990–2001)	Percentage point reduction in dollar a day poverty rate (1990–2001)
Bangladesh	3.1	–0.1
Cambodia	1.8	3.7
China	4.0	15.1
Indonesia	1.9	10.7
India	3.0	11.9
Nepal	2.6	n.a.
Sri Lanka	1.7	–2.8
Vietnam	4.2	35.6

Source: World Bank (2006).

poverty reduction while those with very strong agricultural performance such as China and Vietnam were very successful in poverty reduction (Table 4.22).

In some cases the aggregate evidence can be a bit misleading. For example, Ravallion and Datt (1996) show that faster agricultural growth in India is strongly related to declines in both rural and urban poverty rates. Also, Ravallion and Datt found that the relationship between poverty and the industrial sector depends on a variety of other factors as well as growth and that the same relationship with poverty reduction does not hold for manufacturing sector growth.

Continuing to focus on India, there has been a decline in productivity growth in agriculture in recent years (Table 4.23). But what is more worrying is that there are now signs of generally declining productivity growth throughout the economy, which, if not addressed, may have adverse implications for the future.

Rural strategies will depend upon local conditions. However, some elements of a successful strategy for reducing rural poverty are common to many countries in South Asia and the Mekong.

Rural industry. Employment generation in rural small-scale and medium-scale industries using labor-intensive technology could be a good way of increasing income and employment for the poor (Foster

Table 4.23 Annual Rates of Growth for Selected Crops in India (%)

Crop	1980–1981 to 1989–1990			1990–1991 to 2000–2001		
	Area	Production	Yield	Area	Production	Yield
Rice	0.41	3.62	3.19	0.63	1.79	1.16
Wheat	0.47	3.57	3.10	1.21	3.04	1.81
Coarse cereals	–1.34	0.40	1.62	–1.84	0.06	1.65
Pulses	–0.09	1.52	1.61	–1.02	–0.58	0.27
Food grains	–0.23	2.85	2.74	–0.20	1.66	1.34
Non-food grains	1.12	3.77	2.31	0.84	1.86	0.59
Oilseeds	2.81	5.20	2.43	0.44	0.66	0.61
Sugar cane	1.44	2.70	1.24	1.72	2.62	0.89
Cotton	–1.25	2.80	4.10	2.21	0.92	–1.26
All crops	0.10	3.19	2.56	0.08	1.73	1.02

Source: Government of India (2002/2003).

and Rosenzweig, 2004). As the industrial sector moves into rural urban centers, the opportunities for employment growth will continue to expand. Foster and Rosenzweig argue that these opportunities will provide a substantial boost to rural employment and income particularly when rural workers have sufficient education and training to be productive members of the formal labor force in labor-intensive industries. The importance of complementarities between education and employment in reducing unemployment is highlighted and this is irrespective of gains in agricultural productivity. In other words rural industries seek out low wage areas which, in many cases, were not the beneficiaries of agricultural productivity growth and which would not have grown without the development of rural industry. Bangladesh was able to make good inroads on poverty in the 1990s, particularly compared with the 1980s, as a result of employment generation in small-scale industries, construction, and services (McKinley, 2005).

Policies that remove barriers to investment in non-farm small scale industries and which promote mobility of labor and capital to take advantage of niche markets for labor-intensive small-scale industries in rural areas will have a direct beneficial impact on rural employment and poverty reduction. It will also have a secondary benefit of being independent of weather factors that have created considerable volatility and uncertainty in monsoon agriculture in Asia.

Research, technology transfer, and extension. Extensive work by IFPRI for the period 1957 to 1987 looked at major food grains and many minor crops to determine the most important contributions to agricultural productivity growth (Evenson *et al.*, 1999). They found that the gains from the green revolution in rice and wheat have increased yields. However, they also found that returns from agricultural research and extension have remained high and recommend that public investment in research and extension can continue to raise agricultural productivity. Policies that remove barriers to technology transfer and which promote a variety of investments in infrastructure would have a positive impact on agricultural productivity and poverty reduction. Khunkitti (2002), Bussolo and O'Connor (2002), and Lipton (2001) found that human capital development improves the conditions for technological adoption and innovation. Yet there is much to explore on the relationship between technology transfer, education, health and agricultural productivity. For example, Craig *et al.* (1997) found that life expectancy had a large and more significant impact on labor productivity in agriculture than commercial fertilizer use, tractor horsepower and research expenditures. Much more research is needed to determine the nature of these interactions and their impact on the poor. New areas such as intellectual property rights and biotechnology also need further investigation. Furthermore, the relationship between technology and macroeconomic policies has to be considered and systematic analysis of the impact of policies on the agriculture system and poverty need to be formulated (Liu, 2002).

Rural finance. Financial liberalization has taken place in many Asian economies in the past several decades. After many years of financial repression where state banks were the only financial institutions and where interest rates, credit allocations, and the number of banks and branches were tightly controlled by governments, an era of financial liberalization was introduced in the 1980s and 1990s. Many of the shortcomings of these liberalization measures were exposed by the financial crisis of 1997/1998. In the aftermath of this crisis, the banking and financial systems of Asian economies have come under additional scrutiny. One of the results of the crisis was a reversal of progress in poverty reduction in several countries and the inability of the poor to cope with slower growth and lack of financial resources. While these difficulties were present during the rapid growth period preceding the crisis, post-crisis financial sector policies have become more cognizant of the financial needs of the poor, particularly in rural areas. In Bangladesh and Cambodia, for example, spreads between lending and deposit rates have widened and the poor have had greater difficulty in accessing credit (McKinley, 2003). The reluctance of banks to lend in rural areas is well known. However, in the face of greater volatility in aggregate economic activity, the impact of these constraints becomes more onerous. The problems are symptomatic in the Mekong countries and in South Asia. One of the conclusions reached by the UNDP team studying these economies (McKinley, 2003) is that private financial institutions still have little incentive to lend in rural areas. Self-finance and loans from state credit institutions are the main source of credit aside from landlords and moneylenders. In China and Vietnam, these institutions are much better developed than in the Mekong or South Asia. China has the Agricultural Bank of China and rural credit cooperatives while Vietnam has the Development Assistance Fund and the Social Policy Bank. There are weaknesses in these institutions including corruption and poor loan supervision and they need to be strengthened. Nevertheless, the rural credit network seems to be on firmer footing than in the Mekong and South Asia. This is not to say that there are no successes in South Asia. The Grameen Bank has been highly successful in Bangladesh and there are a number of successful rural cooperatives in India.

Rural infrastructure. Work by Mellor and Mudahar (1992) suggests that increased production in Asia during the green revolution was primarily the result of higher yields on irrigated land. Rain-fed agriculture contributed much less to gains in productivity. These trends have continued and have been reinforced by the introduction of more high value-added crops on irrigated land. These findings suggest that extension of irrigation to more rain-fed land would provide a significant boost to agricultural productivity. Is this possible and would it help the poor significantly? Data comparing the share of irrigated to total cropland in the late 1980 and late 1990 show a significant increase in this share in only a couple of countries, namely Bangladesh and Vietnam. Increases in other countries have been more modest (see Table 4.24) and agricultural output growth reflects this stagnation in the extension of the irrigation network.

Can irrigation be extended further in countries where it is languishing? The era of large dams and extensive irrigation schemes that accompany them seem to be over for a number of reasons including lack of suitable river systems to be dammed up as well as environmental concerns (3 Gorges project in China, for example). Tube well technology has advanced and the number of new wells has increased. However, this kind of irrigation may not be sustainable because of insufficient recharge. Nevertheless, selective extension of the irrigation network, better maintenance of existing irrigation facilities, and more efficient allocation of water can improve the efficiency of existing irrigation networks.

Work on India by Bhattarai and Narayanamoorthy (2003) suggests that poverty and irrigation are closely related over the long run period from 1952 to 1989. They estimate the elasticity of output with respect to irrigated areas at 0.32; 1% increase in cultivated area has brought

Table 4.24 Growth in Agricultural Output and Extension of the Irrigation Network — 1990s

Country	Average annual growth in agriculture (%) — 1990–2001	Irrigated land as percent of cropland — 1998/2000 and 1989/1991 (in parentheses)
Bangladesh	3.1	48(17)
Cambodia	1.8	7(6)
China	4.0	40(45)
Indonesia	1.9	15(16)
Nepal	2.6	38(23)
Sri Lanka	1.7	35(28)
Vietnam	4.2	41(26)

Source: World Bank (2006); McKinley (2003).

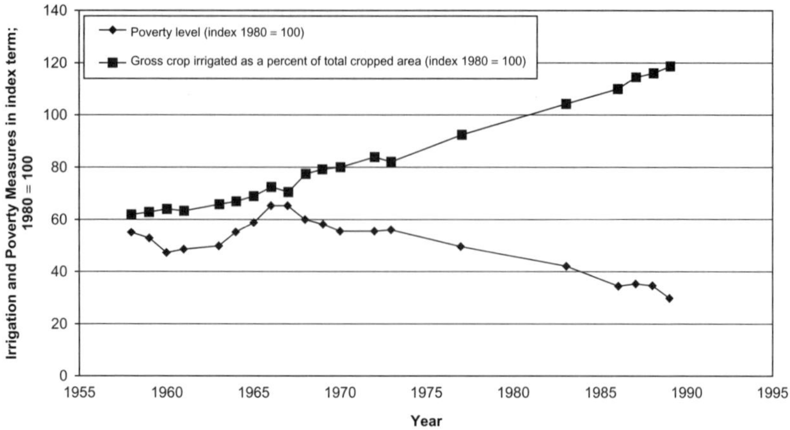

Figure 4.1 Irrigation and Poverty in India

Source: Bhattarai and Narayanamoorthy (2003).

about an increase in total factor productivity of 0.32% and both the poverty level and poverty gap have been falling as the irrigation index has been increasing since the mid-1960s. Furthermore, it does not look like there has been any slowdown in the upward trend in irrigation since the mid-1970s (see Figure 4.1). Therefore, at least in India and perhaps in Bangladesh and Pakistan as well, the scope for continued expansion of the irrigation network is possible although the recharge from aquifers would have to be considered in a longer term strategy.

Health, Education and Population

We have mentioned the key role of health and education in reducing poverty in South Asia and the Mekong. We have also highlighted the shortfall that is likely in meeting Millennium Development Objectives unless there is a dramatic improvement in the delivery of health and education to the poor in the next decade. Even if the countries of South Asia do not achieve universal primary education by 2015 it is still important to focus on education and health. Historically, countries in South Asia have devoted more resources per capita to tertiary education than to primary education. This is a different pattern from the practices of countries in East Asia and even in Southeast Asia where there has been greater balance among the three tiers of education.

We noted above that interrelationships between health and education and technology transfer are important. Lin (1999) shows that greater education and schooling is more related to farm decision-making such as crop mix and the mix of farm and non-farm activities. With more education, farmers have greater awareness of choices in terms of activities and a greater understanding of how to maximize returns from different activities. As economies develop and the size of off-farm employment opportunities grow, education provides a way of raising incomes by offering farmers choices of agricultural technology and alternative employment.

> A possible policy implication is that given the choice between investing in new crop varieties tailored to the conditions of the poor and investing in their education, preferences should be given to the latter (Bussolo and O'Connor, 2002, p. 20).

Research has demonstrated that population and labor force growth interact with educational and health indicators over time in reducing poverty. We know that building up human capital is a powerful way to increase income earning potential and raise families out of poverty. We also know that high population growth rates are negatively related to the level of average educational attainment and literacy as well as health indicators such as life expectancy and infant mortality (UNESCAP, 2002).

Countries that have already experienced the demographic transition from high to low death and birth rates have had a much better record in reducing poverty, increasing literacy and average level of education as well as lowering infant mortality and increasing life expectancy. Women's education and health play a critical role in this process. As more women become educated they tend to take better care of their families in terms of providing good health care and nutrition and are making sure their children attend school. The rate of family formation also slows as women's labor force participation rises and educational attainment increases. These observations are confirmed by the data in Tables 4.21 and 4.25.

Table 4.25 Education, Population, and Poverty Reduction for Selected Asian Countries

Country	Education index (component of HDR index 2002)	Population growth rate (decade of 1990s)	Average annual reduction in 2 dollar a day poverty 1990–2001 (%)
Bangladesh	40	2.1	0.4
China	80	0.9	3.9
India	57	1.5	0.9
Indonesia	79	1.2	2.1
Malaysia	80	1.8	4.0
Nepal	48	2.3	none
Pakistan	42	2.6	0.6
Philippines	91	1.9	1.4
Sri Lanka	84	0.9	none
Thailand	84	0.8	2.8
Vietnam	84	1.3	3.6

Source: UNESCAP (2001), United Nations (2002), and author's estimates.

Furthermore, countries where fertility rates have already fallen will have less pressure to educate children at the primary level and will be able to devote more resources providing for secondary, technical, and tertiary levels of education and also to improve the quality of education. For this reason it is critical for countries to make the demographic transition as quickly as possible. This process can be accelerated by focusing more on women's education and also by enhancing reproductive health services and information including access to birth control methods. Additional measures would include reducing discrimination against women in the workplace and providing opportunities for women to enter the labor force.

Empowerment of the poor and civil society

Civil society is a term that has been coined in recent decades to describe interest groups that are not associated with the government or the business community that have been organized and developed to further their own agendas. They have been called NGOs or more broadly civil society organizations (CSOs). In the context of poverty reduction we will focus on the activities of these organizations to reduce poverty. How effective have they been and how effective can they be in the future? There are many anecdotes that relate to how NGOs and CSOs have been helpful to the poor. Proponents of putting greater emphasis on civil societies point to the structural and institutional difficulties that keep resources from being allocated and delivered to the poor. We have cataloged these difficulties, including discrimination, control of resources by urban elites, a lack of a voice in government institutions where budget allocations are made, and a general feeling of powerlessness. The themes of alienation, lack of inputs into the development agenda, and lack of attention to the perceived needs of the poor are often voiced by NGOs and their supporters.

The work of Asia Pacific Civil Society Forum summarized in its background paper and presented to the ESCAP Committee of Poverty Reduction provides a useful summary of the issues raised by NGOs and other proponents and active members of civil societies (ESCAP, 2003). The key to making the breakthroughs targeted for 2015 is to actively involve the poor in the development process. There are many NGOs in Asia working with governments to formulate development objectives and meet development targets. And, NGOs are independent enough to reflect alternative critical perspectives. They can be active advocates of different development objectives and serve as forces for implementing change within society as well as a source of constructive pressure on governments.

Three themes run through this ESCAP document and agendas proposed by others in the civil society movement. They are summarized as follows:

Better access, information and interaction with policy makers

Civil society organizations have few opportunities to participate in the formulation of policy with national and local governments. They believe that multilateral institutions have the ear of government and there is little room for citizens to participate in policy dialogue and to be active in shaping public policy. These civil organizations are representatives of the poor and by excluding them from policy dialogue the voices of the poor are lost. Whatever interactions there are between government and civil society are often characterized by a lack of trust and there is a narrow base of representation as only well-organized lobbying groups are represented, and smaller and less well-organized groups such as farmers and fishermen, women's rights, migrants, and indigenous people are not fully represented. At the same time, governments seem more interested in growth and greater trade and do not seem to be open to a poverty reduction agenda. There also is often a disconnect between government efforts to implement MDG objectives and the efforts of civil society to achieve similar objectives. Furthermore, the MDG objectives are not concrete or specific enough for the poor to claim ownership. It is a top-down approach that does not have local support or endorsement.

Lack of respect for local governments, their objectives, accomplishments and plans

Aid conditionality and adherence to donor policy suggestions undermine domestic authority and create resistance to foreign aid programs. Such approaches can also stop critical examination and dialogue regarding the impact of neo-liberal policies such as structural adjustment programs and capital market liberalization that may, in certain circumstances, be contributing to poverty rather than reducing it. Imaginative alternatives to reducing poverty and ensuring food security that have worked at the microlevel might possibly be ramped up to be applied at a broader national level. Bilateral and multilateral lenders are not aware of these possibilities and/or not willing to explore them. Furthermore, by approaching poverty as a technical rather than a human issue, poverty reduction becomes another business for aid agencies and their consultants.

Concentration on targets and macroeconomic objectives misses the point

Reducing poverty requires the active involvement of the poor and a more fundamental and comprehensive understanding of the causes of poverty and how to deal with it. By focusing on simple poverty line approaches that may be useful at the macrolevel and for comparison among different countries, more fundamental issues are often ignored. These include issues such as women's poverty, the poverty of migrant and displaced workers and of children, the aged, the extremely poor, and minorities. Furthermore, unless the poor themselves are involved, poverty programs will fail to attract local support. Spending needs to be reoriented to address the problems of the poor. Only a government that represents these poor adequately can achieve this objective.

Many of the suggestions put forward by civil society advocates echo the ideas of previous sections of this chapter including progressive taxation and access to water, education, social services and health care. In addition, the importance of those employed in the informal economy and the rights of workers and their safety should be a priority of a government responsive to the needs of the poor.

Urban Poverty

Much of our discussion up to now has focused on rural poverty, primarily because the incidence of poverty is higher in rural than in urban areas and the majority of the poor still live in the countryside, particularly in South Asia, where poverty is most oppressive. However, there have been substantial shifts in population to the cities and these trends are predicted to continue. The urban population of Asia is projected to nearly double between 2000 and 2030 to over 1 billion from around 650 million to 1.2 billion. The annual rate of increase in the last 25 years of close to 4% per annum is exceeded only by that in Sub-Saharan Africa. Over the next generation, the urban increment alone will account for 130% of the total population growth in the region, as there will be an absolute decline in rural populations (World Bank, 2003). The size of urban areas will increase as populations increase along with the interconnected of different urban regions. The number of cities with over 1 million residents will increase as well as the number of megacities with populations over 10 million. It is useful to group the challenges facing the poor into three interrelated aspects which ESCAP describes as poverty of money, poverty of access, and poverty of power (UNESCAP, 2000).

Poverty of money

The urban poor face many of the same issues as the rural poor. They earn little because their skill base is low and they have limited job security. They are generally unable to accumulate physical assets for themselves, to avail of further education or to educate their children beyond elementary school. The majority of the urban poor work in the informal sector in unskilled occupations in small-scale industries, where productivity is low and competition is high. More often than not there is no social safety net and the poor are highly vulnerable to the vagaries of business cycles. They have limited access to credit and if they can borrow to even out their income stream in bad times it comes at high interest rates from moneylenders. Because urban markets are generally more integrated than rural markets, those who are cut off from the flow of goods and services in the formal sector operate in an informal setting living day to day. Goods and services are more expensive per unit since they are purchased in small volume. Refrigeration and bulk buying is not common and they are often at the mercy of the police and government bureaucrats who extort bribes for doing business and obtaining licenses.

Poverty of access

Many poor families live in urban slums with limited access to urban services. These informal urban slums are located along river banks, railway lines, steep slopes, and areas close to garbage dumps, which serve as source of income for scavengers who gather and sell items they can salvage from these dumps. These slum residents are often in violation of urban ordinances and are not provided with basic infrastructure services such as piped water, electricity, waste water disposal, and solid waste collection by the government. There are few schools, and health services are minimal and often charge high rates as part of the pattern of discrimination against the poor. Life expectancy, infant mortality, and incidence of morbidity are high as a result. The lack of education for the young and poor helps to perpetuate the cycle of poverty from generation to generation.

Poverty of power

Like their poor rural counterparts, the urban poor are cut off from the decision-making processes that determine the allocation of government resources. They have no voice in this decision-making process and are viewed with distrust by the rich who often believe that they would make poor decisions if given power. They suffer more in societies where the level of ethical discourse is limited and decision-making is made based on connections, bribes, and opaque processes rather than openness, transparency, and the rule of law. Because they are marginalized in so many ways the urban poor are acutely aware of the gap between their standards of living and their rich urban neighbors. Income distribution in the cities is much more unequal than in the countryside. Furthermore, the poor are constantly reminded of these disparities as they live their lives in the urban ghettos in close proximity to power centers and government resources. Populist leaders such as Estrada in the Philippines have built power bases among the urban poor. The challenge is to leverage this large urban political power base into a coordinated action force to deal with urban poverty problems. In the Philippines, Estrada exploited them but did little to lift living standards. In Thailand, by contrast, the Thaksin regime made progress in reducing rural poverty while neglecting its constituent base in poor urban areas.

Final Thoughts

Strong and sustained economic growth over the next decade or so will help to bring poverty levels down throughout the Asian region. In South Asia and parts of the Mekong region, additional measures at the micro and human development level have to be undertaken to deal with poverty. In India, more needs to be done to address discrimination against scheduled castes and tribes. In tribal areas and other bypassed areas in India and other South Asian and Mekong countries, discrimination and lack of trust have also been rooted in local traditions for many years. These social attitudes are long-standing and have strong historical roots.

Combined with a culture of corruption they reinforce cultural, political and social divisions that keep the poor from accessing better education, health and sanitation. They also contribute to the continuation of a social divide where the poor have no voice in decision-making. As a result, the poor have very limited ownership of public policies that affect them. Nor do they have any control over how public resources are allocated to address their economic, social and political interests. Lacking political sophistication, the poor are sometimes taken advantage of by unscrupulous politicians and government officials.

Reducing these kinds of social distance, integrating the poor into the apparatus of decision-making, and establishing a respectful and trusting relationship between the poor and the non-poor that involves civil society as well as the government are long-term goals for reducing poverty. Better data would also help to identify how to develop programs. Who exactly is poor? What are the levels of education and health of the poor? Where do they live? How extensive is the depth and duration of their poverty? By assembling a better database, a more coherent and comprehensive program of poverty reduction can eventually be achieved.

Bibliography

Alam, A., M. Murthi and R. Yemtsov (2005). *Growth, Poverty and Inequality: Eastern Europe and the Former Soviet Union*. Washington, DC: World Bank.

Asian Development Bank (ADB) (1992). *Asian Development Outlook*. Manila: Asian Development Bank.

Asian Development Bank (ADB) (1994). *Asian Development Outlook*. Manila: Asian Development Bank.

Asian Development Bank (ADB) (1995). *Escaping the Poverty Trap*. Manila: Asian Development Bank.

Asian Development Bank (ADB) (2000). *Asian Development Outlook*. Manila: Asian Development Bank.

Asian Development Bank (ADB) (2002). *Asian Development Outlook*. Manila: Asian Development Bank.

Asian Development Bank (ADB) (2004). *Poverty in Asia*. Manila: Asian Development Bank.

Bajpai, N., J. D. Sachs and N. H. Volavka (2004). Reaching the Millennium Development Goals in South Asia. Center on Globalization and Sustainable Development, Columbia University, Working Paper No. 17.

Bhattarai, M. and A. Narayanamoorthy (2003). Irrigation Impact on Agricultural Growth and Poverty: Macro Level Analyses in India. Water Policy Research Highlight. http://www.lk.iwmi.org/Assessment/Publications/conference_papers.htm [Retrieved 1 October 2006].

Bourguignon, F. and C. Morrison (2002). Inequality among World Citizens: 1820–1992. *American Economic Review*, 92(4), 727–744.

Bussolo, M. and D. O'Connor (2002). Technology and Poverty: Mapping the Connections. In *Technology and Poverty Reduction in Asia and the Pacific*. OECD and ADB, Development Centre Seminars. http://www.adb.org/Documents/Conference/Technology_Poverty_AP/default.asp [Retrieved 14 October 2006].

Craig, B. J., P. G. Pardy and J. Roseboom (1997). International Productivity Patterns: Accounting for Input Quality, Infrastructure and Research. *American Journal of Agricultural Economics*, 79(4), 1064–1076.

Deininger, K. and L. Squire (1996). A New Data Set Measuring Income Inequality. *World Bank Economic Review*, 10(3), 565–591.

Dollar, D. and A. Kraay (2001a). Growth is Good for the Poor. Policy Research Working Paper Series No. 2587, The World Bank. [Published as: Dollar, D. and A. Kraay (2002). Growth is Good for the Poor. *Journal of Economic Growth*, 7(3), 195–225.]

Dollar, D. and A. Kraay (2001b). Trade, Growth, and Poverty. Policy Research Working Paper Series No. 2615, The World Bank. [Published as: Dollar, D. and A. Kraay (2004). Trade, Growth, and Poverty. *Economic Journal*, 114, 493–515.]

Dowling, J. M. and M. Konchoro (2006). Income Distribution in Indonesia and the Philippines. Mimeo.

Dowling, J. M. and R. Valenzuela (2004). *Economic Development in Asia*. Singapore: Thomson Learning.

Dowling, J. M. and G. Wignaraja (2006). Central Asia's Transition After Fifteen Years: Growth, Regional Cooperation, and Policy Choices. OREI Discussion Paper No. 2, Office of Regional Economic Integration. Manila: ADB.

Evenson, R. E., C. E. Pray and M. W. Rosegrant (1999). Agricultural Research and Productivity Growth in India. Research Report 109, IFPRI.

Food and Agricultural Organization (FAO) (1998). FAO Press release PR 98/31. Rome: UN FAO.

Foster, A. D. and M. R. Rosenzweig (2004). Agricultural Productivity Growth, Rural Economic Diversity and Economic Reforms: India, 1970-2000. *Economic Development and Cultural Change*, 52(3), 509–542.

Gaiha, R. and A. Deolalikar (1993). Persistent, Expected and Innate Poverty: Estimates for Semi-Arid Rural South India, 1975–1984. *Cambridge Journal of Economics*, 17(4), 409–421.

Gajwani, K., R. Kanbur and X. Zhang (2006). Patterns of Spatial Convergence and Divergence in India and China. Paper presented at the annual bank conference on development economics, St Petersburg, Russia.

Garbaccio, R. F. and S. J. La Croix (1998). Regional Convergence in Output and Consumption in China: Evidence from the Maoist and Reform Periods. Mimeo. Honolulu: University of Hawaii, Department of Economics.

Government of India (2002/2003). Economic Survey. Ministry of Finance.

Hasan, R. and M. G. Quibria (2002). Poverty and Patterns of Growth. ERD Working Paper Series No. 18. Manila: Asian Development Bank.

Indo-Pacific Fishery Commission (1994). Socio-Economic Issues in Coastal Fisheries Management: Proceedings of the Indo-Pacific Fishery Commission. Bangkok, Thailand, 23–26 November 1993. *RAPA Publication*, Vol. 8, 442, Bangkok: FAO Regional Office for Asia and the Pacific.

International Fund for Agricultural Development (IFAD) (2002). Assessment of Rural Poverty: Asia and the Pacific. http://www.ifad.org/poverty/region/pi/PI_part1.pdf [Retrieved 10 October 2006].

International Fund for Agricultural Development (IFAD) (2005). Rural Poverty in Vietnam. http://www.ifad.org/operations/projects/regions/PI/factsheets/vn.pdf [Retrieved 10 October 2006].

Khunkitti, S. (2002). Technology and Development Policy in Poverty Reduction: The Case of Thailand. In *Technology and Poverty Reduction in Asia and the Pacific*. OECD and ADB, Development Centre Seminars. http://www.adb.org/Documents/Conference/Technology_Poverty_AP/default.asp [Retrieved 14 October 2006].

Klasen, S. (2004). Malnourished and Surviving in South Asia, Better Nourished and Dying in Africa: What Can Explain This Puzzle? Discussion Paper No. 214, Department of Economics, University of Munich, Germany.

La Croix, S. J., R. Garbaccio and M. Southichak (1998). Regional Economic Development in Asia: Progress, Pitfalls and Policy. In *Sustainable and Balanced Development in the Asia Pacific*, EDAP Joint Policy Studies No. 4, United Nations Development Programme, Korea Development Institute.

Lin, J. Y. (1999). Technological Change and Agricultural Household Income Distribution: Theory and Evidence from China. *Australian Journal of Agricultural and Resource Economics*, 43(2), 179–194.

Lindert, P. and G. Williamson (1985). Growth, Equality and History. *Explorations in Economic History*, 22, 341–377.

Lipton, M. (2001). Reconnecting Agricultural Technology to Human Development. Background Report for Human Development Report 2001, UNDP.

Liu, Y. (2002). Poverty Alleviation in the People's Republic of China's Rural Areas: Problems, Strategy, Policy and the Role of Science and Technology. OECD and ADB, Development Centre Seminars. http://www.adb.org/Documents/Conference/Technology_Poverty_AP/default.asp [Retrieved 14 October 2006].

Manasan, R. G. and S. Chatterjee (2003). Regional Development. *The Philippine Economy*, March, 342–381.

McKinley, T. (2003). The Macroeconomics of Poverty Reduction. Mimeo. New York: UNDP, Bureau for Development Policy.

McKinley, T. (2005). MDG-Based PRSPs Need More Ambitious Economic Policies. Discussion Paper. New York: UNDP, Bureau for Development Policy.

Mellor, J. W. and S. Gavian (1999). The Determinants of Employment Growth in Egypt — The Dominant Role of Agriculture and the Rural Small Scale Sector. Mimeo. Abt Associates, Inc.

Mellor, J. W. and M. S. Mudahar (1992). Agricuture in Economic Development: Theories, Findings and Challenges in the Asian Context. In M. Lee (ed.), A Survey of Agricultural Economics Literature, Vol. 4, *Agriculture in Economic Development.* Minneapolis: University of Minnesota Press.

Mollett, H. (ed.) (2002). *Tackling Poverty in Asia.* London: BOND.

Navaneetham, K. and S. Jose (2005). Poverty, Malnutrition and Mortality in South Asia: A Review of Issues and Options. CICRED Seminar on Mortality as Both a Determinant and a Consequence of Poverty and Hunger, Thiruvananthapuram, India, 23–25 February, pp. 1–21.

Quibria, M. G. (2002). Growth and Poverty: Lessons from the East Asian Miracle Revisited. Asian Development Bank Institute (ADBI) Research Paper No. 33. Tokyo: ADBI.

Ravallion, M. and S. Chen (2004). *How Have the World's Poorest Fared Since the Early 1980s?* Washington, DC: World Bank.

Ravallion, M. and G. Datt (1996). How Important to India's Poor is the Sectoral Composition of Economic Growth? *World Bank Economic Review*, 10(1), 1–25.

Ravallion, M. and B. Sen (1994). How Land-based Targeting Affects Rural Poverty? World Bank Policy Research Working Paper No. 1270. Washington, DC: World Bank.

Sachs, J. and A. Warner (1995). Economic Reforms and the Process of Global Integration. *Brookings Papers on Economic Activity*, 1995(1), 1–118.

Save the Children (2006). http://www.savethechildren.org/ [Retrieved 21 October 2006].

Sen, A. (1992). Missing Women. *British Medical Journal*, 304(6827), 587–588.

Srinivasan, T. N. (2004). China and India: Economic Performance, Competition and Cooperation. Mimeo, Yale University.

Srinivasan, T. N. and J. Bhatwati (1999). Outward–Orientation and Development: Are Revisionists Right? Mimeo, Yale University.

Timmer, P. (1997). How Well Do the Poor Connect to the Growth Process? CAER Discussion Paper No. 17, Harvard Institute of International Development.

Transparency International (2004). Transparency International Annual Report 2004, Corruption Perceptions Index. http://ww1.transparency.org/in_focus_archive/AnnualReport04/annual_report04.htm [Retrieved 10 July 2006].

United Nations (various years). *Human Development Report.* New York: United Nations.

United Nations (2006). UNDP Millennium Development Goals MDGs. http://www.undp.org/mdg/basics.shtml [Retrieved 21 July 2006].

United Nations Economic and Social Commission for Asia and the Pacific (UNESCAP) (2000). Urban Poverty Alleviation Paper presented at the Regional High-level Meeting, Hangzhou, People's Republic of China.

United Nations Economic and Social Commission for Asia and the Pacific (UNESCAP) (2001). *Economic and Social Survey of Asia and the Pacific.* Bangkok: United Nations.

United Nations Economic and Social Commission for Asia and the Pacific (UNESCAP) (2002a). *Economic and Social Survey of Asia and the Pacific.* Bangkok: United Nations.

United Nations Economic and Social Commission for Asia and the Pacific (UNESCAP) (2002b). Population and Poverty in Asia and the Pacific. Background paper presented at the Fifth Asian and Pacific Population Conference, Bangkok, United Nations.

United Nations Economic and Social Commission for Asia and the Pacific (UNESCAP) (2003). Committee on Poverty Reduction: The Millennium Development Goals (MDGs) and the Eradication of Extreme Poverty and Hunger. Discussion Paper. Bangkok, Thailand: Asia Pacific Civil Society Forum, United Nations.

United Nations Economic and Social Commission for Asia and the Pacific (UNESCAP) (2004). *Economic and Social Survey of Asia and the Pacific.* Bangkok: United Nations.

United Nations Economic and Social Commission for Asia and the Pacific (UNESCAP) (2006). Income Generation and Poverty Reduction: Experiences of Selected Asian Countries. Development Papers No. 26. New York, United Nations: UNDP, Poverty and Development Division.

UNESCAP-UNDP-ADB (2005). *A Future within Reach.* Bangkok and Manila: UNESCAP and ADB.

Winters, A. (1999). Trade and Poverty: Is There a Connection? In Dan en-David Håkan Nordström and L. Alan Winters (eds.), *Trade, Income Disparity and Poverty.* Geneva: WTO.

Wolfe, M. (2004). There's a Proven Way to Rid the World of Poverty. *Straits Times* (Singapore), 6 May.

World Bank (1993). *The East Asian Miracle: Economic Growth and Public Policy.* New York: Oxford University Press.

World Bank (1998). *Reducing Poverty in India: Options for More Effective Public Services.* Washington, DC: World Bank.

World Bank (2002). *World Development Report.* Washington, DC: World Bank.

World Bank (2003). Urban Poverty in East Asia: A Review of Indonesia, the Philippines and Vietnam. East Asian Urban Development Working Paper Series No. 11. Washington, DC: World Bank.

World Bank (2006). *World Development Indicators Online*. Washington, DC: World Bank.

Wu, X. and J. M. Perloff (2004). China's Income Distribution over Time: Reasons for Rising Inequality. Department of Agricultural & Resource Economics, UC Berkeley, Working Paper Series No. 977.

Chapter 5

Corruption, Ethics and Governance

As I have said, the first thing is to be honest with yourself. You can never have an impact on society if you have not changed yourself... Great peacemakers are all people of integrity, of honesty, but humility.

— *Nelson Mandela (Nobel Prize for Peace), 1993*

Corruption in Asia has begun to receive a considerable amount of attention in recent years among economists, government planners, ordinary citizens and the international community. A global survey of observed opinions conducted by Transparency International in 2003 showed that political corruption was the number one problem followed by public contracting and judicial and law enforcement.[1]

In order to develop possible outlook scenarios for corruption and governance in the region, it is important to review recent developments in the region. This will enable us to develop an understanding of the forces that determine the nature, level, and extent of corruption in the Asian region as well as the costs of corruption that are borne by the community at large.

Corruption

Corruption and other illegal practices impose considerable costs on society in terms of the misallocation of resources that results from corrupt practices, changes in incentives that draw resources away from productive work, risks to health from sale of illegal substances, human rights violations from illegal immigration, and the dangers that arise

[1]All survey results and indices compiled by Transparency International are based on observer opinions.

from roads, buildings, plants and infrastructure that have been constructed by corrupt contractors. As much as $1.5 trillion dollars is paid in bribes each year, according to ongoing research at the World Bank Institute (WBI) on a worldwide basis in both developing and developed countries. This is about 5% of world GDP which is estimated at 30 trillion dollars. This figure, as large as it may seem, could also be underestimated since it may miss embezzlement of public funds such as the estimated embezzlement of between $15 and $35 billion by former Indonesian leader Suharto and $5 billion by Ferdinand Marcos in the Philippines (as reported by Transparency International). Furthermore, these estimates do not include the extent of fraud within the private sector, but only bribery transactions between private and public sectors.

A calculation of the total amounts of corrupt transactions is only part of the overall costs of corruption. Corruption constitutes a major obstacle to reducing poverty, inequality, and infant mortality in emerging economies. Economic research at the World Development Institute of the World Bank suggests that countries that tackle corruption and improve their rule of law can increase their national incomes by as much as four times in the long term (see World Bank (2006) for further details).

How much of this corruption occurs in Asia? Transparency International's Corruption Perceptions Index shows that of 146 countries rated, 9 Asian countries were in the top half and 15 in the lower half of the distribution (Table 5.1). This suggests that a good share of the world's total corruption occurs in Asia. By population size both China and India (with more than 2 billion people between them) have relatively high scores.

Nature of Corruption

The most widely used definition of corruption is a case where public officials use their office and influence to divert money and resources for private gain, that is using public power for private gain. There are three general categories where this occurs — revenue raising, revenue spending, and regulation.

Table 5.1 Corruption Perception Index 2004 — Transparency International

Country	Corruption Perception Index ranking (out of 146 countries worldwide) — Low scores good, high scores bad
Singapore	5
Hong Kong	16
Japan	24
Taiwan	35
Malaysia	39
South Korea	47
Thailand	64
Sri Lanka	67
China	71
Mongolia	85
India	90
Nepal	90
Papua New Guinea	102
Philippines	102
Vietnam	102
Kazakhstan	122
Kyrgyzstan	122
Pakistan	129
Indonesia	133
Tajikistan	133
Turkmenistan	133
Azerbaijan	140
Myanmar	142
Bangladesh	145

Source: Transparency International (2004, pp. 8–9).

When revenue is raised by the government, corruption can take place when internal revenue service workers and tax department officials reduce taxes due in exchange for a bribe. This can be for personal and corporate income taxes, sales taxes and duties, customs duties, and other tax liabilities. It can be for driver's licenses, passports, visas, import and export licenses, access to credit, and a myriad of other laws and regulations that control the access to resources and restrict the actions of private parties. There is a joke in many Asian countries that businesses keep several sets of books depending on who needs to see them. Where markets exist, bribery and corruption help to clear the markets by allocating fixed resources to those offering the highest bribe not the highest bid.

The second avenue for corruption is through influence in the way that government resources are spent. Private companies obtain contracts for construction of public projects and other spending by the government by bribing officials that award these contracts. After getting the contract supervision of construction is often lax, building codes are ignored, and inferior materials used. The bigger the bribe the more the contractor will be tempted to cut corners in order to make a profit. Lawmakers also channel government projects to their districts and their contractors, friends, or relatives in exchange for votes and kickbacks.

Finally, regulators, including police, the military and those in the judicial system, become part of the corruption process when they receive protection money, accept bribes to influence verdicts, or blackmail bankers into making uncollateralized loans to themselves, family members, and cronies.

Each of these corruption relationships involves a principal and an agent. The principal is the top level of government interested in seeing that there is no corruption and that the system is honest and transparent. The agent is the lower level official charged with enforcing the laws. Corruption and bribery occur when the agent does not act as directed by the principal. The task of anticorruption actions and legislation is to make sure that the agent does the task assigned by the principal without bribery or other corrupt practices. This supervisory task is complicated by the possibility that many other agents are also involved in corrupt practices which may also extend to other principals as well.

Moving outside of the sphere of governmental corruption, a considerable amount of corruption and bribery are involved in the perpetuation of illegal activities. In Asia the major offenders are drug trafficking, prostitution, arms smuggling, human rights abuses as part of smuggling rings, and illegal gambling. These illegal activities, often controlled by criminal syndicates, are protected by the military and police departments through a network of bribery and corruption.

These abuses result in considerable human suffering as well as having an adverse impact on overall well being and living standards of the society as a whole.

While some economists have defended small-scale bribery of public officials on the grounds that it increases efficiency, this is only because the incentive to bribe is the result of stalling and complex bureaucratic procedures that can be skirted by the bribe. If the system were efficient to begin with without bribes then there would be no increase in efficiency. Furthermore, if low-level corruption is tolerated what is to stop it from escalating and it also undermines the legitimacy of the government and its ability to uphold the law. If bribes do serve to improve efficiency the fees should be made public and posted. A market based on illegal payoffs is not efficient.

The focus of corruption can also differ depending on the strength and honesty of the political regime. Wedman (1997) argues that there are three basic types of corruption. First, when central governments are highly corrupt, looters use governmental powers to systematically plunder public funds and extract bribes. Second, if corruption is less pervasive rent-scraping involves manipulation of licenses and the systematic scraping off of these rents by public officials. Third, where corruption is even less pervasive dividend-collection involves a transfer of a percent of profits earned by private companies to public officials as a bribe for awarding the contract. Wedman believes that looting has the worst overall impact on growth and efficiency and dividend collection the least.

Corruption and the Government

When studying corruption, it is important to understand existing institutional arrangements, particularly between the private and public sectors. For example, it is critical to study the kind of government, the size of government and the relationship between the bureaucrats that administer government policy and the politicians who make policy. A considerable amount of information has been accumulated about these variables and relationships. This section provides a short summary of this research as it pertains to the Asian region.

Cross section evidence on corruption suggests that the amount of corruption is positively correlated with the size of government (Goel and Nelson, 1998). When the government bureaucracy is extensive with interests in a variety of economic activities, then the scope for corruption is wider and the amount of corruption increases. The relationship between corruption and government size seems to hold generally across a broad set of countries including developed and developing alike. Within Asia the relationship also seems to hold although the range of both corruption and size of government are not as wide as they are in the global economy. Untangling the relationship between the size of government and the level of corruption is complicated by the interrelationship between corruption, the level of income, the degree of competition and openness to trade, and institutional characteristics of government bureaucracies such as civil service pay scales.

The organization of corruption within the government has widespread implications for the amount of corruption and the impact on corruption "efficiency". Corruption that is organized by a ruling oligarch such as Marcos or Suharto, so called "grand corruption" (Moody-Stewart, 1997) is highly organized, controlled, and supervised by a centrally located and powerful government office. It benefits government officials who are bribed by private firms. In one sense, this network of corruption is efficient in its collection capacity (see Shleifer and Vishny, 1993). It also fits into Wedman's looter category. Once Marcos and Suharto were overthrown, the number of bribers increased, creating a kind of corruption "free for all" which may have increased the total amount of corruption and bribery. However, it falls into Wedman's second and third categories.

Much of high level corruption is focused on big contracts and bidders for such contracts. Bidders offer bribes to get contracts and then pay additional bribes to either get fat contracts that guarantee high profits, or to officials who allow them to use lower quality materials and construction methods. Because infrastructure and defense procurement projects with a short time horizon offer the quickest and often the most lucrative source of corruption, a corrupt administration will tend to stress such projects at the cost of less capital-intensive and longer gestation periods such as school and curriculum development, public health, and poverty reduction. The social costs of such misallocation of resources is substantial, let alone the extra costs of construction and the risks of having substandard roads, bridges, and buildings produced by corrupt contractors.

Concessions for timber, oil and gas, and other natural resources are also subject to the same short-term orientation. In Malaysia and Indonesia, for example, logging concessionaires routinely accelerated clearing and exceeded their concession allotment of logs, and power companies erect small and costly power stations on barges to minimize infrastructure spending and write off capital costs quickly (Rose-Ackerman, 1999; Schwartz, 1994).

Privatization can also provide fertile soil for corruption. Assets of companies to be privatized can be undervalued and this information leaked to contractors (for a price) who are willing to make a higher bid based on this information. After the fact both the government and the contractor benefit; the former by appearing to have made a good deal because they got a high offer based on faulty values and the latter because they probably still bid less than the true market value of the public company. Furthermore the privatized company may continue to take advantage of monopoly power as part of an under-the-table agreement made with the government.

Aside from contracts and concessions, government departments where officials grant licenses, assess duties, and levy taxes are often corrupt. This is particularly true when time delays are costly to those who are applying for the license or having their goods processed. Ports and other transport check points are often corrupt, and efforts to clean them up are met with stiff opposition. Port handling times in Asia are notoriously slow compared with efficient ports in Europe (Yusuf and Evenett, 2002). Indonesia subcontracted many of its port operations and duty collection activities to a Swiss firm, only to later return it to the hands of its own bureaucracy. In addition to large-scale corruption, there are numerous opportunities for petty corruption. Policemen extort bribes from small businesses, operators of small vans, buses and private vehicles, sometimes on the pretext that laws have been violated. Tax authorities take bribes from businesses and private citizens to reduce tax liabilities, and small businesses avoid sales and other taxes by offering bribes to tax collectors.

Determinants of Corruption

There have been a number of empirical studies of corruption using cross-section information from a wide array of industrial and developing countries. These studies are based on a number of hypothetical relationships between corruption and a variety of explanatory variables. There are several economic, political, judicial and social variables that are important determinants of corruption (Treisman, 2000; Ades and di Tella, 1999; Kaufmann, 1997; Seldadyo and di Haan, 2005; and Serra, 2004). These are reviewed below.

Economic variables

Economic variables include the level of per capita income, the relative level of pay for government officials, the openness to international trade and a high level of oligopoly and economic rents in the domestic economy. Income is the most powerful economic determinant of corruption. Other things equal, richer countries are much less corrupt than poor countries. Table 5.2 shows the results of a simple regression between corruption and per capita income for 26 Asian economies in 2002. Income per capita is highly related to the level of corruption. Nearly 90% in the variation in corruption is explained by the best fitting regression. The elasticity at the mean shows that a change in income reduces corruption between 0.21 and 0.27 standard deviations, depending on whether PPP or atlas income series are used.[2]

Higher pay serves as an incentive for government officials to eschew corruption. However, it is also possible and even likely that corrupt regimes pay their civil servants well resulting in simultaneity problems in determining cause and effects. Openness is important because it reduces rents and introduces competition, both of which

[2]Kaufmann and Kraay (2002) measure corruption as a standard normal variable with mean zero and variance one.

Table 5.2 Regression Results — Income Per Capita and Level of Corruption

Dependent variable	Income — PPP	Income — Atlas method	R squared adjusted
TI		0.00018 (9.94)	0.802
WB		0.00076 (7.82)	0.707
TI	0.0000157 (13.16)		0.882
WB	0.0000086 (8.92)		0.773

Notes: Regression coefficient are displayed in the table and *t* values displayed in parentheses, Data sources TI — Transparency international corruption index 2004, WB — Kaufmann and Kraay (2002). Atlas method income is obtained using local exchange rate to US \$, PPP is purchasing power parity level of income.

reduce corruption. The same holds true for domestic monopoly power which is also reflected by the positive relationship between corruption and the size of the mining (including oil and natural gas) sector. Mining leases and contracts for extraction of mineral resources are controlled by the government and are often subject to bribery and corruption.

Social, political and judicial variables

Whether the country had a long period of British rule is the one social variable that has had a significant negative impact on corruption. The impact of this British tradition, in Britain itself and in its former colonies including North America, Australia, India, and many other countries in Asia and Africa is probably the result of a system of justice that held corruption down. Anglican and other Protestant religious traditions show a lower level of corruption than Catholic, Moslem, Hindu, Buddhist or other religious traditions. It is speculated that the reason for this is that Protestantism instills a distrust of state institutions and this is helpful for exposing corruption. Other religions have a hierarchical structure that often works in concert with the state.

British rule is somewhat correlated with the common law judicial tradition that also prevailed in these countries and which offered protection against official abuse of power. However, the importance of common law in reducing corruption has an impact over and above British heritage. This derives from the fact that many other countries have adopted the common law tradition where law evolves based on precedents rather than codes drawn up by central governments. Because of this common law is more flexible and arguably more in tune with the will of the people that emphasizes procedures rather than fixed codes. Consequently it is more likely that residents of countries with these traditions are more likely to see corrupt procedures as violations of the legal system. Democratic institutions tend to reduce corruption but only when such institutions have been in place for many decades.

Governmental variables

The economic analysis of Becker and Stigler (1974) and others working in this field suggests that corruption is a positive function of incentives such as higher pay and better chances for promotion as well as being a negative function of the risk of being caught and punished. We saw that countries observing common law traditions probably provide a more credible threat of punishment and this threat seems to deter corruption. On the other hand, evidence for higher salaries reducing corruption is more difficult to determine because of the simultaneity problem discussed above. Nevertheless, economic logic would suggest that higher salaries for government officials would reduce corruption; if they are caught they lose a higher paying job.

Turning to the structure of government, federal governmental structures tend to be more corrupt than governments run by a central administration although the evidence is not overwhelming. This is probably because corruption becomes more diffused with the devolution of power from the central government. High level corruption gives way to more petty corruption and bribery. Furthermore, when local officials have more autonomy they are less under the control of central authorities who might otherwise keep them from exercising their own ability to develop a network of corruption.

We can conclude from this short review of the determinants of corruption that bribery and corruption are endemic in many societies and intertwined with the social, institutional, and political fabric. Certainly economic growth, reducing rents, and opening up to trade will help to reduce corruption. Yet we observe that many Asian countries, including China, India, Indonesia and Philippines, are open and growing fast, but also have high corruption scores. Hopefully, corruption has begun to fall as a result of these positive economic developments. However, we do not yet have adequate time series data to see whether this is true.

Trends in Corruption

Drawing inferences on the time path of corruption from the cross-section data, we would expect to find an upward trend in income associated with a reduction in the level of corruption over time. Systematic surveys of time series data on corruption is available from the files of Transparency International since the early 1980s, a period of 25 years or so. Looking at this evidence reported in Figure 5.1, where high values reflect less corruption and more honesty, there does not seem to be a downward trend in indices of measures of corruption for many countries. If anything, the trend is up, not down. Among the three largest economies, corruption has been trending upward in India and China, and down in Indonesia. In the NIES, Singapore has an exemplary record of low corruption and this record has improved slightly from high to very high levels of honesty. Corruption has been low in the other NIES as well, remaining rather flat in Taiwan and Hong Kong with a slight dip in the early 1990s and late 1980 and improving slowly in Korea up until the Asian crisis.

In Southeast Asia there is also a reduction in corruption in Thailand in the last decade compared with the 1980s and early 1990s. On the other hand, corruption in the Philippines has increased since the Asian crisis after falling for nearly two decades while there has been a slight increase in Malaysia. There are not enough data to draw conclusions for Vietnam. In South Asia, aside from the increase in

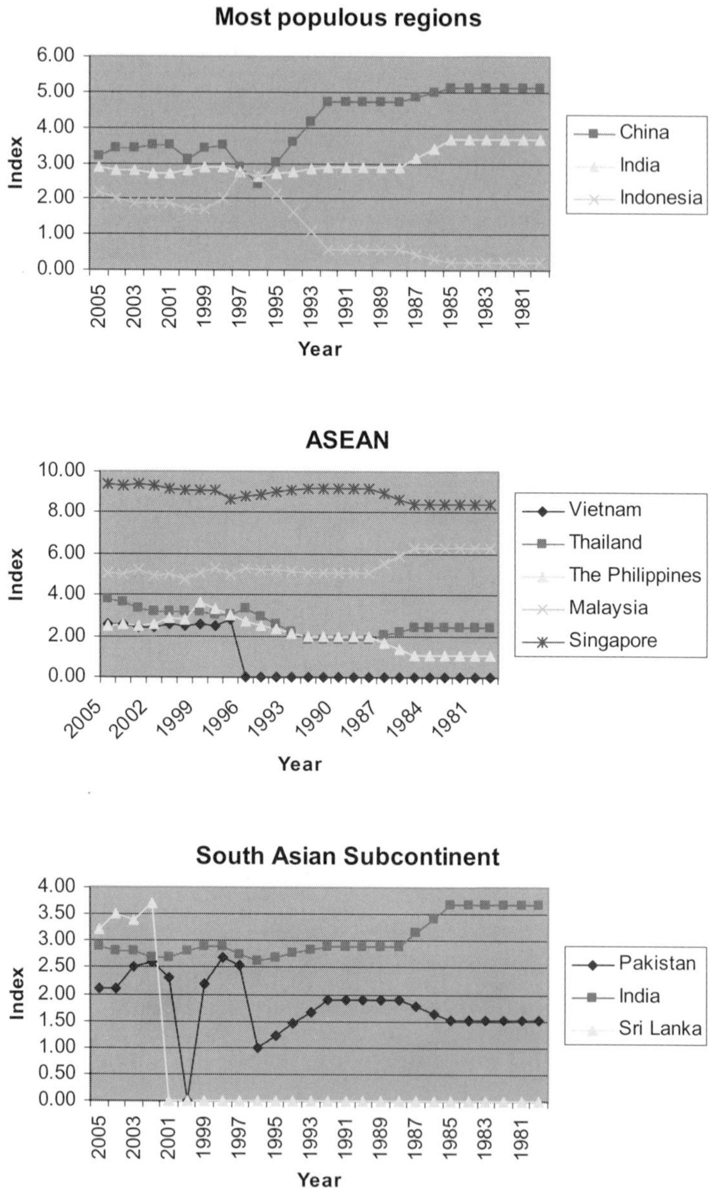

Figure 5.1 Corruption 1980–2005 — Transparency International

Source: Transparency International.

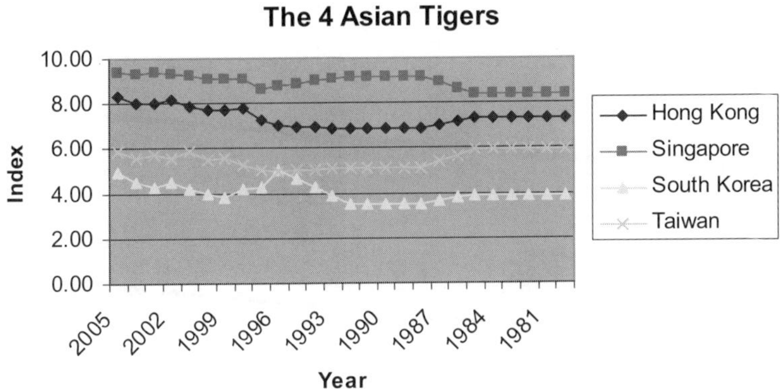

Figure 5.1 (*Continued*)

corruption in India, the experience in Pakistan and Sri Lanka is hard to interpret because of a lack of a continuous time series.

Table 5.3 shows the level of corruption from Transparency International and the World Bank corruption indicators for selected years from 1996 to 2005 and also the opacity index constructed by the Kurtzman group. Only a few countries show a reduction in the level of corruption and some of these, such as Singapore, already have low corruption levels.

If we carry the conclusions drawn from analysis of Figure 5.1 and Table 5.3 forward, we might be observing a divergence in corruption leading to a bimodal distribution. Rather than rising incomes eventually resulting in low levels of corruption in many countries, we are seeing that a large group of poor countries are caught in a low level corruption trap at the same time that others at higher levels of income are becoming more and more honest.

There may also be a correlation between broad measures of governance and corruption. A World Bank study reported on more fully in a subsequent section of this chapter compiles a broad set of governance indicators, including corruption, for a number of countries over the period 1998 to 2004. Of the Asian countries mentioned in this survey in five different categories (see discussion of governance

Table 5.3 Trend in Corruption — 1996–2005

Country	WB (1996)	WB (1998)	WB (2000)	WB (2002)	WB (2004)	WB (2005)	TI (1998)	TI (2002)	TI (2005)	OA (2004)
Azerbaijan	–1.03	–1.08	–1.13	–1.07	–1.21	–1.01		2.0 (95)	2.2 (137)	
Cambodia	–1.00	–1.34	–0.79	–0.90	–0.98	–1.12			2.3 (130)	
China	–0.00	–0.20	–0.38	–0.40	–0.59	–0.69	3.5 (53)	3.5 (59)	3.2 (78)	50
Bangladesh	–0.49	–0.47	–0.68	–1.02	–1.29	–1.18		1.2 (102)	1.7 (158)	
Hong Kong	1.64	1.67	1.43	1.45	1.56	1.68	7.8 (16)	8.2 (14)	8.3 (15)	20
Japan	1.34	1.10	1.28	1.21	1.22	1.24	5.8 (29)	7.1 (20)	7.3 (21)	28
Kazakhstan	–0.90	–0.93	–0.92	–1.08	–1.16	–0.94				
S. Korea	0.61	0.33	0.33	0.33	0.12	0.47	4.2 (43)	4.5 (40)	5.0 (40)	37
Kyrgyz	–0.84	–0.76	–0.93	–0.86	–1.01	–1.06		2.3 (88)	2.6 (107)	
Lao PDR	–1.00	–0.77	–0.88	–0.94	–1.04	–1.10			3.3 (77)	
India	–0.32	–0.24	–0.31	–0.41	–0.38	–0.31	29 (66)	2.7 (71)	2.9 (88)	48

(Continued)

Table 5.3 (*Continued*)

Country	WB (1996)	WB (1998)	WB (2000)	WB (2002)	WB (2004)	WB (2005)	TI (1998)	TI (2002)	TI (2005)	OA (2004)
Indonesia	–0.49	–1.03	–1.05	–1.19	–0.96	–0.86	2.0 (80)	1.9 (96)	2.2 (137)	59
Malaysia	0.57	0.67	0.21	0.33	0.29	0.27	5.3 (29)	4.9 (33)	5.1 (39)	
Mongolia	0.41	–0.34	–0.28	–0.08	–0.52	–0.56			3.0 (85)	
Myanmar	–1.25	–1.37	–1.30	–1.43	–1.60	–1.44				
Nepal	–0.29	–0.66	–0.63	–0.44	–0.70	–0.71			2.5 (117)	
Pakistan	–1.04	–0.82	–0.94	–0.85	–1.06	–1.01		2.6 (77)	2.1 (144)	45
Papua N.G.	–0.27	–0.77	–0.92	–0.82	–0.94	–1.08			2.3 (130)	
Philippines	–0.41	–0.34	–0.53	–0.55	–0.58	–0.58	3.3 (55)	2.6 (77)	2.5 (117)	50
Singapore	2.38	2.43	2.44	2.35	2.42	2.24	9.1 (7)	9.35 (5)	9.4 (5)	24
Sri Lanka	–0.23	–0.30	–0.16	–0.20	–0.17	–0.31		3.7 (52)	3.2 (78)	

(*Continued*)

Table 5.3 (*Continued*)

Country	WB (1996)	WB (1998)	WB (2000)	WB (2002)	WB (2004)	WB (2005)	TI (1998)	TI (2002)	TI (2005)	OA (2004)
Taiwan	0.82	0.80	0.63	0.71	0.63	0.63	5.3 (29)	5.6 (29)	5.9 (32)	34
Thailand	−0.33	−0.32	−0.37	−0.32	−0.24	−0.30	3.0 (61)	3.2 (64)	3.8 (59)	35
Tajikistan	−1.76	−1.18	−1.12	−1.08	−1.20	−1.08			2.1 (144)	
Uzbekistan	−1.05	−1.04	−0.87	−1.04	−1.16	−1.07		2.9 (68)	2.2 (137)	
Vietnam	−0.76	−0.81	−0.71	−0.78	−0.67	−0.68	2.5 (74)			

Notes: WB signifies World Bank data, TI signifies Transparency International data, and OA signifies the Kurtzman opacity index. World Bank data from Kaufmann *et al.* (2005a). The scores are normally distributed variables with zero mean and standard deviation of 1 (*Z* variables). The Transparency International scores range from 1 to 10, higher scores indicate less corruption. Rankings are shown in brackets for the Transparency International data. The Kurtzman opacity index (see http://www.kurtzmangroup.com/ for details) is based on a number of indicators including World Bank statistics, Transparency International corruption index, and the International Country Risk Guide.

below) all but one experienced a significant decline in the effectiveness of governance over the period.

Examining this evidence more carefully we can make several observations. The first is that a decade or two might be too short a time period to observe a downward trend in corruption. It may take many decades to reduce corruption as income increases. The second is that the observation period includes the years of the Asian crisis, a time when many countries in Asia were under stress. As a result the Asian financial crisis may have resulted in temporary setbacks in the fight against corruption in these crisis countries. This is certainly reflected in the figures for Philippines, Korea, and perhaps Malaysia. A third explanation is that the data on corruption are not accurate enough to capture small changes in corruption over a few years. For example, the Transparency International data for the early 1980s are compiled as average over five-year period and are based on a smaller sample base of observations within each country than in subsequent years. As a result the time series evidence should be viewed with some skepticism.

In any event, the data compiled by Transparency International and the World Bank do demonstrate the difficulty in dealing with corruption as a short-term problem. Concerted efforts over a significant time period are required to make a dent on the psychological, economic, and institutional environments that spawn corrupt institutions and the mentality of the people. If these efforts are not made, the "vicious cycle of poverty" could be joined and reinforced by a "vicious circle of corruption".

Impact of Corruption on the Economy

Market clearing and bribery

Governments that control the markets for goods and services often participate in bribery and corruption. The way this works is through a dual market system. Goods are supplied by the government to selected buyers at subsidized prices while the general public pays much higher prices. When the government has control of many markets this can be a rich source of corruption for government officials who get payoffs to get access to products at state-subsidized prices. This was, and continues to be, a common practice in centrally planned economies. In China, market and subsidized prices were sometimes very far apart. "The market price of many commodities were 2 to 7 times higher than subsidized prices" (Rose-Ackerman, 1999, p. 10; see also *New York Times*, 1998). In financial markets, including access to bank credit and preferred rates of interest on loans, bank officials also routinely accept bribes from borrowers to secure access to subsidized funds. Furthermore, high-profile public officials, military officers, and cronies of the leaders of the country often borrow from banks on the basis of a letter to the bank officials without any credit check and often without any plans to repay. Memo lending in Indonesia and large numbers of bank loans from Chinese banks are largely responsible for the high level of non-performing loans in these two countries (Dowling and Yap, 2004; *People's Daily Online*, 2004).

In Pakistan estimates put the cost of the banking crisis of 1996–1997 at 10% to 15% of GDP as a result of NPLs provided to friends of the regime (Burki, 1997, p. 9). Issue of export and import licenses is another way that patronage and bribes arise (see Dowling and Yap, 2004 for the situation in Indonesia).

In cases where bribers and non-bribers compete for services, the briber pays for faster service. However, this process does not act in the same way as a market where those willing to pay higher prices bid up the market price. Rather, because the bribery mechanism is secretive, the bribe price is secret information and those who are unwilling to bribe never know the going bribe price. Even those who offer bribes that are refused also do not know the market clearing bribe. Aside from the welfare losses sustained by losers who refuse to bribe, markets with bribes are less efficient than honest markets, because considerable time is spent in keeping bribery market information secret.

When bribery involves sequential use of resources such as upstream and downstream water, those at the end of the queue often get very little while those at the head of the queue may waste what they are allocated. When bureaucrats can ration both the quantity and quality of the service provided, bribes are also inefficient market clearing mechanisms. Public services like obtaining a passport, getting a telephone installed, obtaining a valid driver's license, qualifying for a profession by taking an exam, obtaining an old age pension, or getting workman's compensation are screened by public officials. Those who bribe receive the service faster or are allowed to circumvent qualification requirements like being able to drive a car, be a nurse, lawyer or teacher! The greater the discretion of the public officials and the more desirous the service to be provided (and the fewer the options) the greater the asking price for the bribe.

Bribes as incentives for bureaucrats and to reduce costs

There are many ways in which bureaucrats create opportunities for corruption aside from expediting the provision of public services. Tax collectors can extort bribes by creating false tax liabilities for businesses and individuals. Tax rates are subject to interpretation by customs and other officials and those who do not pay bribes can be charged above the legal rate. The uncertainty created by such systems act as an impediment to foreign businesses and they also provide a strong incentive for all businesses to keep several sets of books (Rose-Ackerman, 1999, p. 17; Prichett and Seeth, 1994). The risk of continuing to condone such systems of "grease money" is that the "grease" cost increases with time as the corrupt become bolder and those paying the bribes do not take any legal action.

Regulatory agencies that are responsible for inspections to ensure workplace safety, public safety, to check to see if building codes have been adhered to, issuance of business permits, and enforce environmental regulations are all possible loci for bribery. In Indonesia small businesses often pay bribes (Robison, 1986); policemen extort bribes from bus drives and jeepney drivers in the Philippines (World Bank, 2000a); and forest concessionaires pay bribes to circumvent cutting requirements and to cut in public lands (Roodman, 1996). Buildings have collapsed in several countries in Asia as a result of poor quality materials, architectural design, and shoddy workmanship (see Rose-Ackerman, 1999 for cases in Korea; and Kidd and Richter, 2002 for other cases).

Organized crime and corruption

Illegal businesses survive by paying bribes to the police, the military, public officials, judges and the information media. Organized crime syndicates exist in many Asian countries. Three Thai economists have exposed syndicates in drug smuggling and distribution, prostitution, gun running and gambling (Phongpaichit *et al.*, 1998). The extent of these activities is estimated to be between 8% and 13% of GDP every year. Organized crime syndicates also operate in other countries. Offering "protection" to legitimate businesses is a favorite activity of organized crime. Gambett and Reuter (1995) suggest that businesses where barriers to entry are low, technology is unsophisticated, labor unskilled, and demand is inelastic are good candidates for organized crime. By intimidating small operators, crime syndicates can extort money from many small businesses with the help of police protection. Garbage collection is a good example. In Asia drug smuggling out of the "golden triangle" of Myanmar, China and Thailand is widespread as are syndicates that control prostitution and smuggling, particularly out of Myanmar and countries in the Mekong region. This entire golden triangle region has been a drug producing area for many years.

Economic Costs of Corruption

The Asian Development Bank (ADB) has analyzed corruption and lists a number of problems brought on by corruption and the cost of corruption. Corruption can add between 20% to 100% to the procurement of government goods and services in a number of Asian countries. Corruption can cost governments as much as 50% of their tax revenues.

Estimated losses due to corruption can total more than a country's foreign debt. Corruption diverts foreign investment, may reduce expenditure on social sectors such as health and education in favor of military spending and large capital projects, and leads to reduced asset life. Corruption undermines the creation of a professional, merit-based civil service. Corruption is a highly regressive tax, and its burden falls disproportionately upon the poor. Combating it is therefore essential for realization of the goal of poverty reduction.

As a number of recent tragedies have revealed, the cost of corruption is often measured not only in dollars and cents but also in human lives. Behind the hundreds and even thousands of lives lost each year to fires, landslides, collapsed infrastructure and ferry sinking, there is often a sordid tale of building codes violated or safety standards overlooked by officials willing to turn the other way. In extreme cases, corruption can contribute to political instability and regime collapse (see ADB, 2001 for further details). In the remainder of this section we look more closely at the impact that corruption has on growth, rents gained by restrictive trade practices and negative effects on investment and on saving.

Corruption retards growth

At the macroeconomic level, corruption and economic growth and development are negatively related. Countries that are high on the corruption list of Transparency International, the Kaufmann *et al.* index, the Business International index, and the Opacity index all have low per capita incomes. Expanding on the income per capita and corruption results displayed in Table 5.2, a regression equation for Asia suggests that a 10% increase in income is associated with between a 2% and 4% decrease in corruption, other things equal (Table 5.4). In a study of over 70 countries during the 1970s and early 1980s, Mauro (1995) estimates that a one standard deviation improvement in the corruption index translates into an increase in the growth of per capita income of 1.3% as well as an increase in the investment rate of nearly 3%.

Corruption is negatively associated with openness

Countries with more open trading regimes have lower levels of corruption. Competition from foreign firms requires domestic firms to be increasingly competitive and competition is not compatible with corruption. Furthermore, competition reduces rents and gives domestic firms role models of efficiency if they are to compete in international markets. Rents are also reduced as markets become contestable and trade increases as a share of income.

Table 5.4 Log Regression of Corruption as a Function of Income, Openness, and Concentration of Mineral Wealth for Asian Economies

Dep. Var Kaufmann	Dep Var TI	PPP Income	Atlas Income	Openness	Resource Dummy	R squared	Sample size
x		0.313		0.122	−0.42	0.78	20
		(5.2)		(1.5)	(2.9)		
x			0.207	0.125	−0.46	0.81	20
			(5.7)	(1.7)	(3.4)		
	x	0.397		0.128	−0.17	0.88	20
		(8.6)		(2.1)	(1.5)		
	x		0.260	0.133	−0.22	0.90	20
			(10.1)	(2.4)	(2.3)		

Source: Data from World Bank and Transparency International, regressions by the author.
Data definition: TI is transparency international corruption index; Kaufmann is from World Bank database constructed by Kaufmann *et al.* (2005b); Openness is sum of exports and imports divided by GDP; PPP Income is World Bank purchasing power parity adjusted income; Atlas income is World Bank unadjusted income; and resource dummy is one when country has significant mineral resources. *t* values in parentheses.

These observations are reinforced by the regression results reported in Table 5.4, particularly with respect to the TI index of corruption, which shows openness is a significant determinant of corruption. While the elasticity is small (1% increase in openness results in one-tenth of 1% decrease in income) the cumulative effects could be large. Greater openness is consistent with more rapid growth and this interactive cycle of more rapid growth, openness, and greater competition serves to reduce corruption even further.

Corruption has serious detrimental impacts on foreign and domestic saving and the level of investment

Less corruption is also positively related to the level of foreign direct investment and the size and robustness of equity markets including stock market capitalization and stock market traded value (Kurtzman *et al.*, 2004; Wei, 1997). The impact of corruption was strong in countries where corruption was highly syndicated and predictable.

The level of corruption also has a negative impact on the level of domestic saving and investment. Countries with low saving and investment rates tend to be more corrupt than those with higher rates of saving (Mauro, 1995; Shleifer and Vishny, 1993; Table 5.5).

Corruption shifts the emphasis of public spending toward projects that lend themselves to corruption such as infrastructure projects and

Table 5.5 FDI, Corruption and Investment in Asian Economies

Country	FDI US$ million (2004)	TI index of corruption (2005)	Investment rate (2004)
Azerbaijan	2311	2.2(137)	40
Cambodia	115	2.3(130)	20.7
China	60600	3.2(78)	45.6
Bangladesh	385	1.7(158)	23.6
Hong Kong	–5702	8.3(15)	23.0
Japan		7.3(21)	33.0
Kazakhstan	3477		
S. Korea	3397	5.0(40)	29.1
Kyrgyzstan	122	2.6(107)	19.0
Lao PDR	17	3.3(77)	22.0
India	4374	2.9(88)	26.5
Indonesia	1046		21.3
Malaysia	2074	5.1(39)	22.5
Mongolia	132	3.0(85)	27.0
Myanmar	na		
Nepal	0	2.5(117)	27.3
Pakistan	906	2.1(144)	18.1
Papua N.G.	na	2.3(130)	
Philippines	57	2.5(117)	17.0
Singapore	5378	9.4(5)	18.3
Sri Lanka	133	3.2(78)	25.9
Taiwan	–5189	5.9(32)	20.7
Thailand	656	3.8(59)	27.1
Tajikistan	na	2.1(144)	
Uzbekistan	115		23.9
Vietnam	1730		35.5

Source: Transparency International and World Bank database. Rank in the world in parentheses.

away from maintenance of existing infrastructure and new social sector projects that raise the level of education and health. Corruption also reduces tax revenue and the effectiveness of industrial policies.

Direct foreign investment is discouraged because corrupt regimes are unreliable, unpredictable, and capricious and their policies are often rearranged to suit certain clients while harming others. Corruption also exacts a higher price on the poor than on the rich. The rich can afford to pay protection and other bribes that help them maintain their monopoly or oligopoly power, while the poor have no recourse but to submit to the bribery demands of the police, licensing bureaus, and organized crime.

In centrally planned economies the vast size of the public sector and the allocation of resources by the government allows for extensive corruption. For example, in the 1980s in China state enterprises operated a two-tier system whereby some resources were sold on the private market and others were allocated at lower prices to select buyers. Li (2001) has estimated that between 1985 and 1989, between 35% and 37% of the production was diverted away from the free market, garnering extensive benefits to the bureaucrats that oversaw the scheme and resulting in a general loss in economic efficiency as resources were allocated according to administered rather than market prices. Other estimates of the size of corruption during the 1980s varied between 10% and 20% of GDP (Li, 2001).

The overall distortions and reductions in efficiency that corruption introduce are difficult to measure. Many observers suggest that they might be significantly larger than estimates of the direct cost of corruption that have been made using convictions for bribery, amounts salted away by Marcos, Suharto, and other oligarchs or other estimates based on tax returns or visible signs of corruption made by looking at diversion of resources and products to corrupt bureaucrats and businesses.

Corruption bears heavily on the poor

Because budget resources are often diverted to lucrative infrastructure projects, rural health clinics specializing in preventive care and primary education are neglected. Where resources are allocated for education materials some of the budget is skimmed off by corrupt officials and students receive fewer resource materials (see example of Philippines below). The poor receive a lower level of social services because budget resources are allocated elsewhere and because they have fewer resources to offer to corrupt officials to help divert resources to their communities. Infrastructure projects focusing on urban projects benefiting the rich will be directed to well-to-do communities through pork barrel schemes to line the pockets of corrupt government officials who conspire with construction contractors. Tax systems which are controlled by the rich tend to be regressive and bear more heavily on the poor. Agricultural marketing and subsidies favor the rich and bear more heavily on the poor because the subsidy programs are designed to help rich farmers more than poor farmers (Rose-Ackerman, 1999).

The Culture of Corruption

Complicity of a wide range of public officials and business leaders is required to sustain a high level of corruption. The general public also has to condone or at least not object strongly to these corrupt practices. One explanation for the strong correlation between corruption and level of development stems from differences in social organization. In market-oriented economies where capitalism has a long history, competitive markets play a critical role in allocating resources and determining the pattern of production and distribution of income and these markets tend to operate efficiently.

When markets and capitalism are not well established, personal connections take their place. This is particularly true when there is general distrust of government. It is essential in such circumstances to have a wide network of friends and acquaintances in a variety of different occupations to facilitate business and to obtain favors. The importance of reputation and trust take on added importance in this setting. Bribery and corruption which are secretive in nature require trust on both sides in order to maintain the bond of secrecy and to provide insurance that neither side will extort or blackmail the other.

The importance of establishing a social network which includes friends, extended family, and business associates is particularly important in Asian economies where there is often a fundamental distrust of government, perhaps dating from colonial times but also reinforced by a history of corruption, bribery and special interests. Oligarchs such as Marcos and Suharto are stereotypical leaders and there are few democratic regimes with a reputation for honesty and probity in the region. Relationships in these social systems are codified and the individual's place in a social hierarchy is established by tradition and reinforced by family values. In Indonesia there is *panchasila*, in the Philippines there is *paki ki sama*, and in China, Hong Kong, Taiwan and Singapore there is *guanxi*. These social, cultural, and political connections are more powerful than money alone. Strong allegiance to family members contributes to cultural values that stress the importance of networks and personal relationships in business rather than objective standards of productivity and efficiency and trust in established institutions. As a result businesses might encourage nepotism and cultivate relationships that build up family networks that connect business with government. Such relationships are characterized by gift giving and mutual favors which can easily deteriorate into corruption.

Hierarchies are also important. In top–down patrimonial relationships where public officials are the patrons and private interests are the clients, a climate for bribery and corruption are easily established. The case of Bangladesh is a good example. Businesses were depending upon the state, particularly after the initial separation from Pakistan and this relationship has continued (Kochanek, 1993). If, on the other hand, business interests are powerful, then they serve as the patron and work through the government to obtain favors. The Philippines is a good example of such a "clientelistic" state, where business interests originating in agricultural monopolies and later transferring to industrial power, have worked in concert with the government for decades (see Kahn (1996) for definitions of different kinds of principal-agent relationships).

In newly emerging democracies such as India the pattern of corruption has not changed much from earlier times when a cozy relationship between technocrats and businesses existed as part of the colonial ruling paradigm and also as a result of having a very large state sector. This involved a combination of ownership of key industries by the government and the power to regulate private enterprise. Tight regulations and need to have various permits to conduct business resulted in the "license–permit raj" (license–permit regime) involving politicians empowered to make decisions, bureaucrats who ran the system on a daily basis, and businessmen who needed permits to get business done.

Lack of competition in domestic markets meant that businesses controlled captive markets, extracted rent, produced inferior products, and failed to adopt new technology. Automobiles continued to be produced with British technology from the 1950s and sold at monopoly prices. This system did not respond to democratic incentives. Indeed political campaign finance was an extension of the reciprocal exchange of favors that characterized the nexus of business, governmental bureaucracy and politics. A sting operation conducted in 2001 by journalists uncovered an intricate web of official corruption running vertically from clerks up through middle men and up to top politicians, and high ranking military officers. It also ran horizontally across the civilian bureaucracy, the military officer corps, and the political elite (Bose, 2002).

As competitive and contestable markets become more prevalent and economies become more open to foreign trade, the level of corruption can be expected to decline, particularly if the governmental administration can be cleaned up and the scope of government interference in the economy reduced. It is particularly important that the scope of government control over the production process and widespread regulation of businesses is reduced.

This has been the general experience of several countries in Asia which were quite corrupt in the early part of the 20th century and even after World War II. The cases of Hong Kong, Singapore and Korea are particularly noteworthy. Their experience, along with other less successful countries such as Indonesia and the Philippines are reviewed in a subsequent section of this chapter.

Controlling Corruption

In the previous section dealing with the causes of corruption we saw that there are a few major determinants of corruption. Turning the analysis around we can begin to analyze how it is possible to affect these variables, thereby increasing the ability to control corruption. There are three important categories. They are (i) economic variables, (ii) administrative and bureaucratic features as well as the justice system, and (iii) political, social and historical factors.

Economic variables

Economic growth seems to have the most significant effect in reducing corruption. Countries in East Asia and Southeast Asia, which currently have high levels of per capita income, have become less corrupt over time. At the beginning of the postwar era there was a high level of corruption in Singapore. There were ample opportunities, law enforcement was lax and civil service salaries were low. Police were involved in gambling and there was widespread bribery in customs, immigration, and internal revenue departments (Quah, 2003). Corruption in Hong Kong was also widespread and continued until the 1970s.

> Corruption was deeply rooted, widespread, generally tolerated and, in some sectors, highly organized. Every part of the public service was infected, especially the disciplined services, namely the police, the customs and excise service, the immigration department, the fire and ambulance services and the prison services (de Speville, 1997, p. 13).

There was also widespread corruption in Korea. Legal sanctions were weak until 1974 when major corruption scandals involving civil servants were exposed and there was increasing public dissatisfaction with bureaucratic corruption. Korea suppliers subverted the US army policy of competitive bidding by colluding to overcharge the US army by 30% or more. Firms also gave gifts to Korean accountants and contractors working for the US army as bribes to increase cost estimates. At the political level vote-buying was commonplace. Estimates by Lee (1990) suggest that the ruling Democratic Justice Party spent between

US$300 million and $1 billion to buy votes in the 1987 presidential election. Polls conducted in Korea show that politicians are considered much more corrupt than any other groups in society — over 70% of respondents named politicians as the most corrupt group in South Korea compared with less than 10% for businessmen and government officials (see Quah, 2003, p. 158 for further details and references). Low salaries for civil servants, the gift giving culture, and other cultural variables have been offered as explanations for corruption in Korea. In other Asian countries also corruption was widespread. The looting of public finances by Marcos and Suharto are well documented and recent exposés of corruption in Thailand have been recorded by Phongpaichit *et al.* (1998).

Over time, however, as incomes have risen and societies have become more open to trade with less corrupt countries in Europe and North America, the level of corruption has begun to decline in some countries. A growing and more socially aware middle class demands higher standards of probity than a society where most of the people are poor, not well educated, and can be easily manipulated by powerful oligarchs. As a result there has been a shift in the willingness to put up with corruption.

As foreign direct investment increases and more markets become contestable it also becomes more difficult to retain an exclusive hold on monopoly rents that derive from exclusive import licenses or other trade restrictions that protect domestic producers and their friends in the bureaucracy. Businesses from industrial countries are also generally less willing to countenance the bribery/corruption culture although there are exceptions such as the Lockheed scandal. Aside from the indirect effects of more rapid growth, higher incomes, more openness, and encouraging competition there is little else that direct macroeconomic policies can do to reduce corruption.

Administrative policies including civil service reform

Civil servants are likely to contact the public in a variety of ways. Each of these contacts presents opportunities for bribery and corruption, from issue of licenses and permits to enforcing regulations to levying customs duties, collecting taxes, and enforcing criminal and civil codes. Local traditions and conditions will also play a role in how the government bureaucrats carry out these duties. Within this context a number of proposals for reform that have widespread, if not universal, appeal. Pay reform is one. If civil service salaries are too low, bribery and corruption may be needed to survive. Furthermore, if pay level differences with the private sector persist, the quality of applicants to the civil service will deteriorate and along with it, the quality of service and the *espris de corps* of the civil service. After a time the civil service will be staffed with either low-skilled/low-productivity workers who cannot get a decent job anywhere else or those who are willing to take bribes to augment their incomes (Besley and McLaren, 1993). Furthermore, recent empirical evidence shows that salary levels of civil servants relative to private sector wages in manufacturing are negatively related to the level of corruption (van Rijckeghem and Weder, 1997).

This evidence suggests that raising civil service salaries is a good way to reduce corruption and bribery. Of course, the richer countries will generally be able to offer higher salaries than poorer countries, so there is a degree of collinearity between higher incomes and higher civil service salaries. In addition to higher pay, civil service reform needs to reduce the number of opportunities and scope for corruption. This involves keeping civil servants and the business interests they serve separate and distinct. Once civil servants go into business on the side or have a private financial interest that conflicts with their public duties it is a golden opportunity for corruption. To avoid conflict of interest, holders of high office should be required to divest themselves of assets earned before being elected (or appointed). Alternatively they can put assets in a blind trust to be managed by an impartial trustee. The civil service also needs to be an efficiently operating meritocracy and not a dumping ground for hacks, relatives, and the unemployable. Nepotism and low productivity are characteristics of many civil services in developing countries. These problems need to be addressed if progress is to be made in reducing corruption in bureaucracies around the Asian region.

Legal reform and corruption

Turning to an analysis of the judiciary and the potential for corruption there are several factors involved. Firstly, it is important to consider the relationship between judicial reform and overall governance. Governance is a much broader category than the judiciary (see the next section.) At the same time, poor governance can have a spillover impact on the judiciary. If there is a widespread breakdown in governance it is easy for a basically honest judiciary to be drawn into corruption.

Where the level of corruption is high, corruption often extends to the legal system. Judges are bribed to hand down verdicts favorable to those accused of corruption and jurors are bribed to throw out verdicts to those same people accused of corruption. Because of its independence the judicial system can feature strongly in the institutional framework to reduce corruption. Furthermore, because it is smaller than other branches of government in the justice system (compared with police) it can be more easily monitored and its integrity increased. However, if it is corrupt it has a cascading impact on the rest of the government. If the police and prosecution agencies as well as the general public lose confidence then there are few chances for corruption to be effectively reduced.

How pervasive is such corruption and what can be done to clean up the judicial systems in Asian countries? There are no systematic and concrete estimates of the degree of judicial corruption in Asia. There is, however, much anecdotal evidence and there are many items in a reform agenda. In Cambodia a survey carried out by the World Bank, 64% of respondents agreed with the statement that the judicial system is very corrupt. In Bangladesh 63% of those involved in litigation had paid bribes to either court officials or the opponents' lawyers and an astounding 89% of respondents were convinced that judges were corrupt. And in the Philippines a large proportion of respondents believed that there was significant corruption and judges could be bribed. In Thailand 46% of respondents believed that the Supreme Court was dishonest (see World Bank (2000b) for evidence about Cambodia; World Bank (2000a) for evidence about the Philippines;

and Phongpaichit *et al.* (2000) for evidence about Thailand). In Thailand indicators of corruption in the judiciary as perceived by the public include delays in execution of court orders, lack of transparency, unfair variances in sentencing for similar crimes, appointment to courts on the basis of political payoff, socializing between judges and lawyers, conflict of interests, and obvious prejudice.

To control corruption in the judicial system, there must first be mechanisms introduced to monitor the performance of the system. A workshop for high level justices convened by the United Nations (2001) made a series of recommendations for developing appropriate indicators of corruption and discovering causes of corruption in the judicial system. Judicial accountability and a course for dealing with bribery and corruption in the judicial system were also recommended. Workshop participants believed that malpractice by other members of the legal profession (outside the judiciary) was largely responsible for perceptions of judicial corruption in the public eye. For example, lawyers on the losing side will presume that their opponents paid bribes to the judge and will make their opinions known. The panel also noted that low pay often results in tips to court staff to facilitate speedy trial for filing of documents. These practices should be stopped as they can lead to corrupt practices. Furthermore, a review of the judicial appointment procedures, the assignment of cases to particular judges, and reduction in trial delays are also important components of a judicial review to reduce corruption.

To facilitate the flow of the justice system it is also important to computerize court files. This will not only speed up the justice process but it will also reduce the chances of files being "lost" or "misplaced" and fees being charged for "finding" these files. Appeals systems should also be introduced to provide a check against corruption in the legal system. There should also be a complaint mechanism whereby a separate panel of judges would investigate complaints and be empowered to suggest criminal prosecution or disbarment.

In some countries where the overall level of corruption in society may be high the judiciary is still highly regarded. In Thailand, for example, a nationwide survey (Phongpaichit *et al.*, 2000) found the police to be the most dishonest of all public officials and the Supreme

Court and ministry of justice to be among the most honest. Since the judicial system is also smaller than the police force and the military, it is also easier to supervise and keep free from corruption, provided effective monitoring mechanisms can be developed. Chances for large-scale corruption in decisions involving large claims by companies or where large government contracts are involved may provide a greater temptation to justices than bribes for petty crimes. Greater care is required to insure that justice is served in these cases.

In centrally controlled economies such as China and Vietnam, where the Communist Party exercises control over the judicial process, the challenges to the justice system in providing fair treatment are many. A series of recent cases in China (*International Herald Tribune*, 2006) suggest that the justice system is in flux as judges exercise independent decisions that are subjected to scrutiny of the Central Committee. As the public becomes more aware of the failures of the judicial system and is able to articulate their grievances the likelihood of reform increases.

Most of the evidence accumulated for the Asian economies suggests that reforms are successfully implemented and enforced through the enforcement of laws and reform of the legal system from within and not through the issuances of anti-corruption edicts or new laws. In the case of the Philippines and Korea there have been many laws passed and yet corruption continues (see section on Asian experience below for details).

Political system and corruption

Is there a way for the political process to help citizens and the political system to fight corruption? So long as the discussion focuses on the reform of bureaucracy and government administration, the role of democratic empowerment to reduce corruption may be marginal. And even if the focus of reform is overtly political it is still possible for democracy to fail (people power revolution in the Philippines is a good example). Treisman (2000) maintains that democracies are only significantly less corrupt after they have in place for several decades. However, Warren (2005) argues that if democracy empowers those affected in collective decision-making then corruption can be fought through the ballot box and other democratic methods. In addition to the substantive harms associated with corruption such as inefficiency and misappropriation of funds and uneven enforcement of rights, corruption also damages the democratic process. Democratic institutions can be energized to deal with corruption by greater use of voting mechanisms to elect public officials, public discussions, development of civil institutions and a civil society, and by decentralizing power. The evidence from a wide range of countries suggests that lower levels of corruption are associated with the presence of a democratic political system as well as higher levels of per capita income

In developing countries and regions this relationship is not as clear. In developing Asia, for example, the countries with the highest Transparency International corruption score and lowest corruption were, in descending order, Singapore, Hong Kong, Taiwan, South Korea and Malaysia. There was a wide gap between these five countries and the rest of the Asian economies. Each of these five countries has high per capita income — the first four have much higher per capita income than the rest of the countries in the region. Yet only South Korea and Taiwan have high freedom indices and an elected democratic government (see Freedom in the World, 2006 and Transparency International, 2006). One reason could be that the desire to get elected or reelected also puts substantial pressure on politicians to resort to corrupt practices such as by vote-buying or passing pork barrel legislation that benefit their constituents. These corrupt practices send a negative signal to the electorate and also reinforce the culture of corruption within the governmental bureaucracy.

We investigate the relationship between corruption and governance in Asia further in the next section.

Corruption Experience in Asia

In this section we review the experience with corruption for several Asian countries based on accounts of informed observers and a review of legislation aimed at reducing corruption.

Philippines

Corruption in the Philippines has existed for many years. It has become an essential feature of the economy and now pervades all levels of government and extends to the private sector. Rule by despotic oligarchs prevailed in the Philippines during the Spanish occupation which ended at the turn of the last century. However, US rule did little to change the entrenched system of government. The transition from agriculture and rural-based production to urban and industry maintained the ownership structure of the means of production and political powers of a few. During the Marcos years, despite a façade of democracy, the power and influence of a group of powerful bureaucrats increased. The people power revolution of 1986 overthrew Marcos and offered the promise of a more widely based and more transparent government. Yet successive regimes have continued in the footsteps of their predecessors. There have been countless scandals and exposure of politicians for corruption and several governments (Marcos, Estrada, and now Arroyo) have been threatened or brought down by public outrage over the magnitude and extent of corruption. Yet there has been little if any progress in stemming the misappropriation and waste of resources that has resulted from such widespread corruption.

There are numerous studies of corruption, surveys of attitudes toward corruption, and estimates of loss from corruption in the Philippines (see, for example, World Bank, 2000a, 2001; Batalla and Malaluan, 2001; de Dios and Ferrer, 2000; de Dios and Esfahani, 2001). Petty corruption has existed for many years in the form of bribes to internal revenue and customs officials and among the police and military in the form of bribes and extortion (protection money). This is the kind of corruption that the ordinary citizen must confront and it also bears heavily on the poor, including small shopkeepers, jeepney drivers, and anyone having to deal with the bureaucracy for a license or permit.

Grand corruption schemes involve contracts from the government which result in large payoffs to government officials. The pervasive nature of this kind of corruption in the Philippines can be gleaned by the estimates contained in Table 5.6. These estimates are also probably on the low side since substantial corruption and bribery remain hidden. Sometimes, the government gets nothing out of it as in the case of the sale of two islands in Manila Bay — the so-called PEA–Amari scam whereby the government was defrauded of billions of pesos by a property developer.[3]

Other big deals included the development of the cultural center where large sums, including from development agencies like the World Bank, USAID were diverted to the Marcos family and IMPSA deal under the Estrada and Arroyo regimes.[4] Grand corruption of a less flamboyant nature are evidenced by the various projects listed in Table 5.6 where contractors and the government conspire to cut corners, raise prices, create fictitious jobs, and other ways to subvert

---

[3]In 1981, Ferdinand Marcos tried to rescue the bankrupt Construction Development Corporation of the Philippines (CDCP) by ordering the sale of several portions of the Manila Bay reclamation area, including the three islands and the site of the ill-fated Film Palace, to the PEA, a government agency he created in 1977. PEA was asked to pay CDCP P1.5 billion for the land and to assume another P1.5 billion of CDCP's debts. Fifteen years later, in the 1990s, PEA found itself the fortunate owner of what had become prime real estate in Manila's hot property market. Although the three islands had by then become a teeming slum marooned in the polluted waters of Manila Bay, developers were quick to see the potential of seafront property just off scenic Roxas Boulevard, adjacent to the international airport, and with a view of a world-famous sunset. On 28 April 1995, the PEA board approved the contract selling the three islands to Amari, which was awarded the Philippines' biggest real estate project ever. On 23 July 1996, Centennial City, a publicly listed company, assumed complete control and ownership of Amari through a stock swap. Centennial then made a killing in the stock market by selling the idea of a new city complete with skyscrapers, parks, a marina, a golf course, and casinos that would rise out of Manila Bay. Not long afterward, two Senate committees investigated the transactions, concluding, after four months of hearings, the government was defrauded of billions of pesos in that deal. But the investigation was unable to find conclusive evidence of the possible involvement of a string of public officials.

[4]The IMPSA case, wherein a power plant contract was awarded to an Argentinean firm with close linkages to the Estrada regime, shows the lack of transparency in the awarding of multibillion-peso government contracts. It also reveals the opportunities for brokering and deal-making that arises from the privatization of the potentially lucrative power sector.

Table 5.6 Estimates of Sectoral Losses from Bribes in the Philippines

Sector	Government agency	Loss estimates
Infrastructure	Department of Highways and Public Works	10% to 30% of total cost of public works projects
	Mount Pinatubo Commission	10% to 30% of total cost of public works projects
Education	Department of Education, Culture, and Sports	20%–65% of total cost of textbooks procured
Health	Department of Health	20%–40% of total cost of medicines procured
Agriculture	Department of Agriculture Rural Fields Unit	10% to 50% of total costs of farm inputs procured
Finance	Burea of Internal Revenue and Bureau of Customs	30%–43% of potential revenue
Environment	Department of Environment and Natural Resources	15% of reforestation costs
Local Government	Philippines National Police	P45 million a year for ghost payroll
Justice	Bureau of Immigration	P200,000 to P300,000 bribes from illegal aliens

Source: Social Weather Stations (2001).

public funds. These estimates are reinforced by estimates from the Social Weather Station's National Survey (2000) (see http://www.sws.org.ph/pr111300.htm) which showed more than 70% of survey respondents saying they thought that there was at least 30% leakage (and more than 50% in some cases) from corruption in road construction, tax collection, textbook provisioning, and purchases of government vehicles and office equipment.

An independent study by the Economist Intelligence Unit estimates that 20% to 30% of the national budget is lost to graft each year. That would be enough to cover next year's projected budget deficit — far more than the struggling economy can afford and enough to provide basic human needs for the millions mired in poverty. Without this kind of investment, the already yawning gap between the rich and the poor will continue to widen — with perilous consequences for social stability (Cordingley, 2001).

Corruption also extends to vote-buying and proceeds from illegal gambling (juteng). President Arroyo has been accused of vote-buying in the election of 2004 and former president Estrada was implicated in illegal gambling as well as other corruption activities. He is now serving time in jail as a result of conviction for these and other crimes.

Observers believe that the extent of corruption has probably been increased by the reduction in power of the central government which came following the demise of Marcos, who held a tight reign on the disposition of the spoils of grand corruption. It has also been influenced by the devolution of political power and patronage to the provinces, where local authorities now have more power to control projects and their implementation.

There have been many attempts to deal with corruption and there are several official bodies that are charged with investigating corruption including the Ombudsman, who has wide investigative powers, and the PCGG, charged with maintaining good governance. The PCGG was originally formed to investigate the illegal activities and return ill-gotten gains of former president Marcos. An abbreviated list of agencies has been compiled by the World Bank and is displayed in Table 5.7.

There are also a number of statutes and many articles written dealing with corruption. Nevertheless, there seems to have been little progress in bringing criminals to justice or in stopping any of the many and varied kinds of corruption. Imelda Marcos, wife of former president Marcos, remains free and is still living a luxurious lifestyle in Manila. Her husband, Ferdinand died in exile in the United States without ever being brought to trial. The evidence of these failed efforts is reflected in continued low tax effort as a result of widespread tax evasion, crumbling infrastructure from poor construction, and failure to complete projects because of lack of funds.

Many capable people have given up and have immigrated to the United States, Canada and other industrial countries. Remittances have increased over the past few years and provided a steady source of income for the poorer segments of the population. To some extent this reduces the pressure for reform of the political system and the strength of the anti-corruption campaign. Furthermore, the vicious cycle of corruption exerts its own powerful nexus which has proved difficult to break. This is compounded by the failure of the political leadership to mount a concerted effort and deep commitment to fight corruption.

Future prospects for reducing corruption in the short run are bleak. Surveys of the Social Weather Station and others suggest that cynicism is widespread, particularly in light of the recent difficulties encountered by the Arroyo administration. Nevertheless, there are some reasons for optimism in the medium run. The business sector is critical of corruption and is likely to support a vigorous anti-corruption campaign to reduce corruption in the public sector. They even expressed a willingness to pay some percent of net income to fund a program of public sector corruption (Social Weather Stations, 2000). Efforts to increase tax revenue have been made and there are signs that government deficit is falling. There have been efforts to tighten money laundering under the anti-money laundering act. There is the potential for acting quickly and effectively to reduce corruption. The Office of the Ombudsman is the constitutionally mandated lead agency for combating corruption and some efforts have been successful in government saving. With the cooperation of other agencies already legally empowered (see Table 5.7) corruption could be reduced across the board.

Thailand

Thailand has never been colonized and it has resisted any efforts of regional powers such as China or Vietnam to increase their economic influence. Nevertheless a culture of patronage and corruption has developed over the decades. Traditionally officials were paid a portion of the money collected as a fee for their services. These practices have continued in the modern era as patronage in the form of gift giving — *sin nam jai* — or gifts of good will. The social structure also stressed the importance of local businessmen, who established power in different localities (see Phongpaichit and Piriyarangsan, 1994 for a detailed discussion of the relationship between gifts and corruption).

Historically Thailand was a monarchy ruled by a king. A constitution was promulgated in 1932 and since then Thailand has been a

Table 5.7 Agencies Charged with Investigating Corruption in the Philippines

Agency	Mandate
Ombudsman — established in 1986	Investigates, prosecutes and adjudicates administrative cases and takes criminal cases to court.
Commission on Audit — established in 1986	Conducts independent audits of agencies' financial position and reports irregularities to Ombudsman.
Civil Service Commission — established in 1986	Sets standards and norms for civil service appointments and establishes punishment for violation.
Sandigabayan — established in 1986	Main anti-graft court that adjudicates criminal cases brought by Ombudsman. Deals only with high-ranking government officials.
Judiciary — established in 1901	Adjudicates law in all areas.
Department of Justice	Acts as government's primary criminal prosecution arm.
Department of Budget	Oversees reform in procurement, management, and streamlining of the bureaucracy
Commission on Elections — established in 1986	Promotes free, orderly, honest, peaceful and credible elections and investigates violations.
Presidential Commission on Good Government — established in 1986	Aims to recover ill-gotten wealth from former presidents Marcos and Estrada.
Bangko Sentral ng Pilipinas — established in 1993	Performs central banking functions. Replaced old central bank formed in 1946.
Securities and Exchange Commission — established in 1936	Oversees registration of securities, evaluation of financial conditions and operations of applications for security issue. Supervises stock and bond brokers and stock-market exchanges. Strengthens corporate governance.
National Bureau of Investigation	Gathers evidence for probable case hearings and files appropriate charges.
Presidential Committee on strengthening and Streamlining Effective Government	Formulates public sector institutional agenda.
Anti-Money Laundering Council — established in 2001	Investigates charges of money laundering and suspicious transactions. Council consists of governor of central bank and chairman of securities and exchange commission.

Source: World Bank (2001) (partial list).

constitutional monarchy. In much of the last 75 years the military has been in control. However, in 1997, a new constitution was ratified and since then democratically elected officials have run the country. Despite all of these political changes the strength of the bureaucracy has been retained and the practice of gift giving has continued. As industrialization has proceeded since the 1970s income growth has accelerated and Thailand has become one of the richer countries in Southeast Asia. Illegal activities, including drug trafficking, prostitution and illegal immigration, gun running, and illegal gambling have thrived along with traditional bribes to public officials.

Data compiled by Transparency International and other international agencies suggests that the extent of corruption in Thailand is not as high as in neighboring Indonesia and the Philippines although higher than in Malaysia. Some of the moderation in corruption may be the result of higher income, income growth, and trade openness, which, as we noted earlier, are all factors contributing to a reduction in corruption. Nevertheless, the scope of corruption and bribery in Thailand is extensive (see also Phongpaichit *et al.*, 1998, for various aspects of illegal activity) and includes all of the government departments where it thrives elsewhere including ports, customs, business licenses, revenue collection, and contracts for government projects.

Some estimates of the extent of corruption have been made by Phongpaichit and Piriyarangsan (1994) and in a reprinted and updated 1998 edition. They estimate the corruption taken by bureaucrats to be between one-half of 1% and 1% of GDP during the period 1960–1990, equivalent to between 4% and 5% of the government budget (page 38 of 1994 edition). This does not include other corrupt activities carried out by the police and the military or the money gained from sanctioning illegal gambling, prostitution, drugs, and gun running. Recent estimates of public sector procurement corruption by Phongpaichit *et al.* (1998) suggest that bribes could be as high as 40% of the value of the contracts. Before contracts are awarded, the purchaser can tailor specifications to favor particular supplier, restrict information to keep out unwanted bidders, and take bribes to ensure contract award. Suppliers can collude to fix prices, water down technical standards, and offer bribes.

There is some evidence that recent efforts by the Thai government to clamp down on corruption is having a positive impact, as reflected by an improved rating in the most recent Transparency International score for 2005 (score of 3.8 is somewhat higher than 3.2 scores recorded between 2000 and 2003). Recent allegations and protests against Prime Minister Thaksin by the middle class in Bangkok, which eventually contributed to his resignation and the eventual replacement of the democratic government by the military, suggest that there may be a growing anti-corruption sentiment that could help to further accelerate the movement toward good government following the return to a democratic government promised in 2007.

Korea

Confucian values and the importance of gift giving in traditional Korean culture created a close relationship between family members and also among the various hierarchies in society. One of the three principles of Korean society establishes the importance of the relationship between the king and his subjects (wives/husbands and sons/fathers being the others) while the five virtues establish loyalty between king and subject and rank order according to age. These cultural values still have an important impact on Korean society and have had an influence on political leaders to amass large campaign funding and corruption money from large companies in recent times.

Several additional factors have had a salutary impact on society to reduce corruption and increase honesty and transparency. The Korean War and its aftermath as well as establishing freedom from the yoke of the Japanese combined with land reform to bring about an egalitarian distribution of land, wealth and income. Furthermore, the bureaucracy was meritocracy-based with good pay and prestige. Finally, the highly competitive nature of evolving Korean industry made it nearly impossible to capture resources for the sole purpose of corruption despite the close connection between the large conglomerates (chaebol) and the government. In Korea competition tended to trump corruption, at least to the extent that resource allocation was not badly distorted by the extensive bureaucratic networks and erection of barriers to entry. Nevertheless there was a significant amount of grand level corruption as the chaebol paid government bureaucrats for official favors. It was what Kang calls a mutual hostage situation (Kang, 2002). The end objective of both government and businesses was to become more highly competitive internationally. Because both principal and agent had the same ultimate objective of higher economic growth for the country, competition and economic efficiency were not adversely affected to a significant extent even though many people got rich. Furthermore, the money from corruption tended to stay in the country and was rerouted to luxury real estate and the stock market rather than flowing overseas into Swiss bank accounts.

Nevertheless, the number and extent of the bribery and corruption scandals must have had a negative impact on the quality of infrastructure that was completed with a smaller budget as a result of the corruption skim and some domestic projects were completed by contractors that used bribery to get contracts rather than efficient production techniques. It would be fair to say that bribery became the rule. A 1999 survey found that nearly three-quarters of multinational executives had been asked to pay bribes in Korea and 50% did so (Ehrlich and Kang, 2002) and another survey of business executives revealed that 60% of respondents said that corruption impeded their business in Korea. The flow of corruption largess was facilitated by the Korean financial system where accounts could be held under false names.

However, as the middle class grew in size the concern for corruption also grew along with the development of a number of civil societies interested in reducing the extent of the corruption network. The Citizen's Coalition for Economic Justice (CCEJ) and the People's Solidarity for Participatory Democracy (PSPD) were among the most vocal. The conviction of several former presidents and disclosures of large sums of bribery helped to reinforce public sentiment and increased the pressure to establish a "real name" system, which was finally established in 1993 by Kim Young-Sam. Other reforms in the 1990s included more open access to government documents, administrative reforms and reforms of political financing, and an anti-corruption law. Former presidents Chun Doo-Hwan and Roh Tae-Woo were arrested, indicted, and convicted in 1995 and 1996 respectively. These convictions

Table 5.8 Number of Bureaucratic Corruption Incidents Reported in Korea by the Media and by Amount of Bribe — 1985 Constant Prices

Size of Bribe	Park regime 1961–1972	Park regime 1972–1979	Chun Regime 1980–1987	Roh regime 1988–1992
1–10 millon won	31	30	56	16
Over 10 million	10	22	21	34
Total	41	52	77	50

Source: You (2005).

reinforced the growing trend toward more honesty in Government. Public opinion toward the acceptance of bribery changed rather dramatically between 1996 and 2003, as reflected by an increase in the strong opposition to bribing policemen which increased from 36% to 62% of respondents, and also to bribing public officials which increased from 45% to 71% over this period of time (You, 2005). There has also been a gradual increase in the Transparency International Index toward more honesty as the index moved from the range of 3.5–3.9 in the 1980s to 5.4 in 2004 and 2005.

Extensive information about the nature and extent of bribery and corruption in Korea is not widely available. Incidents are reported in the media and they have been assembled by You and reported in Table 5.8. Throughout the entire period the Chaebol continued to make significant political contributions of between 20 and 50 billion won between 1988 and 2004.

The major scandals are further summarized in Table 5.9.

China

Offering of gifts to elders and those in positions of power and influence (*Guanxi*) has been a long tradition in China. Before economic reforms began in 1978 China was a centrally planned and controlled economy with no private property and no entrepreneurs. The country endured considerable hardship as a result of policies introduced during the Cultural Revolution and the great leap forward. Combined with the *Guanxi* tradition and the opportunities that

Table 5.9 Major Scandals in Korea After 1992

Name and date of scandal	Characteristics of scandal	Estimated loss from bribery/corruption
Hanbo Scandal — January 1997	Large conglomerate went bankrupt. Received special treatment from Kim Young-Sam and large loans from banks. President's son was involved.	$6 billion
Furgate or clothes bribery scandal — May 1999	Wives of senior government officials spent bribery money on furs and jewelry	$165 million
Lee Yong-ho scandal — May 1991	Bribery of 20 senior prosecutors and politicians	
Financial supervisory commission scandal — November 2000	Bribes to financial regulators by owners of finance companies	$300,000 or more
Daewoo scandal — July 2001	Collapsed under $80 billion debt. Auditors concealed extent of debt. Bribes of 470 million won.	Difficult to ascertain. At least $1 billion

Source: Ehrlich and Kang (2002); Blechinger (2000).

opened up for entrepreneurship and accumulation of income and wealth as privatization spread through the economy it is not surprising that corruption and bribery were used whenever opportunities presented themselves. While moderated to some extent by Confucian values and traditions of honesty and integrity, the rise of corruption corresponded closely with the spread of market forces.

In the first phase of the reform process there were fewer opportunities for corruption. The *household responsibility system* was developed in agriculture, and extended later on to retail stores, small factories and even public schools under the *contract responsibility system*, whereby farmers were assigned publicly owned land to till for their families while paying a fixed rent to the government. Such leasing arrangements provided incentives to farmers, and businesses and market forces were allowed to operate in order to achieve efficient allocation of resources, particularly for the goods brought to market after resources allocated by the state were paid for at fixed prices. These leasing arrangements were eventually joined by other market mechanisms that encouraged the growth of the non-state sector which now accounts for much of the total output of the Chinese economy.

The two-track system whereby goods were distributed on state-regulated and market tracks amounted to between 20% and 25% of GDP between 1981 and 1988. This two-track system provided further opportunities for corruption and allowed officials to smuggle goods and control inputs from one track to the other. Furthermore, decentralization gave local officials the ability to allocate resources and provide services for businesses operating in their jurisdictions. Firm managers maintained good relations with government officials to ensure the provision of water and power and other support services that are required to maintain efficient business operations not to mention the timely delivery of raw materials and intermediate goods. The cozy relationship between local bureaucrats and the business community enabled firms to obtain smooth and effective cooperation with the local community and maintain efficiency in operation. The bureaucrats are rewarded by nepotistic favors (*tanwu shouhui*) and the appropriation of public property or public funds for personal benefit (*nuoyong gongkuan*) (Fan and Grossman, 2001; Sun, 1999, 2004). Large state-owned enterprises and the financial system were not subject to these leasing arrangements and the strong inflow of foreign direct investment, particularly in the coastal regions, was determined by contracts between foreign firms and local governments.

Given the fact that obtaining objective and systematic data on corruption is difficult, external agencies such as Transparency International have been publishing figures on corruption in China since the early 1980s. These figures, shown in Table 5.3, indicate that corruption has been increasing throughout the last 25 years, from a level of 5 down to 3. As capitalist forces became stronger so also did the level of corruption. Chow (2005) suggests that this is a result of developments that occurred during the reform process. Restructuring of state-owned enterprises followed the Party slogan "Retain control of the large and let loose the small". Small-and medium-sized state-owned enterprises were transformed into share-holding companies and the shares were sold to the management and staff of the enterprise. This generally improved performance and effectively resulted in the privatization of these enterprises, although shares were not generally publicly traded. There was scope for corruption in these small and medium enterprises including smuggling between Guangdong and Shenzhen Special Economic Zone and Hong Kong. There was also ample room for corruption when licenses were granted to foreign companies to operate in Shenzhen and in other special economic zones. As the wealth of these provinces increased so did corruption — as measured by figures reported by Damon (2003).

The scope for corruption was much greater in large-scale enterprises which were also transformed into share-holding companies in the late 1990s, but where shares were held primarily by the government and administered by government officials. Many of these large-scale enterprises continued to exercise monopoly power in strategic industries such as oil, steel, cement, and in public utilities. In these large-scale industries corruption is more widespread as managers and their staff take bribes from customers, use enterprise resources for their personal use, and allocate resources to maximize corruption rather than profit and increase economic efficiency. Regulations and corporate governance provisions have not reduced corruption in large-scale industries (Chow, 2005; Sun, 2004). Furthermore, Sun argues that the large-scale corruption by the state-owned enterprises creates more detrimental impacts on resource allocation and economic efficiency than the smaller-scale bribery that characterized the earlier period. It also helped to legitimize corruption within the society as a whole.

Looking at all prosecuted crimes nationwide in the 1990s, Sun (2004, Table 1.2, p. 39) shows that embezzlement, bribery, and misappropriation of funds accounted for around 80% of all criminal cases prosecuted. Embezzlement is usually uncovered by audit or other accounting supervision and the same would be true of misappropriation of funds, although the latter would be more difficult to be traced. Bribery, on the other hand, is more difficult to uncover since there would not necessarily be a traceable paper trail. As a result bribery is probably under reported more than embezzlement and misappropriation of funds. Profiteering from the dual track pricing system of the 1980s also resulted in a number of cases brought to court. Profiteering fell dramatically once the two-track system was eliminated.

The rank of offenders being prosecuted shifted over time as well. In the earlier period of the 1980s average bribes were small and minor officials were implicated. In the 1990s more high ranking officials were implicated and the size of the bribes also increased. This is consistent with the argument that corruption shifted to large state-owned industries in this period as China entered a period of high level corruption. Throughout the reform period most corruption occurred where state property was involved. The private sector has only become involved very recently as foreign companies have become more important and as the economy has opened up further. As the two-track system of allocation was phased out and replaced by market mechanisms and public investment projects were subcontracted to the private sector, bidding and competition for these projects increased, along with corrupt practices including bribery to get contracts and to avoid building inspections as well as to cover up the use of inferior materials. For instance, there was a number of severe building collapses between 1991 to 1996. In addition, a number of buildings that were inspected in the second half of the 1990s were found to be defective including 98% of Bejing's housing construction projects (Sun, 2004).

In the financial system there were many opportunities for corruption since banks are state-owned and faced no outside competition. Bribes could easily be exchanged for the access to loans. Furthermore, bank officials were not used to scrutinizing loan applications, having been used to a system where they passively granted loans to state enterprises. As a result there has been a gradual build up of non-performing loans by the state banks and the old habits of poor management and loan supervision remain. In addition, high ranking military officers and other high ranking officials still use their positions to force banks to lend them money without collateral and with little hope of repayment. Land developers and building contractors also began to bribe public officials. New policies introduced in the early 1990s allowed the sale, lease, and development of land and properties. This involves bribery to get access to the land along with further bribes to get proper zoning, building permits, ownership transfer, and financing for the improvement of the land. Taxes on newly developed land and for small businesses were also subject to negotiation and bribery as the economy grew rapidly and the private sector expanded in the 1990s and in the new millennium.

Redressing income inequality between the Western Region and the coastal provinces has also been subject to corruption. The effectiveness of infrastructure spending on highways, railroads, power, and education has been reduced by bribery and corruption of local officials who have subverted funds and even refused to pay local workers. Illegal taxes were levied on farmers by local governments until quite recently and immigration to cities was also controlled and bribes paid to obtain appropriate documentation to move. These attempts to redistribute income to the poorer eastern regions are compounded by corruption of local officials that have historically had a large amount of autonomy. Abuse of funds appropriated for anti-poverty programs has been widespread, being diverted to slush funds or other illegal uses. Abuse of money allocated for resettlement is also widespread. The corruption experience of Guangxi Autonomous Region is a good case in point. Many officials from this province were indicted for corruption and Li Chenglong, a former party secretary and mayor was executed in 1999 for taking 16 million yuan in bribes and other illegal income (see Sun, 2004 for further discussion).

The judicial system also follows corrupt practices including missing tax collection, reduced duties on traded goods, building permits without proper inspection, sale of licenses to operate businesses, extortion by police and military officials, complicity in smuggling by police and the military, bribery to buy votes, and payoffs to judges.

One of the anomalies of corruption in China is that the spread of market forces since the early 1990s has resulted in more, not less corruption. SOEs now have to compete with the private sector and have to resort to corrupt practices to maintain their market control, influence government decision-making, and control bureaucratic processes.

In the context of the general discussion of corruption by Shleifer and Vishny (1993) and Rose-Ackerman (1999), this means that the decentralization of the decision-making processes resulted in

> more decision makers, more discretionary and more decision processes that are involved in decentralized structures, leading to more bribes required to get things done. Moreover, outcomes of bribes are also more uncertain in more decentralized structures (Sun, 2004, p. 214).

India

Corruption has not only become a pervasive aspect of Indian politics but has also become an increasingly important factor in Indian elections. The extensive role of the Indian state in providing services and promoting economic development has always created the opportunity for using public resources for private benefit. As government regulation of businesses was extended in the 1960s and corporate donations to political candidates were banned in 1969, trading economic favors for under-the-table contributions to political parties became an increasingly widespread political practice. During the 1980s and 1990s, corruption became associated with the occupants of the highest echelons of India's political system. Rajiv Gandhi's government was rocked by scandals, as was the government of P. V. Narasimha Rao. Politicians have become so closely identified with corruption in the public eye that a *Times of India* poll of 1,554 adults in six metropolitan cities found that 98% of the public is convinced that politicians and ministers are corrupt, with 85% observing that corruption is on the increase (*Times of India*, 2006).

The prominence of political corruption in the 1990s is hardly unique to India. Other countries also have experienced corruption that has had a severe impact on their political systems. What is remarkable about India is the persistent anti-incumbent sentiment among its electorate. Since Indira Ghandi's victory in her 1971 "*garibi hatao*" ("abolish poverty") election, only one ruling party has been reelected to power in the central government. In an important sense, the exception proves the rule because the Congress (I) won reelection in 1984 in no small measure because the electorate saw in Rajiv Gandhi a "Mr. Clean" who would lead a new generation of politicians in cleansing the political system. Anti-incumbent sentiment is just as strong at the state level, where the ruling parties of all political persuasions in India's major states lost 11 of 13 legislative assembly elections held from 1991 through Spring 1995. (The preceding two paragraphs draw heavily on http://www.indianchild.com/corruption_in_india.htm the website for the Indian government corruption agency. Along with http://cvc.nic.in/ these two websites provide detailed evidence of corruption on India.)

Bureaucratic corruption depends on the widespread reach of the government and the seemingly infinite number and variety of government departments that monitor and regulate economic activities. There are 20 million bureaucrats in central and state governments and in quasi-government institutions and over 200,000 government offices in the country. Corruption extends across all sectors and occupation and from petty to grand corruption.

Endless discussion of corruption in India in professional journal and the media has done little to reduce the level, scope, and size of corruption. Outside observers who compare corruption in different countries such as Transparency International continue to rate India as one of the most corrupt countries in the world. What are the sources of this continued high level of corruption? Certainly the size and extent of the bureaucracy is an important factor. The nature of the bureaucracy is another. It is secretive, hierarchical and there is a lack of trust between government departments or among employees in the same department. Decision-making is convoluted and often shifted from one department to another. As a result it is not easy to hold civil servants accountable and

> non-performance has never been grounds for disciplinary action. The result is a psychology of evasion whenever possible. There are even year-long delays in the implementation of major decisions (Dwivedi and Jain, 1988, p. 209).

It is not hard to understand why patrons of the bureaucracy would resort to bribery to get things done. Certainly there is strong evidence that in situations where processing of claims can be sped up, the rate of bribery and corruption falls (see Hunt, 2006). The class structure of Indian society reinforces the culture of corruption. The powerful dominate the powerless and this pecking order continues from top to bottom of society. As a result whenever someone with higher status can exercise control over someone with lower status he or she will exercise that power. Sometimes it is physical, sometimes emotional and sometimes economic. Bribery and corruption fit nicely into this scheme.

Transparency International did an extensive survey of petty corruption in India in 2005. Over 14,000 people were interviewed from 20 different states. They were asked about their experiences dealing with several government departments including needs-based services such as rural financial institutions, police, income tax assessments, judiciary, and land assessment including property taxes. Other basic services such as schools, water supply, electricity, and hospitals were also included. The major findings are shown in Table 5.10. The police, land tax department, and the judiciary were judged the most corrupt. Around a half to 80% of respondents had a direct experience of bribing someone in these three departments. Other departments were less involved with corruption. TI also made some recommendations

Table 5.10 Bribery by Department in India

Department	Direct experience of bribing — % of respondents
Rural banks	19
Individual tax assessment	20
Municipalities	23
Judiciary	47
Land Administration (taxes)	48
Police	80
Schools	18
Water supply	9
Electricity	20
Government hospitals	27

Source: Transparency International (2005a).

based on their findings, including introducing more competition where possible so that consumers can have other options, alloting funds to departments based on performance (for example, to schools on the basis of exam results and dropout rates), simplifying procedures so that citizens can fill in forms without looking to middle men or other facilitators, introducing greater transparency so that allocations and spending are a matter of public record, outsourcing where possible, using technology to reduce the need to visit government offices, increasing staff training to help them manage stress, keeping up with new technology, and providing better service.

Recent initiatives include the formation of the Central Vigilance Commission (CVE) which is to enforce the Prevention of Corruption Act. It is relying on the public by using the internet and publicly disclosing names of corrupt officials and encouraging whistle blowers. The CVE cannot deal with political campaign and other corruption, being confined to dealing with bureaucratic corruption only. It is also dealing with some of the agenda items suggested by TI including simplifying rules and procedures and it is also attempting to clean up procurement fraud and corruption by requiring greater transparency in the bidding process.

Prospects and Policies for the Future

The experience of corruption in Asia reviewed above highlights the difficulties that are faced by the countries in the Asian region in reducing corruption in the future.

East Asian experience and prospects

The countries in East Asia are doing well economically and are managing to deal with corruption effectively. This pattern is expected to continue and perhaps even strengthen in the future. Their success is due to a combination of factors. Both Hong Kong and Singapore have had the benefit of British colonial rule and, as entrepot trading centers, had to maintain open trading environments. Since trade and foreign investment were so important to both of these economies there was little scope for corruption if they were to remain competitive and attract foreign investment. Furthermore, they put a significant amount of effort and resources into monitoring and policing corruption.

For example, Hong Kong's ICAC independent anti-corruption agency has more than ten times as many staff per 100,000 residents as the Philippines and 100 times more investigators to investigate corruption. Furthermore, the ICAC has the confidence of the people and has developed a three-pronged strategy of investigation, prevention, and education. It also ensures confidentiality to those reporting corruption offenses. Quah (2003) stresses the commitment of the government to anti-corruption and the confidence in the ability of the government to control corruption and maintain the cooperation and trust of the people. This is evidenced by the proportion of complainants who identified themselves increasing from one-third to two-thirds over a decade (Quah, 2003, p. 145). Salaries of civil servants have also kept pace with comparable levels in the private sector so that the civil service attracts quality candidates. The strength and good reputation of anti-corruption efforts are also good publicity for Hong Kong and help it to continue to attract foreign investment and improve economic efficiency. Efforts to continue the work of ICAC should strengthen anti-corruption enforcement and serve as a model for other countries in the region.

In Singapore, the Prevention of Corruption Act (POCA) was adopted in 1960 along with the Corrupt Practices Investigation Bureau (CPIB) empowered to enforce it. To deal with corruption, which was rampant during the colonial period, several interrelated strategies were followed. The People's Action Party was committed to curbing corruption and initially it strengthened anti-corruption powers of CPIB as well as provided more resources to it (it still remains small relative to Hong Kong's ICAC). It receives and investigates complaints, investigates malfeasance of public officials, and examines procedures in the public service to minimize opportunities for corrupt practices. Justice was swift and punishment was severe. Eventually, as per capita income in Singapore increased rapidly, salaries of public officials were raised; until Singapore now has one of the highest paid civil services in the world (see Quah, 2003 for further details of the Singapore experience). By adopting a dual strategy of reducing both the opportunities and incentives for corruption, Singapore has been able to achieve and maintain a healthy and honest bureaucracy. Political will remained strong and was a critical factor in this success. In the future the government is expected to continue a strong anti-corruption stance and work to maintain Singapore's reputation as the most honest country in Asia. It is one of the few countries where the corruption index of Transparency International has continued to improve, albeit from already high levels. The government also realizes the public relations' potential to maintain Singapore's good reputation and is expected to maintain and even further strengthen its reputation and enforcement of anti-corruption measures.

In Korea, anti-corruption efforts have focused on political corruption as well as corruption involving the chaebol and government officials. Significant progress has been made in establishing a more transparent environment by establishing the real name system for financial transactions and prosecuting a number of officials involved in bribery schemes. However, some of the politicians convicted have been freed, tending to undermine public attitudes toward the commitment of the government. The fundamentally egalitarian nature of Korean society, a favorable income distribution and a highly educated population have made the task of controlling corruption easier. There is a low level of petty corruption and many of the government bureaucracies are highly professional. Furthermore, the number of corruption cases reported by the Anti-Corruption Citizen's Coalition has increased as well as the number of officials reprimanded for corruption. The proportion of citizens reporting that corruption is a serious problem has also declined, although still around 75% of those polled (Quah, 2003, p. 169). Nevertheless, the public still has a skeptical attitude about the commitment of the government to reduce corruption. Korea still lacks an anti-corruption culture and a coordinated network to deal with corruption. Public officials still seem to lack anti-corruption discipline and a bureaucracy that is thoroughly a meritocracy. Nevertheless, the trend toward establishing a more honest society has begun and is not expected to abate. The examples set by Hong Kong and Singapore should also provide further impetus for Korea to increase its efforts to reduce political corruption related to campaign finance and bribing of public officials to obtain government contracts. Public outrage and press coverage should also reinforce these positive trends.

After Singapore and Hong Kong, Taiwan has the highest honesty rating in Asia. Characterized by more small-and medium-scale industries it does not have the traditional problems of large-scale industries bribing bureaucrats and presidential hopefuls that have characterized the Korean and Japanese experience. It also has a very even distribution of income and an egalitarian social structure as a result of the land redistribution and influence of the United States following the communist revolution in China. The historical experience of Taiwan shows how the spread of democracy has brought about a reduction in corruption and an increase in bureaucratic openness. For more than 50 years the Kuomintang (KPT) party ruled Taiwan. It amassed a real estate and business empire, and set up monopolies for its own companies and became a large business syndicate and one of the richest political parties in the world (Fell, 2003).

Local corruption was widespread and vote-buying and political patronage and bribes by contractors were common. With the rise of political opposition in the 1990s and accusations of corruption, there has been a dramatic shift in the corruption landscape. Vote-buying has decreased and political donation laws have been strengthened. While more needs to be done, Taiwan has made significant progress in cleaning up corruption through the democratic process (*Taiwan Times* interview with Transparency International, 2004; Fell, 2003). These efforts are expected to continue in the future. Reducing corruption and demonstrating its honesty in political and economic affairs is also a strong card for Taiwan to play in its continuing battle with China. By further establishing its position of honesty and integrity it can hope to take the high ground in the international political arena, which China continues to have a reputation for corruption and dishonest dealings.

There are good reasons to believe that East Asian NIEs of Hong Kong, Korea, Taiwan and Singapore will be able to maintain pressure to reduce corruption and bribery even further in the future. Trends in Transparency International indices of corruption have showed a downward trend in corruption in the past few years in all of these economies and the passage of new laws in Korea and Taiwan can be expected to increase the ability of the regulatory agencies to reduce corruption even further. The virtuous cycle of honest and integrity already established in Hong Kong and Singapore can be expected to continue. They should be joined by Korea and Taiwan as the NIEs further build on their already established status for economic efficiency, adding to it a reputation for integrity and honesty.

China, Southeast Asia and South Asia

Success in reducing corruption in the rest of the Asian region has been spotty. The culture of corruption is ingrained in many of the economies of Southeast Asia and South Asia as well as China. In China, the interest in the experience of Hong Kong and Singapore is particularly relevant. In these small, basically Chinese-run economies, there was widespread corruption at the end of the colonial era. Yet they have been able to effectively control corruption through a centrally run and administered bureaucracy.

What lessons can China learn from their experience? Is it possible to reduce the plunder of the SOEs through the imposition of more central controls? In what other ways can the Singapore and Hong Kong experiences be useful for China as it moves toward a more democratic, yet still centrally controlled economic and political system? Initially, rising income will provide the wherewithal for the central government and the provincial governments to raise the salaries of government bureaucrats. A more open trading environment should pave the way for a reduction in the scope of rent seeking and corruption as more markets become contestable and domestic firms have to compete with foreign firms which are less willing to enter the corruption culture by bribing local officials. Furthermore, rapid development is creating a growing middle class that is less willing to put up with widespread corruption. This should provide additional impetus for the government to crack down on corruption. Furthermore, membership in WTO and the responsibilities that membership entail should also add weight to the anti-corruption fight. As more private banks are approved to do business in China under the terms of joining WTO it is possible that financial reforms will become more binding and corruption will be reduced in the future. Genuine privatization of state-owned banks would also help.

In addition to these macroeconomic developments it is worthwhile to stress the importance of increasing civil service salaries in China. Reforms must ensure a honest judiciary, stiff penalties for wrongdoers, raising public awareness, and continued political commitment. Within the military and the party hierarchy it may be more difficult to reduce the scope of "high level" corruption. The privilege to corrupt is still heavily ingrained and the party hierarchy and the military need to be unified if they are to maintain control of the country. If corruption at the highest levels is to be reduced, both the party and military leaders have to agree on this objective. In the short run reform in the judicial system is expected to continue. In 2006 several changes have already been made that suggest more legal transparency will continue. It is still too early to tell whether other reforms in the financial sector and within the bureaucracy will continue to gain momentum. There is still resistance within the Party apparatus even though there have been a number of executions of public officials convicted of corruption. As the legal system is strengthened and becomes more open to international scrutiny these draconian measures can be expected to diminish. Whether or not this will reinforce the general anti-corruption drive is problematic. Certainly, growing sentiment for a more open and honest society among the educated and the young will have to be dealt with, either by force or by gradual removal of barriers to free speech and a continuous fight for honest and transparent government.

In Indonesia the recently established government of Susilo Bambang Yudhoyono has taken a number of steps to deal with corruption. Recent surveys by Transparency International and the Political and Economic Risk Consultancy (PERC) suggest that these measures are having some impact on corruption, particularly public perception of corruption. The TI scores recorded in Figure 5.1 show that Indonesia made steady upward progress in reducing corruption until the Asian crisis. Thereafter, the index dipped for a few years and is now improving again, although it still remains low. This trend in corruption has dovetailed with devolution of responsibility from the central government to the provincial authorities. Even greater autonomy is possible in the future as Aceh has become more independent. Other provinces, particularly in the eastern region, are also gaining greater control over budgets. However, experiences of other countries suggest that the opportunities for corruption multiply as power devolves from the center to the periphery and this will present an ongoing challenge for the government. The most probable outcome is that there will be a period of consolidation of gains in fighting corruption as the newly elected regime gains more confidence in dealing with the various forms of corruption. Recent changes in the management of Pertamina and the signing of new oil exploration contracts with overseas partners is a good sign.

In Thailand the corruption index has also been improving since the Asian financial crisis. The combined efforts of anti-corruption political forces along with the academic community and journalists have exerted significant pressure to reduce the level of corruption and bribery. The works of Phongpaichit and Piriyarangsan, whose two books (1994 and 1998) have helped to publicize the nature and extent of corruption, are good examples of how journalistic and academic expertise have come together to pressure the government to take action against corruption. Recent public demonstrations which resulted in the accusation of the prime minister of corruption in the sale of company assets, are additional evidences to prove that anti-corruption drives have taken root in the awareness of the middle class in Bangkok. Nevertheless, high level corruption continues. Thailand will be able to reduce corruption further only if anti-corruption forces can marshal enough public opinion to increase the enforcement of anti-corruption laws. A key to this drive will be the continued participation of the middle class, the journalistic community, and vocal academics who can continue to publicize corrupt officials.

In the Philippines anti-corruption has made little headway and there is nothing to indicate that the situation will change in the near future. Political and military forces banded together to dislodge

President Marcos and then President Estrada. Now there are strong efforts to oust President Arroyo. Each time a president has been ousted the forces of anti-corruption have claimed a victory for honesty and integrity. However, little substantive progress has been made aside from the Ramos presidency, when indices of corruption got better (see Table 5.2) and the reputation of the government improved both internally and externally. During the Ramos presidency political situation stabilized because of the president's reputation and standing with the military. Since then corruption has returned to high levels. Petty corruption is still as strong as ever and high level corruption saps the efforts of the government to build a more modern infrastructure network and to improve the educational system. Low tax effort is the result of tax evasion and budget deficits have increased despite new tax levies and efforts to increase tax compliance. Corruption scores compiled by Transparency International and other agencies such as the World Bank and PERC show continued high levels of corruption.

Aside from Singapore, Malaysia is the least corrupt of the Southeast Asian economies. It also has the highest standard of living in Southeast Asia. However, it has been unable to break free of the graft and corruption nexus that has characterized the relationship between contractors and the government bureaucracy. Petty corruption is probably not as pervasive as in Thailand, the Philippines, or South Asia, yet there is still corruption in the government bureaucracy. In the early days following independence Malaysia was a very clean country, having adopted the bureaucratic esprit of the British civil service. However, with adoption of the new economic policy and the transformation from a primarily agriculturel based economy into a newly industrializing economy Malaysia developed a wide ranging bureaucratic apparatus and many new agencies to supervise all of the programs connected with the new economic policy (NEP). An anti-corruption agency was introduced to deal with corruption in 1967 and continues to be the focus of anti-corruption efforts. Additional agencies were added in the 1990s. Nevertheless, there is still a substantial amount of corruption involving different government departments, contractors, and other business enterprises. The challenge for the future is to improve the enforcement of anti-corruption measures by building on the already strong reputation of the civil service, which is ranked only behind Hong Kong and Singapore in terms of efficiency by a recent poll (Aziz, 2005). This will require additional initiatives since Transparency International index has been stuck at virtually the same level since 1987. The new government of Dato Badawi, who became Prime Minister in late 2003 may provide such impetus. Since assuming office the Prime Minister has launched a number of anti-corruption initiatives spearheaded by the National Integrity Plan. Several corruption schemes have been uncovered including a kickback scheme in the Ministry of Public Works and investigation of corruption in the police force. Work on cleaning up campaign finance, a more independent judiciary, and access of information and a free press are also on the agenda. If there is follow through on these initiatives Malaysia could quickly move up on the rankings and become much less corrupt in the near future (Transparency International, 2005b).

Corruption in South Asia has remained high for the past two decades. The review of the corruption situation in India presented above is typical. Steps to cut down corruption suggested by TI and CVC will have to be extended and strengthened. The CVE has outlined a plan to eliminate illegal campaign finance. They suggest an amnesty period where such illegal funds can be claimed, and by paying back income taxes (Vittal, 2001). In addition the divide that separates the poor from the rich needs to be reduced. Petty corruption weights much more heavily on the poor and they are treated as second-class citizens when they visit government departments. They are also the subject of bribery in election campaigning.

General policies for the future

It has been emphasized in this chapter that corruption tends to increase the number of less urgent investment projects in developing countries and inflate their costs, both original capital expenditures and subsequent repairs and maintenance expenses. Misallocation of scarce capital and over-investment in "white elephant" and pork barrel projects are a serious drain on the economic resources of a country. It slows overall growth and development, and reduces economic efficiency. However, the recent Asian economic crisis may be viewed as a silver lining in fighting corruption in many Asian countries, as both the public and private sectors are increasingly determined to introduce major reforms and institutional changes to increase business transparency and institutional reforms. In this sense, the 1997 accord by OECD member countries to fight foreign bribery can serve as a critical milestone in fighting corruption worldwide.

In order to control corruption, Asian countries need to adopt the following policy measures. First, there should be drastic deregulation and liberalization of the economy. Where there is less government regulation, there is less scope for corruption. Simultaneously, further government downsizing is called for with a much smaller civil service force. With the saving in the personnel costs, the salary levels of the civil servants should be made comparable to those in the private sector as in Singapore. Second, strong measures need to be taken to enhance business transparency and to improve corporate governance in Asia, thereby promoting the competitiveness of Asian business firms through effective management rather than through political patronage. Third, financial sector reforms are critical to allocate scarce capital through market forces than by bribery-influenced political and bureaucratic intervention. Fourth, governments have to begin to function more like the private sector. Efficiency is important in the public sector. To achieve this, nepotism has to be reduced, along with policies that make the government the employer of last resort. Furthermore, civil service salaries have to be raised at the same time efficiency is increased and labor shedding adopted. Fifth, a paper trail and better auditing along with computerization of accounts is needed to reduce public waste and to control corruption. This entails computerized banking and check clearing, education of civil servants in the use of modern technology, and computerization of essential government services such as drivers and business licenses, income, excise and other taxes. Finally, political and election reform is essential to lessen the need for politicians to depend upon huge political contributions, often provided in the form of bribery from the business sector.

Governance

Governance has a number of different dimensions and is a much broader concept than corruption. The World Bank suggests five dimensions to governance. These include

- voice and accountability, which measures political, civil and human rights;
- political instability and violence which measures probability of violent threats to or changes in government including acts of terrorism;
- regulatory burden which measures the incidence of regulations that impede the working of market forces;
- rule of law which measures the quality of contract enforcement, effectiveness of the police and the legal system, as well as the probability of violence in the society;
- control of corruption which measures the exercise of public power for private gain including a wide range of possible corrupt activities.

From Figure 5.2 we note that good governance is loosely associated with income per capita. The table shows that richer countries are more law abiding. Different authors estimate the causal interaction of governance and income in their own way and yet the estimates are quite similar. Most observations fall on a straight line constrained to go through the origin.

Observing this relationship, it is no wonder that since the end of World War II the avowed purpose of the United States and to a lesser extent European countries, has been to spread democracy and political freedom around the world. The United States believes that democracy is the only system of government that ensures political freedom. Democracy gives citizens the right and privilege of electing their own leaders rather than being ruled by an authoritarian dictator or regime. As the strongest, richest, and most powerful military power these democratic values have been spread by the United States in various ways, including military forces in Korea, Vietnam, Iraq, Guatemala, and failed attempts in Cuba.

How effective has democracy been in ensuring political freedom? And what is the relationship between democracy and corruption and is

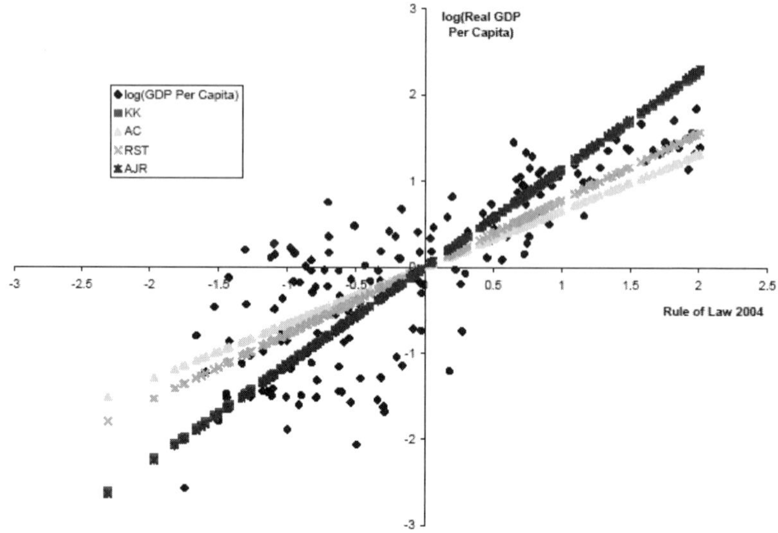

Note: The per capita GDP data has been standardized to have mean zero and standard deviation one. The four fitted lines correspond to four different sets of estimates of the causal impact of governance on incomes in the long run: AJR – Acemoglu, Johnson and Robinson (2000), KK – Kaufmann and Kraay (2002), AC – Alcala and Ciccone (2004), RR – Rodrik and Rigobon (2004). All scenarios were based on our 2004 KK estimates of Rule of Law.

Figure 5.2 The Development Dividend of Good Governance

Source: Kaufmann *et al.* (2005b).

democracy that closely related with good governance? Over the past few decades many countries have adopted democratic systems. In Asia there are 10 democratically elected governments according to Transparency International. Table 5.11 compares the Gastil freedom index including civil and political liberties, with the Transparency International index of political freedom. Visual inspection suggests that the relationship is not strikingly obvious.

This relationship between corruption and political freedom is further explored by fitting a regression equation where the TI corruption index is regressed on several measures of political freedom. Using the data in Table 5.11 we ran a simple regression with the Transparency International index of corruption (TI) as the independent variable and the indices of civil liberties (CL), political liberties (PL) and freedom (FREE) as the three independent variables.

Table 5.11 Corruption, Political and Civil Liberties, and Freedom

Country	TI corruption index — 2005	Index of political liberties	Index of civil liberties	Freedom index
Bangladesh*	1.7	4	4	2
Cambodia	2.3	6	5	3
China	3.2	7	6	3
Hong Kong	8.3	5	2	2
India*	2.9	2	3	1
Indonesia*	2.2	2	3	1
Kazakhstan	2.6	6	5	2
Kyrgyzstan	2.3	5	4	2
Laos	3.3	7	6	3
Malaysia	5.1	4	4	2
Mongolia*	3.3	2	2	1
Nepal	2.5	6	5	3
Pakistan	2.1	6	5	3
Papua New Guinea*	2.3	3	3	2
Philippines*	2.5	3	3	2
Singapore	9.4	5	4	2
South Korea*	5.4	1	2	1
Sri Lanka*	3.2	3	3	2
Taiwan*	3.9	1	1	1
Tajikistan	2.1	6	5	3
Thailand*	3.8	3	3	2
Turkmenistan	1.8	7	7	3
Uzbekistan	2.2	7	7	3
Vietnam	2.6	7	5	3

*Countries have democratic elections held with a multiparty system.

Source: Transparency International; Table 5.3.

Notes: Freedom Index has three rankings, free (1), partially free (2), and not free (3); Political and civil liberties are ranked on a scale from 1 to 7 with 1 being most free and 7 being least free.

The results are reported in Table 5.12. The fit is not tight. Corruption and freedom are not that closely related, even though the relationships have the correct *a priori* sign, that is, less corruption is associated with greater freedom and more civil and political liberties. Only one of the three coefficients is significant at the 10% level.

Table 5.12 Regression of Transparency International Index of Corruption (dependent variable) on Indices of Civil Liberties (CL), Political Liberties (PL), and Freedom (FR) for 24 Asian Economies

Independent variable	Intercept	Regression coefficient	*t* value
CL	5.2	−0.45	−1.90
PL	4.1	−0.15	−0.76
FR	4.9	−0.52	−1.31

Source: Data from Table 5.11.

What are the reasons for this? First the relationship between democracy and corruption can involve significant time lags. The democratically elected governments in industrial countries have been in place for a long time and the political institutions that underpin democracy have also been developed over many decades. Furthermore, newly installed political systems that have a democratic foundation can easily be subverted by corruption. Elected officials have to run for office and do what they must to attract the voters. Political campaigns need to be funded and this requires campaign contributions from their constituents and the general electorate and the community at large. This often results in corruption as campaign contributors ask for favors once the candidate is elected. Those running for office can also bribe voters to vote for them. Finally legislation is passed to ensure that those in office are reelected and this can involve government spending on "pork barrel" projects in the representatives' district. The contracting for these projects can also involve corruption. These kinds of political corruption are commonplace, not only in Asian countries but also in other regions including industrial countries.

Vote buying and corruption

Vote buying has been an issue in recent elections in many Asian countries including Philippines, Indonesia, Malaysia, Korea, India and Thailand (see Thailand, 2006; Taiwan, 2006; Indonesia, 2004; http://www.pathfinder.com/asiaweek/magazine/2000/0407/sr.main3.html for Asia in general; and http://www.ned.org/forum/

asia/june01/session2.html for experience in Philippines, India, Indonesia and Thailand).

Despite the headlines about vote-buying in elections it is difficult to draw strong conclusions about trends in vote buying. Journalistic reports suggest that there is less violence surrounding elections than there used to be. Still many candidates have been killed or injured in the jostling for political advantage and the spoils that come to those who get elected. However, spending on vote buying does not seem to have abated, even in countries where elections have been ongoing for many years. Incentives for change require a more educated, richer and more politically aware electorate in many countries. One analyst puts it this way:

"Why would anyone want to tinker with this system?", Jong Wan Kim asked after hearing the myriad problems of these four countries (Philippines, India, Indonesia, and Thailand) (http://www.ned.org/forum/asia/june01/session2.html) Politicians need money to win elections; business firms have money to donate in return for favors from government; and voters can be bought off for trifling sums from the politicians. Where, he wondered, is the incentive for change?

One of the problems in achieving the aims of a democratic society interpreted in western terms is to understand the intellectual and political divide between the rich and the poor in the poorest countries in Asia. An object lesson from the Philippines is a good example. Philippine society is divided into five categories which are well understood by most educated people and even by the poorer classes. The rich and upper middle classes comprise about 15% of the electorate. The remainder belongs to the three other classes. In the election and eventual impeachment of Joseph Estrada, the poor and the rich had diametrically opposed views of Estrada. The poor loved him because he was considerate and kind to them. He honored them and treated them with dignity. It did not matter to the poor that he was corrupt. They could not understand what corruption meant. The rich on the other hand viewed him as an unprincipled and thoroughly corrupt politician. Schaffer (2002) summarizes it well:

> Good politics for many in the upper and middle classes is, in sum, the clean politics of transparency; while bad politics is the dirty politics of rag-like corrupt politicians (p. 17).

For those poorer people, good politics is often a politics of personal dignity, a politics in which politicians treat the poor with kindness and compassion while bad politics is a politics of insult and rudeness. This is why Estrada was loved by the poor and reviled by the rich. How will this political divide be resolved in the Philippines and, indeed, in many other poorer countries of the region where the same divide is observed? In Thailand the threat to Prime Minister Thaksin was from the rich and upper middle class. He was widely supported in the countryside. The agenda for clean election reform is the project of the rich. Unless institutions are further strengthened and the populous has sufficient education and experience to appreciate the process of democracy and the importance of voting as an expression of political views, democracy could well undermine economic reforms and contribute to further corruption. The development of new political parties in India catering to the needs and aspirations of poorer segments of society may be a sign that political maturity is possible. However, it also suggests that there is a strong correlation between the rise of party competition for votes and political corruption (Banerjee and Pande, undated).

Is there a "Honesty Divide"?

In the literature of economic development there has been much discussion of ways in which industrial and development countries are separated. The "development divide" argues that poor countries find it difficult to overcome poor education, low income, poor health, and the vicious cycle of poverty in their quest to accelerate growth, raise income per capita, and reach industrial country status. The "digital divide" suggests that poor countries are lacking in information and computer technological skills and infrastructure, including the internet, to keep up with industrial countries and thus, they are continuing to fall behind. These two development divides interact to keep poor countries from becoming part of the dynamically evolving global community with access to modern technology, better education and medical treatment, and greater economic efficiency. As a result there is evidence that the world distribution is deteriorating. Poorer countries are falling further and further behind the rich (Table 5.13).

Table 5.13 Global Distribution of Income, 1960–1998

	1960	1970	1980	1990	1998
Share of poorest 20%	2.3	2.3	1.7	1.3	1.1
Share of richest 20%	70.3	73.9	76.3	82.8	86
GDP per capita for OECD countries PPP US dollars	$6,000				$17,000
GDP per capita for East Asia and Pacific	$1000				$2500
GDP for least developed countries	$400				$600

Source: UNDP (1999); http://www.redefiningprogress.org.

The experience with corruption in Asia suggests that we may be able to add another divide to this list, an "honesty divide". The Transparency International indices of corruption for Asia displayed in Table 5.2 suggests that corruption in East Asia (Korea, Hong Kong, Singapore and Taiwan) has fallen in the past two decades while corruption in the rest of Asia, with perhaps the exception of Indonesia and Thailand in recent years, has either remained stagnant or has gotten worse.[5] Unless this trend is reversed, this third "divide" threatens to interact with the other "divides" to create additional impediments to the viable economic development of the poorer countries in the Asian region. Dealing with the corruption issue becomes even more critical when we realize that the three large economies in the region, particularly India and China, are realizing their growth potential to generate rapid increases in living standards for their people. Sustaining such gains in income per capita will require higher labor productivity and

[5]World Bank figures which rank Asian countries in percentiles viz-a-viz all countries show an even more dramatic difference between the NIEs and the rest of Asia. All four NIEs are in the top two quintiles of the distribution (Singapore 99.9, Hong Kong 90.6, Korea 62.1 and Taiwan 73.9). Aside from Malaysia whose ranking is just above Korea, the rest of Asian economies surveyed are in the bottom two quintiles and five (Cambodia, Indonesia, Laos, Myanmar and PNG) are in the bottom quintile. To achieve a modicum of honesty these economies will have to leapfrog over the middle quintile, from 40 to 60, a difficult task given their collective history of corruption and weak governance.

enhanced economic efficiency. To attain these goals it is also essential that corruption be reduced and honesty and integrity restored.

Bibliography

Acemoglu, D., S. Johnson and J. A. Robinson (2000). The Colonial Origins of Comparative Development: An Empirical Investigation. NBER Working Paper No. 7771.

Ades, A. and R. di Tella (1999). Rents, Competition and Corruption. *American Economic Review*, 89(4), 982–993.

Alcala, F. and A. Ciccone (2004). Trade and Productivity. *Quarterly Journal of Economics*, 119(2), 612–645.

Asian Development Bank (ADB) (2001). Third ADB/OECD Conference on Combating Corruption in Asian and Pacific Economies. http://www.adb.org/Documents/Events/2001/Combating_Corruption/default.asp.

Asiaweek.com (2000). How Asia Works: Democracy for Sale. 7 April. http://www.pathfinder.com/asiaweek/magazine/2000/0407/sr.main3.html [Retrieved 23 April 2004].

Azfar, O. and T. Gurgur (2001). Decentralization and Corruption in the Philippines. Mimeo.

Aziz, T. A. (2005). International Case Study: Stamping Out Corruption in Malaysia, 113th International training course visiting experts' papers. http://unpan1.un.org/intradoc/groups/public/documents/APCITY/UNPAN019124.pdf.

Banerjee, A. V. and R. Pande. Parochial Politics: Ethnic Competition and Political Corruption in India. Mimeo, undated.

Batalla, A. and A. Malaluan (2001). Yellow Paper II Beyond EDSA: The Post –ERAP Reform Agenda, Governance and Corruption in the Philippines. Mimeo.

Becker, G. and G. Stigler (1974). Law Enforcement, Malfeasance and the Compensation of Enforcers. *Journal of Legal Studies*, 3(1), 1–19.

Besley, T. and J. McLaren (1993). Taxes and Bribery: The Role of Wage Incentives. *Economic Journal*, 103, 119–141.

Blechinger, V. (2000). Report on Recent Bribery Scandals. Transparency International Workshop paper on Corruption and Political Party Funding in La Pietra, Italy.

Bose, S. (2002). A Culture of Corruption? Democracy and Power in India. Global Policy Forum. http://www.globalpolicy.org/nationsl.corrupt,2002/0624sum.htm.

Burki, S. J. (1997). Governance, Corruption and Development: The Case of Pakistan. Paper presented at Workshop on Governance Issues in South Asia, Yale University.

Chow, G. C. (2005). Corruption and China's Economic Reform in the Early 21st Century. Center for Economic Policy Studies, Working Paper No. 116, Princeton.

Cordingley, P. (2001). The Cost Of Corruption. *Asian Week*, 19 October.

Damon, F. Y. (2003). Corruption in Mainland China Today: Data and Law in a Dubious Battle. In J. B. Kidd and F. J. Richter (eds.), *Fighting Corruption in Asia*. New Jersey: World Scientific Publishing.

de Dios, E. and F. D. Ferrer (2000). Corruption in the Philippines: Framework and Context. Mimeo prepared for Transparent and Accountable Governance Project.

de Dios, E. and H. S. Esfahani (2001). Centralization, Political Turnover and Investment in the Philippines. In J. E. Campos (ed.), *The Boom and Bust of East Asia Corruption* (Chapter 5). Quezon City: Ateneo de Manila University Press.

de Speville, B. (1997). Hong Kong: Policy Initiatives against Corruption. Paris: OECD.

Dowling, J. M. and C. F. Yap (2004). Indonesian Economic Development — Mirage or Miracle? In S. LaCroix (ed.), *Miracles and Mirages in East Asian Economic Development*. University of Hawaii Press.

Dwivedi, O. P. and R. B. Jain (1988). Bureaucratic Morality in India. *International Political Science Review*, July, 205–214.

Ehrlich, C. P. and D. S. Kang (2002). Independence and Corruption in Korea. *Columbia Journal of Asian Law*, 16(1), 1–68.

Fan, C. S. and H. I. Grossman (2001). Entrepreneurial Graft in China. *Providence Journal*, 3 May, B5.

Fell, D. (2003). Democratization and Political Corruption in Taiwan. Mimeo, School of Oriental and African Studies, London. See also http://www.soas.ac.uk/taiwanstudiesfiles/workingpapers/fell1.pdf.

Freedom in the World (2006). http://www.freedomhouse.org/uploads/pdf/Charts2006.pdf.

Gambett, D. and P. Reuter (1995). The Mafia in Legitimate Industries. In F. Gianluca and S. S. Peltzman (eds.), *The Economics of Organised Crime* (pp. 116–139). Cambridge: Cambridge Press.

Goel, R. K. and M. A. Nelson (1998). Corruption and Government Size: A Disaggregated Analysis. *Public Choice*, 97, 107–120.

Hunt, J. (2006). Why are Some Public Officials More Corrupt than Others? In S. Rose-Ackerman (ed.), *International Handbook on the Economics of Corruption*. Edward Elgar.

Indonesia (2004). http://www.businessweek.com/magazine/content/04_28/b3891432.htm.

International Herald Tribune (2006). 8 and 9 March.

Kahn, M. (1996). A Typology of Corrupt Transaction in Developing Countries. *Institute of Development Studies Bulletin*, 27(2), 12–21.

Kang, D. S. (2002). *Crony Capitalism: Corruption and Development in South Korea and the Philippines*. Cambridge Studies in Comparative Politics, Cambridge University Press.

Kaufmann, D. (1997). Corruption: The Facts. *Foreign Policy*, 107(Summer), 114–131.

Kaufmann, D. and A. Kraay (2002). Governance Indicators: Aid, Allocation and the Millennium Challenge Account. Mimeo, World Bank.

Kaufmann, D., A. Kraay and M. Mastruzzi (2005a). Governance Indicators for 1996–2004. World Bank Policy Research Working Paper Series No. 3630.

Kaufmann, D., A. Kraay and M. Mastruzzi (2005b). Governance Matters IV: New Data, New Challenges. World Bank.

Kidd, J. B. and F. J. Richter (eds.) (2002). *Corruption and Governance in Asia.* London and New York: Palgrave Press.

Kochanek, S. (1993). *Patron-Client Politics and Business in Bangladesh.* New Delhi: Sage publications.

Kurtzman, J., G. Yago and T. Phumiwasana (2004). The Global Cost of Opacity: Measuring Business and Investment Risk Worldwide. The Kurtzman Group. http://www.opacityindex.com/.

Kurtzman Group website (2006). http://www.kurtzmangroup.com/ [Retrieved 26 May 2006].

Lee, M. (1990). *The Odyssey of Korean Democracy: Korean Politics 1987–1990.* Westport, Connecticut: Greenwood Publishing Group.

Li, W. (2001). Corruption During the Economic Transition in China. http://faculty.darden.virginia.edu/liw/papers/zif-c3.pdf.

Mauro, P. (1995). Corruption and Growth. *Quarterly Journal of Economics*, August, 681–711.

Moody-Stewart, G. (1997). *Grand Corruption in Third World Development.* Oxford: Worldview Publishing.

National Endowment for Democracy (2001). The Democracy Forum for East Asia: Political Finance and Democracy in East Asia: The Use and Abuse of Money in Campaigns and Elections. http://www.ned.org/forum/asia/june01/session2.html [Retrieved 26 February 2004].

New York Times (1998). 6 March.

People's Daily Online (2004). http://english.people.com.cn/200403/11/eng20040311_137239.shtml. 11 March [Retrieved 25 June 2005].

Phongpaichit, P. and S. Piriyarangsan (1994). *Corruption and Democracy in Thailand.* Bangkok, Thailand: Silkworm Books.

Phongpaichit, P., S. Piriyarangsan and N. Treerat (1998). *Guns, Girls, Gambling, Ganja: Thailand's Illegal Economy and Public Policy.* Bangkok, Thailand: Silkworm Books.

Phongpaichit, P., N. Treerat, Y. Chaiyapong and C. Baker (2000). *Corruption in the Public Sector in Thailand: Perceptions and Experiences of Households.* Chulalongkorn University, Bangkok: Political Econony Center.

Prichett, L. and G. Seeth (1994). Tariff Revenue and Tariff Reform: Some New Facts. *World Bank Economic Review*, 4, 1–16.

Quah, J. S. T. (2003). *Curbing Corruption in Asia: A Comparative Study of Six Countries.* Manila, Philippines: Eastern Universities Press.

Redefiningprogress.org website (2006). http://www.redefiningprogress.org [Retrieved 28 February 2006].

Robison, R. (1986). *The Rise of Capital.* Sydney, Australia: Allen and Unwin.

Rodrik, D. and R. Rigoban (2004). Rule of Law, Democracy, Openness and Income: Estimating the Interrelationships. NBER Working Paper No. 10750. Washington, DC: National Bureau of Economic Research.

Roodman, D. (1996). Paying the Piper: Subsidies, Politics and the Environment. Paper 133. Washington, DC: Worldwatch.

Rose-Ackerman, S. (1999). *Corruption and Government: Causes, Consequences and Reform.* Cambridge, UK: Cambridge Press.

Schaffer, F. C. (2002). Disciplinary Reactions: Alienation and the Reform of Vote Buying in the Philippines. Paper prepared for delivery at the International Conference on "Trading Political Rights: The Comparative Politics of Vote Buying", Center for International Studies, MIT, Cambridge, 26–27 August.

Schwartz, A. (1994). *A Nation in Waiting: Indonesia in the 1990s.* Boulder: Westview Press.

Seldadyo, H. and J. di Haan (2005). The Determinants of Corruption: A Reinvestigation. Paper Prepared for EPCS-2005 Conference, Durham, England.

Serra, D. (2004). Empirical Determinants of Corruption: A Sensitivity Analysis. Discussion paper. UK: Global Poverty Research Group, Economic and Social Research Council.

Shleifer, A. and R. W. Vishny (1993). Corruption. *Quarterly Journal of Economics,* 1083(August), 599–617.

Social Weather Stations (2000, 2001). National Surveys. http://www.sws.org.ph/pr111300.htm.

Social Weather Stations (2000). Business Opinion Survey. August–October.

Sun, Y. (1999). Reform, State and Corruption: Is Corruption Less Destructive in China than in Russia? *Comparative Politics,* 32(1), 1–20.

Sun, Y. (2004). *Corruption and Market in Contemporary China.* Cornell University Press.

Taiwan (2006). http://www.atimes.com/china/DF29Ad01.html.

Taiwan Times interview with Transparency International (2004). http://www.taipeitimes.com/News/editorials/archives/2004/03/29/2003108221.

Thailand (2006). http://www.asiademocracy.org/content_view.php?section_id=1&content_id=338.

Times of India (2006). http://timesofindia.indiatimes.com [Retrieved 20 May 2006].

Transparency International (2003). *Annual Report. Corruption Perceptions Index.* Berlin, Germany: Transparency International.

Transparency International (2004). *Annual Report. Corruption Perceptions Index.* Berlin, Germany: Transparency International.

Transparency International (2005a). India Corruption Study to Improve Governance. New Delhi, India: Transparency International India.

Transparency International (2005b). Report Toward a Corruption Free Malaysia. http://www.transparency.org.my/documents/ TOWARDS a CORRUPTION FREE MALAYSIA.

Transparency International (2006). *Global Corruption Report.* Berlin, Germany: Transparency International.

Treisman, D. (2000). The Cause of Corruption: A Cross-National Study. *Journal of Public Economics*, 76, 399–457.

United Nations (2001). Strengthening Judicial Integrity Against Corruption. Office for Drug Control and Crime Prevention. Vienna: Center for International Crime Prevention.

United Nations Development Programme (UNDP) (1999). *Human Development Report.*

van Rijckeghem, C. and B. Weder (1997). Corruption and the Rate of Temptation: Do Low Wages in the Civil Service Cause Corruption. IMF Working Paper No. 97/93.

Vittal, N. (2001). Corruption and the State. *Harvard International Review*, 23(3). http://hir.harvard.edu/articles/925/ [Retrieved 23 January 2006].

Warren, M. E. (2005). Democracy against Corruption. Paper presented at Conference on the Quality of Government, Quality of Government Institute, University of Goteborg.

Wedman, A. (1997). Looters, Rent-Scrapers, and Dividend-Collectors: Corruption and Growth in Zaire, South Korea and the Philippines. *Journal of Developing Areas*, 31, 457–478.

Wei, S. J. (1997). How Taxing is Corruption on International Investors? Working Paper No. 6030. Cambridge, MA: National Bureau of Economic Research.

World Bank (2000a). Report No. 20369-PH Combating Corruption in the Philippines. http://www.worldbank.org/eapsocial/library/corruption.htm.

World Bank (2000b). Cambodia, Governance and Corruption Diagnostic: Evidence from Citizen Enterprise and Public Official Surveys. http://www.worldbank.org/wbi/governance.

World Bank (2001). Report 23687. Combating Corruption in the Philippines: An Update. http://www.transparency.ca/Readings/TI-F01.htm.

World Bank (2006). The Costs of Corruption. http://web.worldbank.org/WBSITE/EXTERNAL/NEWS/0,,contentMDK:20190187~menuPK:34457~pagePK:34370~piPK:34424~theSitePK:4607,00.html [Retrieved 5 January 2006].

You, J. S. (2005). Embedded Autonomy or Crony Capitalism? Explaining Corruption in South Korea, Relative to Taiwan and the Philippines, Focusing on the Role of Land Reform and Industrial Policy. Paper prepared for delivery at the Annual Meeting of the American Political Science Association, Washington, DC, 1–4 September.

Yusuf, S. and S. J. Evenett (2002). *Can East Asia Compete? Innovations for Global Markets.* Washington, DC: World Bank.

Chapter 6
Environment

It is your human environment that makes climate.

— Mark Twain (American humorist and writer), 1835–1910

In order to develop scenarios for environmental developments in the Asian region it is necessary to review recent history. Economic growth in Asia has been very strong in recent years, particularly in the past couple of decades. This growth has brought prosperity and higher living standards to the region. However it has also resulted in a significant acceleration in environmental damage and degradation. This chapter reviews the record of environmental impacts and policies and then goes on to develop possible scenarios for the next 20 to 30 years.

We recognize that the environment is inextricably intertwined with overall economic activity in the region and is responsive to environmental, economic and social policies. For this reason, we will be looking at the broader basis for environmental policy as we examine the recent past and chart possible courses for the future.

Recent Environmental Developments in Asia

The approach taken here is to review the current state of the environment and the impacts of population and economic growth; the development of freshwater and ocean water resources; air quality; land and forest resources; hazardous wastes; and global warming.

Population growth

Population growth and economic growth combined put increased pressure on the environment. We single out population growth to highlight the impact that higher population density alone has on the resource base of Asian economies. Over the past few decades the rate of population growth in Asia has fallen as a result of a variety of factors. The demographic transition from high to low birth and death rates has resulted in a dramatic abatement in population pressure in the richest countries on East Asia (Japan and the newly industrialized economies (NIE) of Hong Kong, Korea, Singapore and Taiwan). China has also made great strides in reducing population growth through its "one child" policy. In Southeast Asia increases in women's education, late marriages, spacing of children, and general family planning policies have also been instrumental in lowering fertility. In South Asia and the Philippines progress has been slower.

The pattern of overall population growth for major countries in the world (past and future projections) is shown in Table 6.1. South Asia and China have been responsible for the bulk of increase in world population and they will continue to be the major sources of population increase in the next few decades. Population projections have moderated over the past few decades as higher incomes, greater educational opportunities for women, and family planning have reduced birth rates and population growth. The panels of Table 6.1 show the incremental contribution by the Asian region to the world population, particularly the extremely high contribution by South Asia. Overall increases in the contribution to global population growth by Asia are projected to continue in the future. However, it is expected to fall dramatically from 2025 to 2050 (from approximately 1.5 billion people from 1950 to 1995 to around 350 million between 2025 and 2050).

As the remaining parts of this section demonstrate, there has been and will continue to be sustained pressure on a wide range of environmental resources, including fresh water and sanitation, clean air, productive arable land, energy and mineral resources, and global warming. Much of the pressure on land and water resources comes from increasing population density in poor rural areas which are unable or unwilling to shift from current environmentally undesirable practices that degrade soil, water, and air quality. Rising

Table 6.1 Population Increase in Top Ten Countries Globally: Past Trends and Future Projections in 100,000 Persons

Country	1950–1995	Country	1995–2025	Country	2025–2050	Country	1950–2050
China	665	India	401	India	173	India	1175
India	571	China	260	Nigeria	81	China	962
Indonesia	117	Pakistan	132	Pakistan	73	Pakistan	318
US	109	Nigeria	127	Ethiopia	63	Nigeria	306
Brazil	105	Ethiopia	80	Zaire	48	Indonesia	239
Pakistan	97	Indonesia	78	China	40	Ethiopia	194
Nigeria	79	US	65	Indonesia	37	US	190
Bangladesh	76	Bangladesh	62	Iran	35	Brazil	189
Mexico	63	Zaire	60	Bangladesh	33	Bangladesh	178
Iran	51	Iran	60	Brazil	23	Iran	153

Source: Adapted from UN Population Division (1996): World Populations Prospects 1950–2050 (Annex I and II).

incomes and continued migration to the cities have ameliorated these difficulties to some extent. On the other hand, increased movements of people along with natural increases in population have created enormous pressure on Asian urban areas. Asia now has many cities with more than a million residents and many of the world's mega cities (see section on Air Quality) with its own environment issues. Additional resources had to be devoted to cleaning up the environmental damage stemming from this rapid and continued growth in population. However, there has been significant time lags between the build up of this damage and public awareness. Meanwhile, government resources to deal with the environmental degradation are insufficient in preventing further damage.

Water availability and water quality

In a poll taken by the Asian Development Bank (1997), water pollution and fresh water depletion were often mentioned as the most important environmental issue by environmental policy makers (see also Brandon and Ramankutty, 1993). It is unlikely that this concern has abated over the intervening decade since this report was prepared. There are several aspects of water resources that are relevant and of concern to policy makers; — first, the availability of water, and second the quality of water. Of all the other environmental issues, the availability of clean fresh water is critical to the maintenance of health. While we defer discussion on the impact of environmental degradation on health to a later section, it is important to maintain awareness of the impact that water policy has on health. It is also important to understand that the use of water is heavily affected by its cost of delivery, particularly in Asia which accounts for 36% of global runoff even though it has the lowest per capita availability of fresh water. The problem is most severe in India, Pakistan and South Korea where high water stress conditions exist. High water stress occurs when the ratio of fresh water used to available water exceeds 40% (see Dowling and Valenzuela, 2004, Chapter 12).

Box 6.1 Water Stress

Malin Falkenmark has suggested a water stress index based on a minimum level of water required per capita to maintain an adequate quality of life in a moderately developed country in an arid zone. She suggests a minimum of 36.5 cubic meters per year per capita to be multiplied by a number between 5 and 20 that reflect the requirements of industry, agriculture and energy. At a minimum, this amount would be somewhere between 180 and 720 cubic meters per year for all uses.

However, experience has shown that when fresh water availability falls below 1700 cubic meters per person per year, the country experiences water stress as water is not available in quantities required to maintain the required lifestyle and for agricultural and industrial use. The higher stress level of 1700 is said to be a warning light to nations where population growth can be expected to put more demands on water resources. Water scarcity exists when the threshold falls below 1000, and when it falls below 500 cubic meters per person, there is absolute scarcity. The 1000 cubic meter benchmark has been adopted by the World Bank as a general indicator of water scarcity.

Sources: Falkenmark and Widstrand (1992); UNEP (1999, Chapter 2); Population Action Council International website (http://www.cnie.org/pop/pai/water-12/htm).

Fresh water withdrawal from rivers, lakes, reservoirs, underground aquifers, and other sources increased more in Asia during the past century than in any other regions of the world, from 600 cubic kilometers per year in 1900 to around 5000 cubic kilometers in the mid-1980s (da Cunha, 1989; UNEP, 1999). To satisfy this increasing demand, a number of large dams have been constructed since the end of the World War II. Irrigation accounts for the bulk of fresh water use at 85% and the balance taken up by households and industries. Currently, water usage exceeds annual runoff and recharge of underground aquifers. Extensive use of irrigation has been responsible for the lion's share of increased water usage, with service to a larger population and industrial development responsible for the rest of the increase. As a result, it is more likely that water shortages will increase unless there is more efficient use of existing water resources. Already, land damaged by over irrigation in India, China and Pakistan, amounting to almost 30 million hectares, accounts for close to half of the world's damaged (AAAS, 2005).

Water development programs for hydropower along with deforestation in important watersheds have reduced water flow in rivers and reduced the area of wetlands. Combined with mismanagement of water resources and increased irrigation, water demand has exceeded annual flow. Water tables have fallen dramatically as aquifers are pumped faster than they can be recharged to make up for the shortfall in the water supply. The overall result has been a reduction in the annual availability of fresh water from 10,000 cubic meters per capita per year in the 1950s to around 4200 in the early 1990s. Availability is much lower in some regions (see Table 6.2). Singapore has to import water from Malaysia which has much more water than the average for the region (21,000 cubic meters per capita per annum) while supplies in Korea, Pakistan and Thailand are considerably less between 1400 and 1900 cubic meters per capita per annum (UNEP, 1999). Water stress also occurs periodically during periods of drought (see Box 6.1).

Proper sewage disposal is a major problem closely linked to water pollution. Pollutants like organic matter, heavy metals and toxic chemicals, sediment, suspended solids and salts contaminate water supplies.

Table 6.2 Renewable Fresh Water Resources in the Asian Region in the Early 1990s (in cubic meters per capita per year)

Australasia and the Pacific	54,732
Southeast Asia	10,909
Greater Mekong	9494
South Asia	2845
Northwest Pacific and East Asia	2414
Asia and Pacific total	4205

Source: UNEP (1999, Chapter 2).

Note: Greater Mekong includes Vietnam, Cambodia, Lao PDR, Myanmar, Thailand, and Yunnan Province in China; Northwest Pacific and East Asia includes Korea, China, Russia, Japan, Taiwan and Hong Kong.

While most major cities in Asia are heavily polluted with raw sewage, industrial effluents, chemicals and solid waste, water pollution is most evident in India and the rest of South Asia. The problem is most pressing in urban areas where disposal of domestic sewage is the major cause of water pollution. In recent years, this has been compounded by contamination from industrial waste. Levels of suspended solids in Asia's rivers have quadrupled since the late 1970s (ADB, 1997) and rivers typically contain four times the world average and 20 times more than OECD members (UNEP, 2001). Access to safe water is lowest in South and Southeast Asia where nearly half the population does not have access to safe drinking water.

Less than half of the Asian population has access to proper sanitation, less than any other region in the world (WHO and UNICEF, 2000). Coverage is even lower in rural areas where only 31% of the population has proper sanitation. In Bangladesh, natural leaching and pollution have resulted in arsenic levels 70 times higher than recommended. In rural India and the rest of South Asia, there is limited access to sanitation facilities in many villages. Dirty water and poor sanitation are responsible for more than half a million infant deaths in the region as well as considerable illness and disability in the general population (ADB, 2001).

The major rivers in the region are either polluted or running dry for part of the year. Portions of the Yellow River, which run through the most important agricultural regions in China, are severely polluted and run dry for much of the year in its lower reaches (ADB, 1997). The flow of the two rivers running into the Aral Sea has been dramatically reduced by irrigation use, causing it to shrink to a fraction of its size from the 1960s. The fouling of waterways and surrounding river basins has resulted in lower incomes for residents and in some cases making them environmental refugees (World Water Commission, 2000). Levels of suspended solids in Asia's rivers almost quadrupled since the late 1970s (ADB, 1997; GEMS, 1996) and sedimentation has also increased.

Aquatic resources have also been adversely impacted by mismanagement and degradation of resources (see Dua and Esty, 1997 for a good summary). World Resources Institute (1996) estimates that all

Pacific fisheries are over-fished and are at risk of depletion. The dramatic increase in the volume of fish landed in the 1980s, particularly from the Chinese, Australian and New Zealand fishing fleets, cannot be sustained (WRI, 1996). Subsidies to build larger boats and the competition between countries to increase yields are largely responsible for the decline in yields. Some countries have turned to aquaculture to make up the shortfall in catch from ocean fisheries. The destruction of coastal mangrove habitats to make way for aquaculture adds to the negative environmental impact. In Thailand and Philippines alone, over 200,000 hectares of mangrove forests have been destroyed in the last 30 years (UNEP, 1999). The destruction of mangroves also increases the risk of tidal swells in low-lying coastal villages, like in the 2005 Tsunami, where damage was exacerbated by the lack of mangrove cover. It also changes drainage patterns causing marine pollution as a result of discharges from rivers and destroying habitats for many kinds of sea creatures that thrive in the tidal basin. Not only are fish stocks threatened, this pollution also threatens coral reefs in Southeast Asia. Pollution also contributes to the greater incidence of so called "red tide", a form of algae growth that results in shellfish poisoning and other aquatic damage.

Air quality

Air quality in Asian cities has deteriorated over the past several decades as the pace of industrialization has accelerated movements of populations from rural to urban areas, and population size, density, and the numbers of vehicles have increased. Along with the growth in industry and greater reliance on motorized transportation, sulfur dioxide and nitrogen oxide emissions have increased dramatically. It is also aggravated by the use of low-quality fuel, inefficient methods of energy production, vehicles in poor conditions and traffic congestion.

Urban pollution is the major cause of poor air quality in Asia. Asia now has 43 cities over 2 million people and half of the world's 31 mega cities with population over 5 million (source: www.mongabay.com/cities_pop_01.htm). Asia ranks as the world's most polluted region — more than five times as high as in industrial countries and Latin America (ADB, 2005). Lead emissions from vehicles are above safe levels and air quality is far below the standards set for air quality by the WHO. Ten of Asia's 11 large cities (except Tokyo) exceed WHO guidelines for particulate matter by a factor of 3 while four exceeded acceptable lead levels and three exceeded acceptable ozone and sulfur dioxide levels (ADB, 1997). Of the 41 cities ranked by WHO, 13 of the 15 dirtiest were in Asia. The burning of fossil fuels is the largest source of air pollution in urban areas and the rapid increase in the number of vehicles contributes to this pollution. The number of cars is increasing by 10% per year in many metropolitan areas and is closely correlated with the growth in income (Hammer and Shetty, 1995).

Commercial energy use is also growing rapidly at around 6% per year, much faster than in the rest of the world where the average growth rate is less than 2% and less than 1% in OECD countries (ADB, 2001).

Burning of wood fuels also contributes to air pollution, particularly in rural parts of South and Southeast Asia. The region accounts for close to half of global consumption of wood fuels. This contributes to indoor pollution in rural areas where wood is used for heating and cooking. The pollution is exacerbated by the use of inefficient and polluting stoves. Women and children bear the brunt of these pollutants which contain sulfur and nitrous oxides as well as arsenic which are byproducts of burning of stubble, dung, and other residues as well as wood (Dowling and Valenzuela, 2004).

In addition to the detrimental effect air pollution has on general health, the cost of lost efficiency at work, lower life expectancy, and the resulting pain and suffering which will be discussed later in the paper, there are two aspects of air pollution that require special mention. The first is acid rain which is the result of a build up of sulfur compounds in the atmosphere brought to earth by rainfall resulting in the destruction of crops and aquatic resources. The second is the presence of a hazy cloud that has spread over much of the Asian region. This haze which contains a mixture of soot, sulfates, nitrates, organic particles, fly ash and dust reduces sunlight by 10% or more. Forest fires in Southeast Asia over the last decade or so, particularly in the 1997 fire in

Indonesia which spread the smoke to surrounding countries, further contributed to this haze. Simulations of global climate models suggest that this haze could have major impacts on monsoon circulation, regional rainfall patterns and temperatures (UNEP, 2001).

The primary cause of the rapid build up of haze and acid rain is the acceleration of the burning of coal in China and India. China is the second largest emitter of carbon pollutants, right behind the United States with 1 billion tons annually. India is in fourth place (Worldwatch Institute, 2006). UNEP estimates that at least two-thirds of acid rain containing large amounts of sulfur is caused by coal-fired power plants with outdated pollution control equipment. Many of the plants are in China and more are coming on line every year. Coal provides two-thirds of China's energy and half of India's energy (Worldwatch Institute, 2006). Coal briquettes used for heating and cooking in homes and small businesses throughout China releases sulfur dioxide and particulates in the atmosphere. Moreover, many of China's and most of India's coal deposits are high in ash or sulfur, creating a greater pollution hazard.

Increased use of coal for industrial use and electricity generation also contributes to air pollution, particularly when newer technologies to reduce pollution such as scrubbers are not used. There is also evidence being accumulated that traces of mercury from coal-fired power plants in China are being spread throughout the Asian region and even as far as the United States. According to the US Environmental Protection Agency (reported in *Financial Times*, 12 April 2006), over 50% of the world's mercury emissions come from Asia and China is a particularly worrisome source. The EPA says mercury levels in China are high enough to threaten lives. In Jiangsu province, for example, heavy metals including mercury, cadmium, and lead were present in 41% of fish species according to the EPA. Furthermore, China's cement kilns, which account for 40% of world production, are major sources of dioxin and furan, and also deadly chemicals that can be transported by air over long distances.

The components of air pollution vary from region to region within Asia. Particulate pollution is particularly high in India and China because of the widespread burning of coal, while sulfur dioxides are highest in

East Asia as a result of greater numbers of automobiles. The use of leaded fuels in Southeast Asia and to a lesser extent in South Asia (ADB, 1997, Table 4.1) keeps lead prevalent in the air.

Nuclear energy is an alternative that is beginning to be considered by Asian economies. There are a few nuclear reactors currently operating although they only contribute a small fraction of the power needs of the region and have not resulted in any significant reduction in carbon dioxide emissions. However, worldwide and particularly in France, Japan, Sweden and South Korea, nuclear energy is responsible for a significant share of electricity generation at 77%, 34%, 44% and 39%, respectively (Uranium Information Center Limited website).

Land and forest resources

Economic growth, population pressure, infrastructure development, logging, soil erosion, and the use of fertilizers and pesticides have resulted in deterioration of land quality, deforestation and land degradation. While the extent of environmental damage depends on the initial starting point, MacKinnon and MacKinnon (1988) estimate that countries in the Asian region had lost 70% to 90% of their original wildlife habitat by the mid-1980s. Further damage has occurred since these estimates were made. Soil erosion, depletion in soil fertility, and water logging causes acidification and salinization resulting in further land degradation. Significant degradation is the result of mismanagement through improper use of fertilizers and irrigation (Oldman, 1994). These problems are widespread in northern India, Bangladesh, Cambodia, Malaysia, Thailand and Vietnam. Excessive extraction of groundwater and faulty irrigation systems that cause water tables to rise have increased soil salinity and water logging. Soil contamination from lead and arsenic contamination occurs in Southeast Asia and South Asia and is further compounded by irrigation with untreated effluents. UNEP (1993) estimates that around 30% of land area in Asia had experienced some degree of land degradation.

Forest cover has also declined. At the end of World War II it is estimated that forest cover in Asia totalled over 15 million square kilometers, 25% of the world's total. Over 79% have been lost and only one-fifth of what remains (a mere 6% of the original forest) can be classified as frontier forest. Furthermore, WRI (1996) estimates that 60% of the remaining frontier forests in Asia are under threat from economic development, and Wilson (2001) estimates that forest cover is declining at the rate of 1% per year. On mainland Southeast Asia most frontier forests are gone with most severe losses in Bangladesh, India, the Philippines, Sri Lanka and Vietnam. Remaining virgin forests in Philippines, Papua New Guinea, Irian Jaya, and both the Malaysian and Indonesian sections of Borneo as well as Sulawesi and Sumatra are being systematically damaged by logging and mining. In the Philippines it is estimated that 90% of mature forest cover have been lost since the 1960s as a result of legal and illegal logging (Hammer and Shetty, 1995). Agriculture and the expansion of farm-land to grow more crops for growing populations as well as shifting cultivation have also contributed to a decline in forest cover and decreasing soil fertility. In South Asia, the FAO, UNDP and UNEP (1994) estimate that 25% of the cropped area is affected by water erosion. India, Sri Lanka and Nepal are the most adversely affected as well as parts of Punjab. When other forms of degradation are included (wind, soil fertility decline, and lowering of groundwater table) as much as 43% of agricultural land suffered one or more forms of degradation (FAO, UNDP and UNEP study quoted in Third World Network's website www.twnside.org.sg/title/land-ch./htm).

Collection of wood for fuel mainly by poor farmers, has also contributed to forest degradation and depletion. Existing total forest cover of 725 million hectares is now only 21% of total land area. Deforestation and degradation threaten biodiversity and ecosystem stability, besides contributing to soil erosion, flooding, and the long-term availability of lumber and other forest products. Lack of property rights have resulted in overgrazing and the growth of shifting cultivation. Soil erosion and degradation are becoming increasingly costly. In the Philippines, the toll of soil erosion on grasslands alone is estimated at over $100 million, whereas losses in Indonesia are at least four times higher (Hammer and Shetty, 1995). Desertification has also contributed to deterioration in the quality of land. UNEP estimates that there are nearly 2 billion hectares of dry lands in Asia and more than half has been affected by desertification. The worst affected is Central Asia although there are significant areas of desertification in South and Northeast Asia (UNCCD, 1998).

Replanting of frontier forests with annual crops or plantations has somewhat offset the problem of deforestation. For example, in 1990, according to the FAO, while there was a decline in natural forest area of 3.9 million hectares, 2.1 million hectares of these areas were planted with agricultural crops (Cropper and Griffiths, 1994). This could account for the weak relationship between population growth and deforestation in the Asian region in the early 1990s.

There have also been substantial losses in soil fertility, erosion and other resource depletion (see FAO *et al.*, 1994). Philippines estimates an annual rate of natural resource depletion equivalent to 4% of the gross domestic product (Cruz and Repetto, 1992). For Indonesia, the inclusion of the loss of timber, oil and soil resources had the net effect of reducing gross domestic produce by about 20%. The annual depletion of soil fertility was calculated as 4% of the value of crop production, or as large as the annual increase in production (Repetto *et al.*, 1989; Magrath and Arens, 1989).

Solid and hazardous waste

As incomes rise and consumers rely more on markets and processed products the amount of rubbish and waste products generated also increase. This occurs as consumers shift from home grown food and locally produced consumer products including clothing and footwear to marketed goods. Furthermore, rising incomes result in a faster turnover of clothing and consumer durables as fashions shift and technological innovations accelerate. Beede and Bloom (1995) estimate that on average 1% rise in income results in about a third of a percent rise in municipal solid waste. Because Asia's incomes are rising so fast and cities (where most of the waste is accumulated) are growing faster than the rest of the region, there has been a commensurate rapid increase in solid waste and collections have not kept up. ADB (1997) estimates that China and India account for 100 million tons of waste each year. Nearly three-quarters of Asia's waste products are dumped into landfills, deposited in vacant lots, or pumped into the ocean. The remaining quarter is treated chemically or incinerated, but often without adequate safeguards (ADB, 1997). The result is an increase in pollution of the land and waterways in and around urban areas as well as further degradation of coastal estuaries and the ocean. Waste collection and management are costly. Municipal governments in Asia spend between 50% and 70% of their budgets on waste management, yet collections remain low. Between 50% and 70% of residents receive service (ADB, 1997; UNEP, 1993). Untreated waste contributes to health problems including rodent and other insect infestations that feed on the waste. In India, an outbreak of the bubonic plague was linked to inadequate waste disposal (UNEP, 1999, Chapter 2).

Hazardous waste is another component of waste disposal. Unlike ordinary waste, hazardous waste contains potentially toxic chemicals including mercury, cadmium and lead. Bacterial and viral infections can also be spread by discarded needles and other medical waste as well as infected animals and insects (for example, malaria and plague). Many of these toxic elements are contained in electronic waste (computers, televisions, cell phones, and electronic toys) which is increasing rapidly throughout Asia. UNEP (2005) estimates that 4 million personal computers are discarded each year in China alone and India generated electronic waste worth $1.5 billion in 2003. Electronic waste is often disposed in landfills, where pollutants can leach into soils and groundwater or burned in incinerators forming dangerous compounds. So far efforts to systematically deal with electronic waste are lagging because it has only recently been recognized as an environmental problem.

Energy

Asia's energy use has been increasing rapidly for the past two decades. China has become the second largest energy user in the world and overall energy in China and India is increasing at an unprecedented rate. For example, Figure 6.1 shows that China's oil consumption registered a fourfold increase in the 20 years from 1986 to 2006 or about 7% increase per year. Total energy consumption

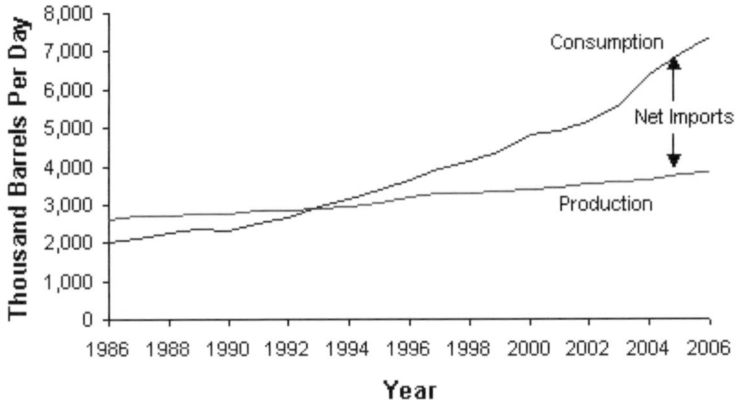

Figure 6.1 China's Oil Production Consumption, 1986–2006 (thousand barrels per day)

Source: US Energy Information Administration (2005).
Note: 2006 data is for January–August only.

increased in China from 4.16 quadrillion BTUs (quads) to 12.8 quads between 1980 and 2001 (over threefold increase), while India experienced a 2.5-fold increase. On a per capita basis these figures are still small — India ranks near the bottom of the list and China is not much higher. However, these rankings are bound to change as income growth and the pace of industrialization and development continue at a rapid pace. Virtually all of the energy needs of these two giant economies and indeed the rest of Asia are based on the burning of non-renewable fossil fuels, primarily oil and coal (nearly 80% of China's fuel consumption is supplied by oil and coal, while India is at 60% of which 56% comes from coal in China and 21% in India) with biomass making up the difference (US Energy Information Administration, 2005). Energy demand in the rest of Asia is also growing rapidly. As a result the Asian region as a whole now produces more carbon emissions than any other region. The environmental impact of this rapidly growing fossil fuel consumption has several components, each of which is taken up separately in other sections of this report. Historically, there are particularly strong impacts on air quality and global warming which will be discussed in this section and in the section "The Way Forward."

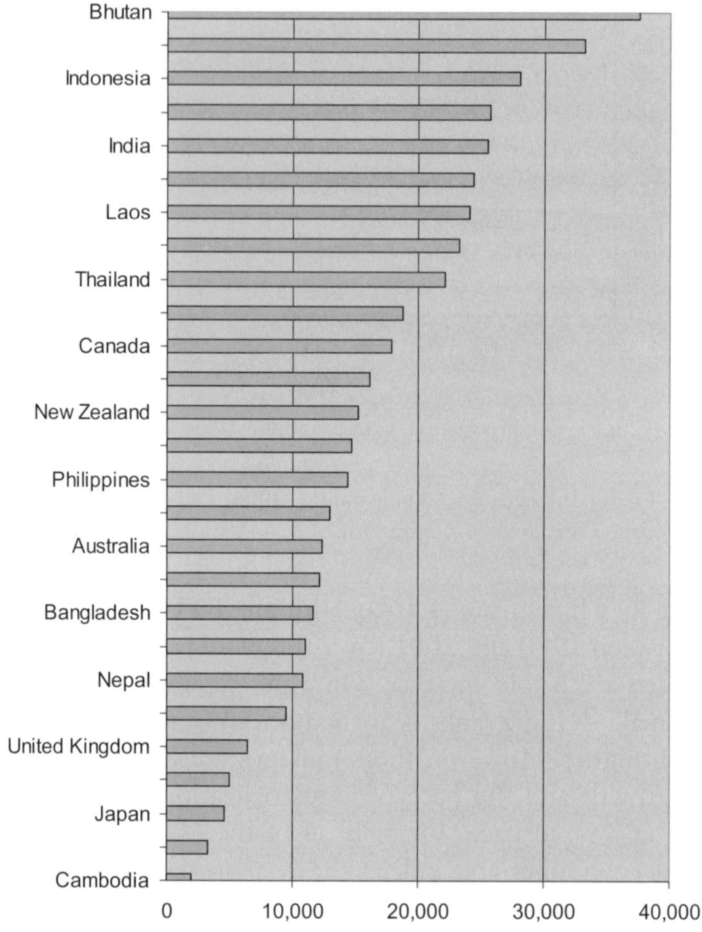

Figure 6.2 World Energy Intensity (Total Primary Energy Consumption Per Dollar of Gross Domestic Product, 2003 Using Market Exchange Rates)

Source: US Energy Information Administration (2003).

Even though many of the poorest countries in Asia use little commercial energy, they are still quite energy inefficient (see Figure 6.2). For example, Indonesia and Bhutan are much more wasteful than Japan and the United Kingdom, requiring more than 30,000 BTUs per dollar of GDP compared with 5000 BTUs for the latter two economies.

Global warming and ozone layer depletion

Global warming is caused indirectly by air pollution, particularly the emission of carbon dioxide into the atmosphere. This is distinguishable from the other components of air pollution not only because of its global aspects but also because its impact is still controversial and occurs over a longer time horizon.

Scientists generally agree that global warming is occurring and many believe that this upward trend in global temperature is accelerating. It is, however, hard to distinguish the relative contributions burning of fossil fuels has over the last couple of hundred years from the natural climate cycle of the earth. Scientists are able to look back and generalize what the climate of the earth was like as far as 2 billion years ago. Figure 6.3 shows that we are now in a period of relatively cold weather compared with the long-term climate cycle of the earth. Furthermore, we are on the upward segment of the curve where temperatures begin to warm. This process takes many thousands of years and is an inevitable part of the dynamics of the earth's climate cycle. We know little about what causes these cycles which are irregular in length. It is also notable that for most of the earth's history it has been much hotter then, than it is now. Furthermore, at the end of the Permian era it is estimated that 99% of all life perished. So the periodic loss of species has been a natural occurrence in the earth's history. This is not to say that we should be complacent in protecting and saving as many of the existing species as possible (see Figure 6.3 for details of historical climate change).

Given this perspective it seems inevitable that average temperatures will get warmer over the next few decades. It may be possible to slow that rate of increase, although the record of past climate changes shows that once the temperature starts to climb, the rate of change will accelerate. Furthermore, there is a systematic relationship between global temperature and the build up of CO_2 in the atmosphere. And the build up of CO_2 in the last 200 years far outstrips the amounts found from exploring core ice samples from the historical record (Gore, 2006).

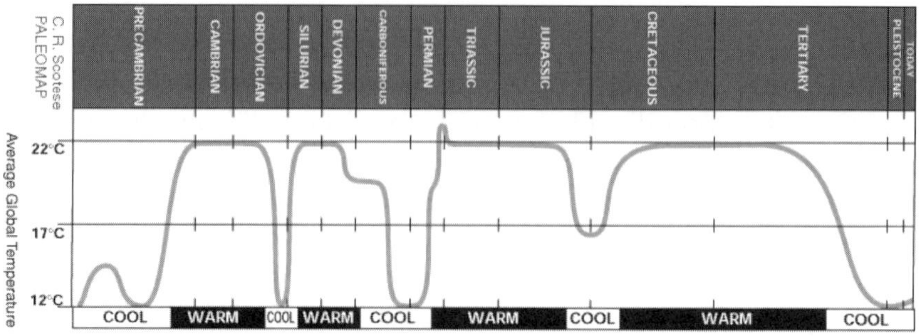

Figure 6.3 Historical Global Temperatures

Source: www.scotese.com/climate.htm.

One troubling aspect of global warming is the possible strong nonlinearities that may arise as the climate changes, further accelerating the tendency for the climate to get warmer. For example, research in Antarctica suggests that changes in the soil's ability to hold carbon may be changing as global warming accelerates. On a global scale the amount of carbon dioxide released by soil invertebrates and microbes is more than 10 times the amount of carbon emissions from fossil fuels. As the other part of the carbon cycle, the oceans and the forests absorb carbon that would otherwise be emitted into the atmosphere. Yet soil holds more carbon than the trees and atmosphere combined. The research in Antarctica and elsewhere suggests that heightened levels of carbon dioxide in the atmosphere hinder the ability of soil to store carbon as a result of increased respiration of microbes in the soil. And this means more carbon dioxide emissions (Heath *et al.*, 2005). Scientists are still unclear about the ultimate impact on climate and whether the results from Antarctica will be confirmed by work with other soils in warmer climates.

How has Asia contributed to global warming? The evidence compiled by international agencies and the scientific community shows that the share of carbon dioxide emissions by developing countries has increased dramatically in the past few years (Table 6.3) as a result of rapid economic growth and reliance on coal for generating electricity and heat. China and India's contributions are

Table 6.3 Carbon Emissions in China, India, Europe, Japan and the United States

Country or region	Carbon emissions (in million tons)	Carbon emissions per capita (in tons)	Carbon emissions per unit of GDP in PPP measures (in tons per million $)	Increase in carbon emissions, 1990–2004 (in percent)
China	1021	0.8	158	67
India	301	0.3	99	88
Europe	955	2.5	94	6
Japan	338	2.7	95	23
United States	1616	5.5	147	19

Source: Worldwatch Institute (2006).

particularly notable. Growth in carbon dioxide emissions has risen sharply for the past two decades but not as fast as income. Between 1990 and 2004 carbon emissions from China and India rose by 67% and 88% respectively (Worldwatch Institute, 2006). At the same time, China's income increased nearly four-fold and India's nearly tripled.

The large hole in the ozone layer that surrounds the earth about 12 to 15 kilometers above the surface was discovered by scientists in the 1980s. This ozone layer, which is sensitive to the release of chlorofluorocarbons (CFCs) commonly used in refrigerators, plastics, aerosol sprays and other applications, is critical in preventing harmful ultraviolet radiation from reaching the earth's surface. A thinning layer is responsible for a greater incidence of skin and eye cancers (including increased risk of melanoma), and cataracts as well as weakening of the immune system. Greater solar radiation is also believed to cause a reduction in crop yields like wheat (5%), potatoes (21%) and soybeans (20%) (UNEP/GEMS Environment Library, No. 7, 1992). Several protocols which mandate a phasing out of ozone-layer damaging products over the next few years have been agreed upon by many countries. As a result, significant progress has been made in reducing the hole in the ozone layer.

Biodiversity[1]

The loss in biodiversity as a result of a larger environmental footprint over the past few decades has been substantial. A recent article by Root *et al.* (2003) in *Nature* estimates that ¼ of all species could be lost over the next 40 years. In Asia these trends are all the more disturbing because the region is one of the most biologically, geographically and ecologically diverse in the world.

The region includes the Himalayas, the second largest rain forest complex, and more than half of the world's coral reefs. The Southeast Asian sub-region is noted as the center of diversity of wild and domestic cereals and fruit species (ASEAN, 1997). Three of the twelve countries designated as being highly ecologically diverse — China, Indonesia and Malaysia are found in the region (McNeely *et al.*, 1990).

> The resources of the region have been increasingly exploited both for international trade and to sustain the growing population. The direct harvesting and export of natural products, particularly timber and fish, the expansion of agriculture into primary forests, wetlands and grasslands, and the replacement of traditional native crops with high-yielding exotic species have had severe impacts on the region's biodiversity. In addition, urbanization, industrialization, pollution, mining, tourism, introduced species, hunting, illegal trade in endangered species and the lack of proper management practices have taken their toll. In the past decade, demand on biological resources increased sharply due to rapid economic and population growth (UNEP, 1999, p. 1).

Increasing habitat fragmentation in Southeast Asia has depleted the wide variety of forest products that used to be the main source of food, medicine and income for indigenous people (MRC/UNEP, 1997). Destruction is less severe in the Mekong basin, particularly along the inaccessible national borders.

A major concern in South Asia, particularly in the Indian subcontinent, has been the loss of biodiversity brought on by long-term

---

[1]The major source for this section is http://www.unep.org/GEO2000/english/0066.htm

pressures on grasslands compounded with rapid growth in human and livestock populations (WCMC, 1995).

Modern agriculture has also reduced genetic diversity. In Indonesia, for example, some 1500 varieties of rice disappeared during 1975–1990 (WRI, UNEP and UNDP, 1992) and similar trends have been observed throughout the region. By 2005, India is expected to produce 75% of its rice from just 10 varieties compared with the 30,000 varieties traditionally cultivated (Ryan, 1992).

Hunting, poaching and illegal trade in endangered species have a widespread impact on biodiversity in many countries. For example, poaching and the illegal harvesting and trade of medicinal plants and animals have increased both in Mongolia and the Republic of Korea, where items are exported to the lucrative black markets of Pacific neighbors (JEC, 1997). In Pakistan, falcons are smuggled to the Middle East, lizards and snakes are killed for their skins, and crocodile hunting is still a popular sport and recreational activity (UNEP, 1999).

While the full extent of habitat change and species loss in the region has not been determined, it is estimated that about two-thirds of Asian wildlife habitats have already been destroyed and 70% of the major vegetation types in South and Southeast Asia have been lost (Braatz *et al.*, 1992; MacKinnon, 1994). Dry and moist forests have suffered 73% and 69% losses respectively, and wetlands, marsh and mangroves have been reduced in extent by 55% (Braatz *et. al.*, 1992). Overall habitat losses have been most acute in the Indian sub-continent, China, Vietnam and Thailand (Braatz *et al.*, 1992).

Remaining tropical forests in the Philippines, peninsular Malaysia, northwestern Borneo, eastern Himalayas, western Ghats in India, southeastern Sri Lanka, and New Caledonia are further threatened by incursion of human settlements and logging activities (UNESCAP/ADB, 1995).

Impact of Increasing Environmental Damage and Degradation

The health, environmental and economic impacts of various kinds of pollution have been estimated by a number of ecologists, environmentalists and economists. We review these estimates in this section. We make no attempt to verify or check the reliability of these estimates nor do we investigate the methodology of damage estimation. In many cases different researchers have arrived at similar estimates of the environmental damage. Differences in the estimated cost of environmental impact result from different assumptions about the degree and extent of damage as well as the time frame and the scope of the damage analysis. Longer time frames and wider scope of damage will obviously result in higher damage estimates. If possible we will try to reconcile different estimates where there are wide differences in the cost estimates of damage. We pool together discussion of environmental, health and economic impacts under the pollution headings introduced in the section "Recent Environmental Developments in Asia".

On an aggregate basis, it is clear that environmental factors are responsible for a significant amount of economic costs and health damages. Between a fourth and a third of the global burden of disease is attributable to environmental factors (Smith *et al.*, 1999) and this figure is probably higher in Asia and other developing countries. Murray and Lopez (1996) estimate that environment-related premature deaths and illnesses account for 18% of the total disease burden in the developing world. Globally, 7% of all deaths are due to inadequate hygienic sanitation and clean water (UNDP, UNEP, World Bank and WRI, 1998). As much as 5% of these deaths were due to air pollution.

Heath impact of air pollution/environmental damage as a result of lower air quality

Polluted air has a direct impact on people's health and also on plants, animals and buildings. The World Health Organization figures (WHO, 2005) show that every year almost 500,000 people in Asia die prematurely because of bad air. The bulk of air pollution is concentrated in large cities, although air pollution has spread throughout the entire Asian region. Pollution in Asia is steadily worsening and is indirectly responsible for the death of thousands in Beijing, Jakarta, Seoul, Bangkok and Manila, according to the World Bank and the

Stockholm Development Institute (SDI). SDI says that sulfur dioxide, ammonia and nitrogen oxides have been rising steadily over the past few decades and ground level ozone concentration has increased. Air pollution in the continent has now surpassed the combined emissions in Europe and North America. It is estimated that 4000 Filipinos die each year because of air pollution in Manila. The mortality figure is the third highest for a city in the East Asian region after Beijing and Jakarta. Bangkok and Seoul were ranked fourth and fifth. Beside deaths, an estimated 90,000 Filipinos in Metro Manila also suffer from severe chronic bronchitis, costing the government 7% of its gross domestic product in terms of health costs according to World Health Organization (WHO) which studied pollution and health studies in 126 countries. However, the air pollution death figures in other big cities are even higher. For example, over 40,000 die in India as a result of air pollution. Aside from an increased fatality rate as a result of air pollution, bad aid results in an increase in respiratory infections including colds, bronchitis and asthma. These respiratory illnesses may also contribute to the spread of tuberculosis, which is a contagious airborne disease. There are an estimated 617,000 deaths per year from TB in Asia, an incidence rate of around 2 per 1000 population or 1,370,000 cases in 2003 (WHO, 2005a).

Urban air pollution also exacts a heavy health toll. Fatalities in Bangladesh, India, Indonesia and Nepal account for about 40% of the global mortality in young children caused by pneumonia (WHO, 1992). It is estimated by the Asian Development Bank (ADB, 1997) that air pollution in South Asian cities causes nearly 100,000 premature deaths per year and over 1 billion work days of lost or reduced productivity. In the PRC's 11 largest cities, smoke and small particles from burning coal are thought to be the primary cause of more than 50,000 premature deaths and 400,000 new cases of chronic respiratory illness every year.

The use of two-stroke engines in inexpensive two- and three-wheel taxis in Asia present an added air pollution hazard. Anumita Roychowdhury, associate director of the Centre for Science and Environment (CSE) in Delhi, India estimates that two-wheelers form a staggering 75%–80% of the traffic in most Asian cities. Two-stroke engines burn an oil–gasoline mix that emit a lot more smoke, carbon monoxide, hydrocarbons, and particulate matter than cars and gasoline-fueled motorcycles.

In Calcutta and Delhi, a study of lung function in urban and rural areas showed a significant difference as a result of poor air. There was impaired lung function in 46% of Delhi adults and 56% of Calcutta adults, but only 21% of those from the Sunderban islands where air pollution is low. A study of traffic intersections in Bangkok, Thailand, found that two-wheelers contributed up to 47% of particulates. When the numbers of two-wheelers were cut back, air quality improved. Similarly, when two-stroke baby taxis were phased out of Dhaka, Bangladesh, in 2002, particulate concentrations dropped by 40%, and carbon monoxide and hydrocarbons concentrations fell significantly.

Air pollution also has detrimental impacts on forests as a result of elevated sulfur content in the air and from deposits of acid rain. Computer models suggest that the impact of sulfur buildup in the atmosphere is having a long-term impact on forests throughout Asia, particularly in China, where the use of coal for fuel is high (Stockholm Environmental Institute, 2003).

Increased traffic congestion from more private cars and motorcycles is an indirect cause of air pollution. A 10% reduction in the number of peak-hour trips in Bangkok would yield an estimated $400 million in time saving for travelers (Hammer and Shetty, 1995). These savings can be replicated many times for other large congested cities in Asia like Jakarta, Manila and Mumbai.

How much does air pollution cost Asian economies? And is it worth cleaning up now? It is difficult to make estimates because of the various economic, environmental and health impacts. The UNDP estimated that the haze of 1997 from forest fires in Indonesia cost Southeast Asian economies $1.4 billion, most of it in short-term health costs as more than 40,000 people were hospitalized with respiratory problems. In general, other estimates suggest that the overall costs of air pollution to the economy, people and forests are quite high. For tuberculosis alone, WHO estimates the minimum cost of treatment at $10 per patient, not including administrative and medical personnel costs (WHO, 2005a). A study by the Singapore Institute of Policy Studies and the World Bank

(Douglass and Ooi, 2000) suggests that air pollution cost Asian cities about 5%–10% of their GDP while the cost of cleanup is only estimated to be 2%–3% of GDP. Some simple remedies like banning all two cycle engines, moving to clean diesel fuel, moving to lead-free gasoline, and installing scrubbers on major industrial polluters would probably reduce air pollution significantly at a small cost.

Health impact and cost of fresh water and ocean pollution

The health impact of poor water and sanitation are extensive. Pierce (2001) estimates that arsenic poisoning may be killing as many as 20,000 people a year in Bangladesh. Worldwide, two billion people — one in every three people on the planet — are at risk from malaria which is responsible for 1 to 2 million deaths annually. There are over 2 million deaths a year from intestinal problems and one out of ten people living in developing countries have intestinal worms. Worldwide, the number of cases of diarrhea is nearly equivalent to one episode per person — over 4 billion cases per year — and likely to be much higher in the developing countries. There are also nearly 200,000 cases of cholera, 100 million cases of dengue fever, and close to 20 million cases of typhoid fever reported (Millennium Ecosystem Assessment, 2005). Loss estimates do not take into account the number of workdays missed from sickness due to water- and sewage-related illnesses.

WHO has introduced the term Disability-Adjusted Life Year (DALY) to measure the economic loss of illness caused by environmental factors associated with the incidence of various diseases. The DALY emerged as a measure of the burden of disease and reflects the total amount of healthy life lost to all causes, whether from premature death or disability over a period of time. The principal suggestion of the 1993 World Development Report (WDR) "Investing in Health" (World Bank, 1993) is to define a package of essential health services and to make extensive use of the concept of the Disability-Adjusted Life Year (DALY).

Poor sanitation, untreated urban sewage, and failure to treat wastewater before it is discharged into river systems are among the leading causes of marine pollution. Overall, some 42% of DALYs are due to water pollution and inadequate sanitation, making it the most important among the major environment-related health risks due to air, water and soil pollution (ADB, 2001, 2005). Diarrhea cost over 62 million DALYs per year at 4 billion cases, while malaria accounts for 46.5 million DALYs worldwide. Together these two waterborne diseases account for over 100 million years of life lost. While the cost of clean-up will vary from country to country, most estimates suggest that clean-up is cost-effective. Realize that with no action the number of DALYs will increase every year as the level of water pollution remains unchecked and population density increases, and that the burden of ill health falls with greater impact on the poor. Life expectancy increases as income rises both across countries and within a particular society. So, environmental clean-up has a secondary benefit of a net positive impact on the quality of life of the poorer members of society.

The costs of overfishing are reflected by the increased costs of fish capture. Larger boats are needed to trawl further away from shore. Some of these costs have been offset by increased efficiency in fishing. Larger boats are able to catch more fish, locate schools of fish with sophistical sonar and other navigational equipment, and process fish on board, thereby reducing the loss from spoilage. However, environmental risks multiply as fish stocks decline and some species near extinction. These risks are exacerbated by the fact that fish populations are subject to wide fluctuations from year to year due to climate changes and other natural factors that are not completely understood. The overall cost of red tide is difficult to estimate, although an occurrence in Manila Bay in 1990 had an estimated loss of $2.4 million to Filipino fishers (UNESCAP/ADB, 1995). Red tides also occur in Hong Kong and China.

Costs of damage to land and forest resources

It is difficult to ascertain the economic cost of changes in land use whereby natural forests are being replaced by urban development, agricultural planting, or plantation of new timber. By not considering the loss of ecological diversity, the negative impact on global ecology and global warming, economists can say that the change in use is justified, i.e., the land is now being used more productively than when it was in its natural state. Urban land is valuable for commercial and residential purposes. Agricultural land is used to raise crops and feed livestock. Tree farming allows the forests to be harvested for timber for dwellings and then replanted again and again. Environmentalists insist that these arguments overlook the impact that deforestation has on the global environment — particularly global warming, and on the loss of ecological richness and diversity that results from the destruction of virgin forests. The costs of losing biodiversity are difficult to measure although some attempts are made and will be discussed below.

As far as the relationship between deforestation and the build-up of greenhouse gases goes, the evidence is clearer. The Intergovernmental Panel on Climate Change estimates that during the 1990s, land use changes resulted in a quarter of the total global build-up in greenhouse gases (Schlamadinger *et al.*, in press). By reducing deforestation, developing countries can contribute to the reduction in global emissions and qualify for entry to a market for carbon emission permits which have been suggested in the Kyoto Protocol. By maintaining forest cover, countries can also conserve existing ecosystems, prevent forest fragmentation and protect watershed. Such a program can also provide income for developing countries by trading emission permits and using the income generated to promote sustainable forest management and reduce illegal logging.

The costs of soil erosion, desertification, water logging, and other kinds of soil degradation are difficult to tally. Nevertheless, FAO made a comprehensive assessment (FAO *et al.*, 1994) of the soil erosion and soil quality in Asia. In South Asia, FAO estimates suggest that if all land in the South Asia region were not degraded, i.e., in conditions that existed before recent population pressure and fragmentation of land, an additional 36 metric tons of cereal would be obtained, equivalent to an increase of nearly 10% over current cereal yields. At current prices, the value of these losses is calculated at $5.4 billion per year.

Water erosion, which accounted for more than half the calculated loss at $5.4 billion, was the primary cause of soil degradation damage.

For the Asian region as a whole, the estimated total losses were equivalent to $10 billion. Water logging and soil salinity were next at $2 billion. Wind and fertility decline made up the balance. The estimates of degradation, if not dealt with in a timely manner, will accumulate over time. Furthermore, these estimates do not take into account forest and rangeland degradation, soil destruction through mining and quarrying, and urban settlements.

Cost of ecosystem and biodiversity loss

Losses in biodiversity and damage to the ecosystem by continued extinction of some species because of air pollution, overfishing, changes in the landscape as a result of deforestation and urbanization, pollution of oceans, and spread of "civilization" to remote habitats, are difficult to quantify. The cost of deforestation and conversion of grasslands to crops have been discussed in the previous sections.

The cost of losses in ecological diversity is more difficult to measure. Ecologists point to the fact that several kinds of medicines are derived from plants and their loss will create a gap in our ability to treat diseases. Many species are introduced to other locations by means of modern transport, either by air, ship, rails, trucks or cars. The intermingling of species causes changes in natural ecology which is not well understood and could result in further loss of biodiversity.

Economists have developed the concept of "option demand" to reflect the utility that arises from simply knowing that natural treasures such as the Grand Canyon are protected. Similarly, the concept of "option demand" can be applied to the continued thriving of a rich ecological environment. However, calculating the amount that society is willing to spend to maintain these natural treasures and the current level of biodiversity is difficult to determine.

Ecologists tell us that the rate of extinction is accelerating. A report from *Environment News Service* (1999) says that:

> The current extinction rate is now approaching 1000 times the background rate and may climb to 10,000 times the background rate during the next century, if present trends continue. At this rate, one-third to two-thirds of

> all species of plants, animals, and other organisms would be lost during the second half of the next century, a loss that would easily equal those of past extinctions.

Given this kind of information, most economists have given up trying to measure the costs of extinction, and have looked more into the cost of maintaining biodiversity, or reducing the rate of loss of species instead. Limiting deforestation, overfishing, and the destruction of coral reefs seem to be the best way to maintain ecological diversity.

Ecologists estimate that deforestation alone has done a significant amount of damage to the ecosystem. A 20-year study has shown that about 12.5% of the world's plant species have become critically rare due to deforestation and the introduction of non-native species. A study suggests that the Amazon damage is worse than previously thought, due to previously undetected types of selective logging and deforestation. The same may be true of damage estimates for parts of Papua New Guinea, Indonesia and East Malaysia.

A 1999 report from the World Commission on Forests and Sustainable Development suggests that major changes in global forest management strategies would be needed to avoid further devastation.

New species of animals and plants are still being discovered in small pockets of land, often within the virgin forests of a developing country. In Papua New Guinea, 44 new species of animals were discovered recently in its forests. Unrestricted logging could adversely affect these animals' habitats, making deforestation an ecological problem.

Evolutionary and scientific knowledge or benefits from a host of vanishing species are lost along with rainforests destroyed by deforestation. India, with 81,000 animal species (some 8% of the world's biodiversity), is under various pressures from globalization and deforestation as are many other biologically diverse Asian regions like Indonesia (WRI, 2005e).

There are various estimates of economic gains and losses from the destruction of coral reefs from coral mining, cyanide poisoning of fish, and dynamite fishing. Although there is short-term benefit to the economy from destroying reef ecosystems — gains of $121,000 from coral mining and $15,000 from dynamite fishing — the estimates of gain in keeping these resources intact are even higher. In the Philippines, it is estimated that an investment of $100,000 to protect coral reef systems would have a net effect of $1.4 million in revenues from increased tourism and sport fishing (ADB, 2005).

Cost of global warming and ozone layer depletion

How much does the build up of greenhouse gases in the atmosphere contribute to global warming and how much does it cost? These are pressing questions and scientists continue to monitor and model the changes in the environment that result from the continued burning of fossil fuels. There is consensus that the earth's climate is getting warmer and that the pattern of global warming is beginning to accelerate. Much of this warming trend is attributed to the burning of fossil fuels and the long term climate cycle of the earth. Some studies also suggest that we are now in an upward cycle of warming that will continue unabated for many thousands of years, regardless of what we do or do not do about it (as shown earlier in Figure 6.3).

So far the cost of global warming seems to be modest since it has only just begun to have a lasting impact on the environment (Nordhaus and Boyer, 2000). The 2005 Tsunami that caused so much devastation to Asia may have been exacerbated by the slight rise in the sea levels in the past couple of hundred years, although most of the destruction was a direct result of the giant waves, the close proximity of human settlements to the sea, and the loss of mangrove forests that would have afforded some protection. Low-lying areas throughout Asia are heavily populated and these people depend, to a large extent, upon the ocean for their livelihood. As the climate warms, more people living in these areas could experience disruption in their way of life.

There is some disagreement about the implications of rising temperatures for sea level rise over the next 100 years. The intergovernmental panel on climate change has been issuing reports since 1999. In its 2001 assessment it estimated that in the next 100 years sea levels would rise globally by at least a few inches and perhaps as much as

3ft, or a little less than a 1m, a catastrophe for low-lying coastal areas and island nations. In its 2007 report (Intergovernmental Panel on Climate Change, 2007) these estimates have been scaled back from the 2001 forecasts. In the worst case scenarios, the interval estimates for sea level rise by 2090–2099 relative to 1980–1999 is an interval from 0.26 m (0.85 ft) to 0.59 ms (1.9 ft). Scientists outside the working group are now saying that these estimates are too low and that the 2001 assessment may also be too low. Some scientists suggest a sea rise of 10 ft (3 ms) on average worldwide. However, such a dramatic increase is deemed to be remote by most of the scientific community and would require a severe break off of the ice cap in Greenland or in the Antarctic and perhaps the melting of the Russian, which would release large amounts of carbon dioxide (Zimov *et al.*, 2006) which would contribute to the sea rise.

In Asia this much of an increase in the sea level would inundate large regions of heavily populated regions in China, Bangladesh and India. Even a 3 ft rise would flood nearly 20% of Bangladesh. Gore (2006) estimates there would be at least 100 million refugees and millions of others would lose their homes and livelihoods from this large increase in the sea level. The economic costs of such a calamity are difficult to imagine. There would also be widespread devastation in the rest of the world. The Netherlands and other low-lying areas in Europe would be covered and Manhattan would be covered up to Harlem. It is important to consider the consequences of such a global warming pattern on the future of mankind and the planet.

How long it will take for global warming to have such disastrous consequences is difficult to predict. However, signs of break up of ice flows in the Arctic, cracks in glaciers in the Antarctic and Greenland are being reported with increased frequency. We also know that the earth has experienced many periods of transition from "icehouse" to "greenhouse" conditions in the past. Glaciers have been much more extensive than they are now in some periods and they have completely melted in others. How the transition takes place, and the speed of the transition is not well understood. It is possible that man's contribution to global warming is accelerating the transition.

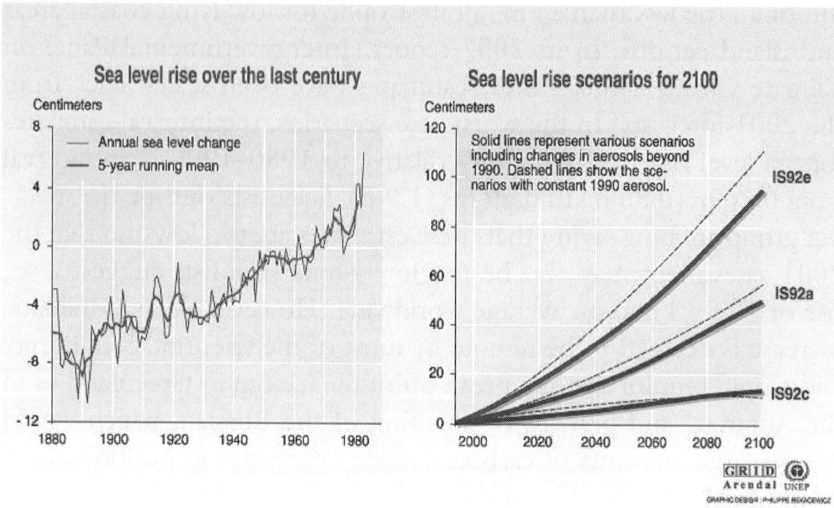

Source: Climate change 1995, The science of climate change, contribution of working group 1 to the second assessment report of the intergovernmental panel on climate change, UNEP and WMO, Cambridge university press, 1996; Sea level rise over the last century, adapted from Gormitz and Lebedeff, 1987.

Figure 6.4 Sea Level Rise Due to Global Warming

The more conservative estimates of the IPCC suggest that the sea level has risen between 10 and 25 cm over the last 100 years and can be expected to accelerate somewhat to rise a further 26 to 59 cm over the next 100 years (Figure 6.4). This scenario would cause less damage although it would cover most of the small islands in the Pacific and would bring widespread damage to low-lying areas in China, Bangladesh, India and Thailand. Other low-lying areas around the globe would also be adversely affected.

Furthermore, climates are likely to become volatile with some regions experiencing extreme weather conditions. The increased number of hurricanes in a number of regions around the world has been noted including Katrina, the hurricane that did so much damage to New Orleans in 2005.

Extreme weather patterns due to global warming is also responsible for exacerbating the loss of biodiversity and increasing the risk of extinction of some species that are already threatened by reduced numbers, habitats and limited climate range.

General estimates of overall loss as a result of environmental damage

In Pakistan, the health impacts of air and water pollution and productivity losses from deforestation and soil erosion were estimated at \$1.71 billion, or 3.3% of GNP, in the early 1990s. In PRC, for example, Smil (1996) estimated that productivity losses caused by soil erosion, deforestation and land degradation; water shortage; and destruction of wetlands amounted to between \$13.9 billion and \$26.6 billion, equivalent to 3.8% to 7.3% of its 1990 GNP. In Jakarta, Indonesia, studies by Ostro (1994) and DeShazo (1996) estimated the annual cost of environmental damage reached \$2.16 billion (equivalent to 2% of GNP), primarily due to the effects of particulates and lead which exceeded acceptable levels of WHO standards. A study by O'Connor (1994) on the effects of the same pollutants in Thailand revealed an annual loss of \$1.6 billion, representing 2% of GNP. The result of a 1993 World Bank Study in the Philippines also showed that health and productivity losses from water and air pollution in Manila during the early 1990s amounted to between \$335 million and \$410 million, or 0.8% to 1% of GNP (ADB, 2005).

Causes of Increasing Environmental Damage and Degradation

Economic activity is all about the production, marketing, sale and distribution of goods and services. This process takes places in a variety of markets. There are markets for labor, services, and perishable and durable goods. In each of these markets, there are buyers and sellers that generally have property rights over what they buy or sell. Some have a marketable skill to sell; others have manual labor. Manufacturers hire labor to make products that are sold under their trademark or through their distribution network to other buyers or retailers. Legal rights govern these arrangements and the legal system adjudicates disputes over these rights.

How is the environment protected in this system? What are its rights? Who owns the environment and how are these owners paid?

Answers to these questions tell us a bit about how the environment differs from other parts of the economic system and why there has been little concern for the environment in most of our human history. The environment is not a player in the market. No one company or group of companies is going to get rich protecting the environment. Those who profit from the environment are more likely to exploit the environment and the resources contained within it. Market mechanisms are designed to integrate the environment into the flow of market decision, and the environment's products (land fertility, natural resources, air and water) are used to increase the amount and variety of goods and services available in our markets.

What characteristics of the environment and environmental goods created this disconnect between conventional economic theory and resource allocation resulting in so much environmental damage in such a short span of history? Fundamentally, it is the failure of the market system to incorporate aspects of resource use and the concept of sustainable development into the equation. There are four related aspects of this failure — externalities, common property resources, property rights, and the role of government. We will consider each in turn.

Externalities

Externalities exist when markets do not reflect the full costs (and potential benefits) of production and consumption. Of particular interest are negative externalities created when production processes which pollute the environment are not reflected in the costs to consumers. As a result marginal social and private costs diverge resulting in market prices that are too low and levels of output too high (Figure 6.5). An example of such a negative externality is the coastal fishermen who use dynamite and cyanide to stun or poison small fish for capture and later sell to retailers and aquarium owners. This practice damages the coral reefs and kills a lot of other marine animals in the process. None of these negative externalities are necessarily accounted for in the prices paid by aquarium owners for the beautiful and colorful fish that swim around in their aquariums. As discussed in the previous sections, air, water and land pollution qualify as negative externalities and can

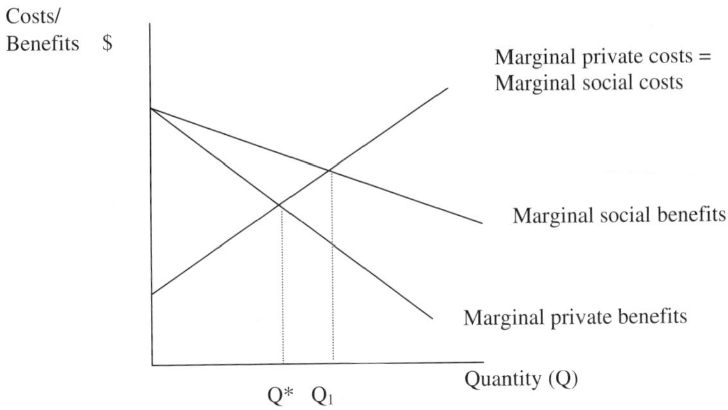

Figure 6.5 Private and Social Costs and Benefits

Source: Dowling and Valenzuela (2004).

be analyzed within a simple methodological framework. Sugar mills dump waste into rivers increasing its toxicity, harming aquatic life, degrading water for downstream users, and spoiling the recreational value of the river; loggers fail to replant causing soil erosion and degradation, reducing biodiversity, increasing silt in streams, and adding to the risk of landslides; and power plants burn high sulfur coal, resulting in high concentrations of particulates that exacerbate respiratory disease and contribute to global warming and acid rain that damage crops and buildings.

Some writers use the term "regular externality" to describe the general sorts of negative externalities like river and stream pollution, poor logging practices, and local overfishing. Theses negative effects are limited in geographic scope and can be dealt with by local authorities. However, other environmental damages, such as global warming, acid rain, and depletion of aquatic fisheries can have ramifications that transcend local and even national boundaries. These "super externalities" need to be dealt with at the international level through treaties that address pollution from a global standpoint. Acid rain from China migrates to Japan and Korea, overfishing increases the risk of extinction of some species, and global warming from burning of fossil fuels threatens the entire global ecosystem.

Developing legislation that deals with super externalities is difficult because individual nations may not view it in their best interest to support legislation and enforcement that harms their economic prospects. International treaties like the Kyoto Protocol and the Law of the Sea have only had varying degrees of success which will be discussed in more depth in the section about the environmental future of Asia.

Another kind of externality, which we call "competitive externality" results when countries compete to attract foreign investment and capture new overseas markets by offering a better climate for businesses without regard to possible damage to their environment. Introduction of environmental regulations could jeopardize these objectives. Countries offering a more relaxed regulatory environment will be able to offer a better deal for potential investors who will be able to manufacture at lower prices if environmental impacts are ignored. This kind of environmental competition is referred to as the "race to the bottom" as competition favors those that ignore or bypass any regulation that will increase costs or have an adverse impact on competitive position. This means jobs for companies and votes for politicians[2].

A final note on externalities relates to positive externalities. Neighborhood improvement schemes, like gentrification of urban areas, create a positive externality for the community at large. Home improvement by a few can result in an increase in property values for the many. In the same way, better schools attract more home buyers interested in better education for their children. Plant a tree programs and other social and nature-based activities also contribute to a sense of community and may also result in lower crime rates.

[2]With reference to government regulation, a race to the bottom is said to occur when competition between nations to attract foreign capital or other kind of economic advantage leads to the progressive dismantling of regulatory standards. It is also used in a general sense when evolutionary trends gravitate to the lowest common denominator such as labor market rules, safety standards, terms of employment, and minimum wages.

Open access and common property

A "pure public good" or an open access resource is defined as a resource where (i) the consumption of a good or service by one person does not reduce the amount available to anyone else, and (ii) no member of the public is excluded from the consumption of the good or service. Using this strict definition, there are not too many pure public goods or services. Clean air, clean streets, parks, highways, lighthouses, a beautiful sunset, and national defense are examples of pure public goods. A pure public good is not "owned" by anyone so there is a very strong incentive to free ride when possible. Those who take advantage of a public good overuse the resource relative to its optimal use and with respect to other goods. Users also tend to consume more in the short run than is optimal.

There are no incentives for members of the general public to pay for public goods and services like clean air, defense or a lighthouse, unless forced to do so by the government. The private sector will not supply them for the same reason. The services of the oceans appear to fall in the category of a pure public good. However, technological advances in tracking fish and increasing fish haul reduce the amount of fish available to different fishing fleets thereby violating condition (i) of our definition of a public good.

Therefore, the use of pure public goods has to be regulated; otherwise these resources will be overused and degraded. Government has to be involved in the difficult task of policing and regulating the use of public goods. Otherwise, overfishing will continue as does air pollution and degradation of natural habitats. Even illegal logging, which is theoretically much easier to control than overfishing, has continued unabated in the Asian region as learned in a previous section.

When there is some control over who uses the resource, or when its use causes a change in the nature of good or service, then we refer to it as "common property resource". Overcrowding of beaches or parks would be an example of how overuse changes the nature of the resource and its use. The term "Tragedy of the Commons" is used to describe the overuse syndrome that arose in the United Kingdom in the early 20th century. This common property resource was degraded when too many came to use the common area and disregarded the rules and regulations on noise and littering.

Theoretically it is possible to regulate the use of common property resource as well as other resources where overcrowding is a problem, like in parks and green areas. However, political will is often lacking and a downward spiral of overcrowding and overuse results. Many beautiful beaches in Asia have been subject to this phenomenon. While the beaches are open use resources, the surrounding land has been so intensively developed that the beaches are teeming with people. In Hawaii, where beaches are open for public use, some beaches have been subject to overcrowding. Popular beaches now require visitors to pay an entry fee.

The key to regulating use of common property resource is in controlling access. Many of these resources should be subject to governmental regulations. In other cases, the regulation of use and entitlements has to be maintained by those with a common interest in the resource. In these instances, the resource in question must have well-defined physical and perhaps biological and social boundaries. The groups of users (and non-users) have to be identified and rules established to define the responsibilities of these users in maintaining resource (see Adhikari, 2001 and Stevenson, 1991 for further details).

The passage of time also plays an important role in the misallocation of resources resulting in environmental degradation. This happens when ecological systems react in a non-linear way to changes in environmental factors. When certain trace elements in a bird's diet are missing, the shells of some bird's eggs are more prone to cracking, thereby compromising the young.

> Because for some environmental problems, the time delay may spread out over decades or even centuries, over time it may be hard to determine, particularly since future citizens are not present today, to affect the decision process by casting their "market" votes (Dua and Esty, 1997, p. 63).

Property rights

In the case of environmental issues, there are many cases where property rights are poorly defined or difficult to enforce even when these rights relate to private property and should theoretically be well defined and protected. In China and Vietnam in particular, property is usually collectively owned and legal rights are not well established with respect to environmental rights and responsibilities. In many parts of rural Asia including Thailand, the Philippines, Malaysia, and other regions in Southeast and South Asia traditions of community land tenure are strong. Furthermore, forest land has been cleared, settled, and cultivated without any ownership rights being established. Sometimes this is because the land was seized illegally and other times because the legal system has been slow in responding to claims for legal title. As a result, tenants out of fear of eviction are reluctant to make improvements to the land that would increase productivity. Their stewardship of the environment is also weak (see Brandon and Ramankutty, 1993 and Lynch and Talbot, 1995 for further details). Furthermore, tenants are loath to leave the land for any length of time for fear that someone else will take over. Without a legal claim this would involve a fight to reclaim the land. Establishing land title can raise the value of the land to farmers with unsecured titles. Having title to the land, farmers are more likely to care for and maintain it. Water rights to a fixed quantity of water per unit time must be awarded by the authority that owns the water which can also be sold creating a market for water that is flexible and allocates water to its highest value use (ADB, 1997). Even when property rights are well established, it is often difficult to evaluate the cause of pollution, assess blame and measure damages. A polluter has strong incentives to fight legislation to limit his level of pollution or tax his company, while the general public is often uninformed, poorly mobilized, and unwilling to fight the polluter in court. NGOs are helping to mobilize public opinion and are willing to fight the major polluters; however their resources are also limited. Most often the government will have to bear the responsibility of framing and enforcing environmental regulations by default.

Role of government

Governments have limited resources and often have limited institutional capacity to enforce regulations or prosecute polluters that break the law. The effectiveness of governmental agencies dealing with the environment is also hampered by the complexity and scope of many environmental issues. In developing countries in Asia, the environment takes a backseat to economic growth, raising living standards and reducing poverty. All of these take resources which often created pollution. As a result, policy makers often see the consideration of environmental issues a roadblock to more rapid economic development. Resources devoted to promoting sustainable development and environmental protection have been generally lacking. Agencies monitoring and developing standards for air and water pollution in Asia are strapped for resources and lack manpower in enforcement. With respect to water resource management, China and ASEAN members lack skilled staff, suffer from shortages of equipment to monitor discharges, and have limited enforcement powers (UNESCAP/ADB, 1995; Dua and Esty, 1997). Industrial pollution goes unregulated in many countries and air pollution is generally uncontrolled. China and India are beginning to shift away from high polluting coal to cleaner fuels such as natural gas but progress is slow. Carbon emissions are still growing rapidly despite these efforts.

In addition to poor and ineffective regulations, many existing policies already in force in Asia are contributing to environmental damage.

Subsidies are one of the biggest contributors to environmental damage[3]. Energy production is subsidized through tax breaks for producers, rebates to consumers, and differential pricing to subsidize the poor. This results in overconsumption of commercial energy, reduced incentives to conserve energy, and higher levels of pollution. Subsidies are provided for logging through lucrative contracts for logging companies to harvest timber without consideration of ecological damage. The fishing industry is assisted by tax breaks on fuel, subsidies that

[3]Subsidies take many forms. The most obvious are direct government payments to producers or consumers. There are also credit subsidies by way of government guarantees, interest subsidies, or subsidized loans. There are tax breaks, subsidized commodities and regulatory subsidies that alter market access or prices paid (Clements *et al.*, 1995).

encourage fishing fleets to expand with bigger and more powerful boats, and income support for artisanal fishermen. These subsidies and competition between countries to catch more fish create overcapacity, encourage overfishing, and threaten ocean ecology. World Wildlife Fund estimates that fishing subsides worldwide are around 50 billion or 20% of the value of total fish catch, with Asia contributing to a substantial part of this total (WWF International, 2005a; Dua and Esty, 1997). Water is also heavily subsidized throughout Asia. FAO *et al.* (1994) estimated that irrigation subsidies amounted to 11.4 billion or 60% of the developing world total. Public irrigation projects do not achieve cost recovery, irrigation water underpriced, tube wells used more and more to draw down water tables; only a fraction of water used is metered and drinking water subsidized. As a result, water tables have fallen, salinization and water logging have spread, while crop yields are adversely affected, and water quality has deteriorated (Whittington, 2003; Dua and Esty, 1997). Agriculture used up to 90% of all water withdrawals, significantly more than other regions. Increasing the cost of irrigation water by 10% would double the amount of water available for residential and industrial users (assuming a price elasticity of −1)[4]. Furthermore, savings can be achieved by increasing the cost recovery for urban water systems, which, according to ADB (1997) recover an average of only 35% of their costs. ADB (1997) estimates that with better pricing and demand management along with some repairs and extension of the distribution system, about half the water supply could be recovered and used to supply people currently without water. Excessive use of fertilizers can cause environmental damage. Eutrophication is caused by deposits of nitrate and phosphate in lakes, ponds, and other water bodies, leading to excessive growth of algae that results in oxygen depletion and killing of fish. Although fertilizer use has increased rapidly in the last few decades, especially in East Asia, fertilizer use is still too low to cause general environmental damage. However, in some areas, such as Java, Indonesia and Punjab, India, nitrate contamination of rivers and lakes may have been caused by nitrogen runoff

---

[4]Bhatia *et al.* (1993) present estimates between −0.4 and −1.5.

(Bumb and Baanante, 1996). Fertilizer subsidies have fallen in recent years; however their use is still substantial and need to be reduced further. Finally, transport equipment is subsidized by publicly funded roads that are not fully paid for by users, and by fuel that does not reflect full environmental costs. This results in overuse and greater air pollution. As income rises, the demand for vehicles rises and the cycle continues unabated.

What are the Options?

There are three major methods for addressing environmental concerns, correcting externalities, and addressing market failure — command and control, taxation and marketable permits.

Many of the environmental regulations adopted by governments in Asia have relied on "command and control" policies. These policies have been widely used in the United States and other industrialized countries. They rely on source-specific standards for pollution and take no cognizance of difference in the costs of pollution by different sources. Some examples are requiring coal-burning utilities to install sulfur-removing scrubbers on their smokestacks, or automobiles to use catalytic converters, the use of lead-free gasoline for all automobiles, or a ban on smoking in public buildings or on airplanes. Command and control policies require the regulatory authority to decide how much pollution is acceptable. These policies are effective when the polluter is easy to identify or in cases where there are many small polluters and monitoring is expensive. When the compliance standard (which usually takes the form of physical standards rather than charges) is high, costs of compliance are also high. Polluters are often tempted to get around the regulations which lead to under-compliance.

Command and control systems are useful in some instances when monitoring costs are high or the environmental damage is great. Littering is a good example of the former. Fines are an effective way to control littering since the costs of monitoring are high and even more difficult to determine the marginal propensity to litter. Toxic waste disposal is another example where costs of environmental damage can be very high. Regulations prohibiting or banning discharge of radioactive wastes are effective measures with a small monitoring cost.

However, command and control policies have proven to be inefficient in cases where more flexible alternatives are introduced. To understand why these more flexible environmental control policies are more effective, consider the cost of abatement at different levels of pollution and the public response to markets for pollution rights. Taxation is one way to account for the cost of pollution by increasing the cost of the good to reflect the added cost of pollution. The most effective taxation method would be to tax the externality itself. When taxing the externality is not possible, a tax on the product itself is a viable substitute. Taxes allow firms to internalize the costs of pollution that were previously ignored. The higher price would also limit the amount of sales of the good in the marketplace. What are the advantages of taxation over regulation? Taxation gives the producer flexibility to decide the structure of production while generating revenue for the government which can be used to fund research on new energy-efficient technologies or to compensate those harmed by the pollution. Examples of pollution taxes are charges for effluent waste water and emission charges for air pollution. These taxes or charges require monitoring and measuring the pollution flows. These taxes can be based on engineering estimates of the flow of pollution or self-reported data verified by random inspections. An alternative tax would be based on the pollution input rather than the pollution itself. For example, taxes on production processes using coal could be based on the carbon content of fossil fuels rather than the amount of carbon dioxide produced. Since these product charges would be based on inputs, monitoring is not required. If the tax structure is well publicized, there is incentive to use cleaner fuels or production processes which would reduce the firm's tax burden. Another tax option is a user charge that helps to defer the cost of environmental damage and reduces use. Road tolls, garbage collection fees, connection fees for water and electricity, and entry fees for access to parks and other recreational facilities are some examples. Deposit and refund systems can also be used for products without a secondary market, like used batteries and pesticide containers.

Finally, marketable permits are another option. These permits would allow the holder to pollute up to a certain amount. This "license to pollute" is issued by the government based on scientific evidence of the environmental impact of the pollution which then determines the amount of regulation that is required. These permits can be traded so that heavy polluters will be allowed to purchase permits from low-level polluters. This type of system gives high cost polluters an incentive to reduce their pollution thereby reducing overall pollution to the environment. By comparison, marketable permits are theoretically more efficient than taxation. A tax system cannot predict the resulting pollution as the tax is levied after the fact. Taxes would have to be adjusted on a trial and error basis in coordination with a monitoring system to ensure that the target level of pollution is not exceeded (see Dowling and Valenzuela, 2004 and ADB, 1997 for details on these different alternatives as they pertain to Asia).

How much money can a more flexible pollution abatement system of pollution controls save? Estimates vary — air and water pollution are probably the most damaging to the environment in terms of impact on health expenditures, losses in worker efficiency due to illness, and general increases in morbidity and mortality. According to some estimates, moving away from command and control systems to more flexible environmental monitoring and control mechanisms could save at least half the existing budget and still achieve the same level of emissions (ADB, 1997, p. 233). The cost savings and improvements in efficiency achieved by a more flexible market-based system could be used to extend the scope of environmental protection and would also enhance industrial competitiveness. Additional benefits would also be obtained by reducing pilferage and system losses through better monitoring. A study by the University of Hawaii and Singapore's Institute of Policy Research estimated that the cost of air pollution in Asia is 5%–10% of urban GDP, while cleanup is estimated to be only 2%–3% (Boyd, 2002).

According to most experts significant increases in prices of carbon emissions will be needed globally to stem the projected rise in emissions from rapidly growing Asian economies of India and China. Some economists suggest that without a demonstration by the United

States and European countries of a strong commitment to reduce carbon emissions it will be doubly difficult to achieve savings in Asian economies. In the United States additional taxes on carbon dioxide emissions have been suggested as well as additional subsidies for development of alternative energy sources. There is, however, no consistent strategy for reducing global warming. The United States did not sign the Kyoto Protocol and it has not developed an alternative plan of its own.

Recent estimates of the cost of combating global warming by reducing carbon dioxide emissions made by a British government report says global warming could have a disastrous effect on the world's economy. Tackling the problem now would require 1% of global GDP, according to Stern (2006), the author of the report. The report says that if we do not act now to reduce global warming the overall costs and risks of climate change will be equivalent to losing at least 5% of global GDP each and every year into the foreseeable future. Nordhaus (2006) argues that these estimates by Stern demand crucially on a zero discount rate. Replacing this assumption with other assumptions about the discount rate reduces the cost estimates and throws open the question of global warming policy. These are questions that, in Nordhaus' view, still remain to be answered.

The Way Forward

In its preparation for a forward looking view of the 21st century, the United Nations Environment Program launched a survey asking scientists and environmental experts around the world about the most pressing environmental issues identified by ICSU's Scientific Committee on Problems of the Environment (SCOPE) (see UNEP, 1999 for details). The survey involved 200 environmental experts, including many research scientists, in more than 50 countries. The survey results suggest that many of the major environmental problems expected in the 21st century are problems that exist now but do not receive enough policy attention. A summary of responses are displayed in Table 6.4. Climate change was the most cited issue in the SCOPE survey although, taken together, water scarcity and pollution ranked higher.

Table 6.4 Major Emerging Issues Identified in the SCOPE Survey by Percentages of Respondents Indicating Most Important Issues

Climate change	61
Freshwater scarcity	29
Deforestation/desertification	29
Freshwater pollution	28
Poor governance	27
Loss of biodiversity	23
Population growth and movements	22
Changing social values	21
Waste disposal	20
Air pollution	20
Soil deterioration	18
Ecosystem functioning	17
Chemical pollution	16
Urbanization	16
Ozone depletion	15
Energy consumption	15
Emerging diseases	14
Natural resource depletion	11
Food insecurity	11
Biogeochemical cycle disruption	11
Industrial emissions	10
Poverty	9
Information technologies	7
War and conflict	7
Reduced resistance to diseases	7
Natural disasters	7
Invasive species	6
Genetic engineering	6
Marine pollution	6
Fisheries collapse	5
Ocean circulation	5
Coastal zone degradation	5
Space debris	4
Persistent bio-accumulative toxics	4
El Nino effects	3
Sea level rise	3

Source: UNEP (1999).

Taking a cue from this list we have focused on a few of the topics that are most relevant for Asia. The background conditions have been discussed earlier in the chapter.

Population growth

Population growth is an important background development that interacts with other environmental variables. If population growth moderates substantially, it will have a beneficial impact on the other environmental drivers. Projections of population increase for the world's most populous nations were presented in Table 6.1. Mid-range estimates project world population to peak at around 10 billion people toward the end of this century. More optimistic estimates suggest that world population could peak at as little as 8–9 billion people by the middle of this century. Much will depend upon progress made in reducing population growth in South Asia, particularly in Bangladesh, India and Pakistan. If population growth can be reduced substantially in these three countries, it is highly likely that the Asian population could peak sometime in the next four decades. To achieve this reduction will require the adoption of additional policies to curb family size including greater commitment to birth control, enhancing the status of women, placing more emphasis on women's education, accessible health care and better employment opportunities. In addition, governments can provide real incentives and tax breaks for smaller families.

Water

Water is not a resource or activity that mankind can control in the same way that forest resources or minerals or pollution can be adjusted, modified, destroyed or enhanced. The amount of water available in the future is determined by the ecological cycle of the planet over which we have no control. Therefore, the focus of this section has to be with policy.

Policies to regulate the use of water through pricing and other mechanisms as well as infrastructure and other civil works to improve accessibility, purity and general cleanliness are required. Other things equal, rising population will put pressure on fresh water resources. While agriculture will continue to use the most water in Asia, fresh-water demand is growing rapidly in urban areas and in the industrial sectors of Asia's fast growing economies. By 2025 India is expected to be water stressed with per capita water availability decreasing to around 800 m^3 per capita per year. China is also expected to reach the water stress threshold of 1000 m^3 per capita per year by 2025 as water usage increases from a current estimated level of around 2300 m per capita per year. Without better management practices, groundwater will be further depleted as water-stressed economies resort to more rapid depletion of groundwater by sinking more tube wells. Government policies are crucial in dealing with water use issues. While the focus and structure of policies will differ from country to country, there are several basic considerations that have to be reflected in these policies.

First, water is not a free commodity. All users have to pay for any amount of water they use for any purpose, whether it is for personal consumption, agriculture or industry.

Second, tariffs have to strike a balance among four objectives — revenue sufficiency, economic efficiency, and equity and poverty alleviation. Cost recovery is important and the revenue stream must be adequate to attract both equity capital and debt financing. Water charges have to include the marginal cost of supplying water or the average incremental costs of supplying water. Similarly, consumers should be treated equally; water bills should reflect the full amount of water delivered. Poverty alleviation is better handled by charging for water used and subsidizing the fixed costs of connection instead (Whittington, 2003).

Evidence from field studies suggests that everyone is willing to pay for reliable water. By pricing water correctly and recognizing the importance of financial viability, water authorities will be able to provide better service to more consumers with greater efficiency. The World Bank estimates that as much as 50% of current water supply in Asia could be saved if sound pricing policies were introduced, implemented and maintained.

Revenue from an improved structure of water tariffs can be directed to improving sanitation, providing wider access to clean water, and building additional infrastructure to deal with the growing needs of urban communities as population increases. Additional challenges of possible climate change and the impact on water availability will have to be addressed along with a careful review of the current levels of groundwater and water tables. Only by maintaining a sufficient level of water tables can we avert the disastrous consequences of dire water shortages when groundwater is exhausted.

Forests

Reforestation and reducing the rate of deforestation are key components of environmental policy in most Asian economies over the medium term. Reforestation reduces the rate of soil erosion, flooding and soil degradation, and also contributes to the reduction of global warming (see below). Population pressure in rural areas and the pressure from the logging industry have been largely responsible for the decline in forested area. Bringing this trend to a halt will require concerted efforts to rebuff the pressures from loggers, increased incentives for reforestation, and development of new plantations, like rubber, palm oil, fruits, and other tree crops. Some of the pressure to cut down forests in the past has been offset by the growth of plantation crops and those trends can be reinforced and encouraged through government policy. Recognizing the external advantages of reforestation, subsidies can be provided for replanting projects in rural areas, particularly where reforestation will compliment conservation efforts and reduce the risk of floods and further soil erosion. It is also imperative that checks and balances are put in place to ensure that these subsidies are not misused by logging interests for other purposes, particularly in areas where local officials are known to take bribes and allow loggers into virgin forests (Cropper and Griffiths, 1994; Perlez and Bonner, 2005). Reforestation is particularly crucial in the tropics since trees grow three times as fast as they do in more temperate climates and are able to absorb much more carbon from the atmosphere. In addition there are other factors that make carbon absorption in temperate climates slower than in the tropics (see http://en.wikipedia.org/wiki/Carbon_offset).

Energy

Energy demand in Asia will be expanding on several fronts. Industrialization will continue at a rapid pace throughout the region and the demand for energy will expand in tandem. Projections by the US Energy Information Administration (www.eia.doe.gov/ooiaf/ieo/enduse/html) suggest that industrial sector energy demand in emerging economies (which are dominated by Asian economies) will grow by close to 4% per annum between 2002 and 2015. Cambridge Energy Research Associates predict that half of total growth in oil consumption will come from Asia in the next 15 years (quoted in Yergin, 2006). China and India will depend primarily on increased use of coal to fuel this expansion, with attendant ecological challenges covered in other sections of this chapter. Demand for transportation will also increase rapidly at around the same rate, or slightly less for emerging economies, although it will be substantially faster in China (6%) and India (4.7%). This is primarily from rapid demand for automobiles by individuals for both work and leisure. The number of cars per capita in both countries is low and there is a substantial scope for expansion in line with rising incomes. Rail transport is well developed in India as a result of investment from its colonial era, although there are still parts that are inaccessible. The rail system is used to overcapacity and in need of upgrades and modernization. Nevertheless the scope for expanding and improving existing public transport in India is much greater than in China which has fewer extensive networks as alternatives to road transport. Some form of river transport system that is possible in both countries will only move a small percentage of the total passengers and freight loads. For the rest of Asia (Southeast Asia and East Asia) energy is also projected to increase at a healthy pace of around 6%. Most of the transport demand will be fueled by oil.

Projections of energy demand take very little notice of downward shifts in demand as a result of lifestyle changes that refocus on mass transit, or the use of clean fuels such as ethanol, or the use of new fuel efficient hybrid cars that use fuel cells in combination with conventional internal combustion engines. This assumption is particularly worrying as we realize that developing Asian economies are going to be responsible for the lion's share of the increase in fossil fuel emissions between now and 2025 and beyond. Carbon dioxide emissions from fossil fuels in Asia (excluding China) increased by 78% from 1990 to 2002 compared with 16.4% globally (UNESCAP, 2005).

Is there an alternative set of scenarios that would result in a slower growth in fossil fuel emissions and a reduced environmental load? One possibility is fuel substitution using ethanol, nuclear energy, wind or geothermal power. Brazil is the world's biggest producer of ethanol, followed by the United States. Brazil plans to export more ethanol in the future — 8–9 billion litres annually. How does it compare with world oil production and exports? 8–9 billion litres of ethanol is equivalent to about 17 hours worth of oil consumption worldwide or 18% of the daily oil consumption requirements of Taiwan, listed as number 20 on the list of the top 20 oil consumers. There are no other countries with an extensive ethanol production capability at the present time.

Recently work in the United States has begun on an alternative ethanol technology called cellulosic ethanol which can use a variety of different feedstocks (Greer, 2005). This technology, which can make use of a wide variety of cellulosic biomass feedstocks including switch grass, agricultural plant waste and paper pulp, also has a higher net energy content than corn ethanol and emits a low level of greenhouse gases. It might bear fruit in the medium term in industrial countries as an alternative energy source, particularly as a substitute for gasoline. It is, however, unlikely to serve as a widespread energy alternative for developing countries in Asia in the next decade.

For coal-burning, countries like China and India have to adopt technologies that convert coal into gas in a process known as integrated gasification combined combustion or IGCC. An advantage of this approach is that contaminants like mercury and sulfur can be easily separated from the gas and disposed of. It is also possible to take the separated carbon dioxide and pump it underground. Unfortunately, IGCC has not been welcomed by the Chinese authorities because of high costs and their desire to keep the technologies that western companies are not willing to share. (Energy prices are still held below border prices to subsidize businesses and consumers and as a result, utility companies are reluctant to spend on foreign equipment.) In the meantime, high sulfur coal continues to be burned and new plants are still using old technology. A similar story can be told for other coal-burning countries in Asia such as India and, incidentally in Europe as well (Bradsher and Barboza, 2006; Landler, 2006).

The outlook for nuclear power is not much better. Nuclear energy share of World energy production is projected to decline between 2002 and 2025 (Figure 6.6). These projections are based on current plans to expand nuclear capacity in France, United States, India, China, and a few other countries. It is unlikely that these plans will be revised to increase nuclear production during the forecast period. Similarly, prospects for other alternative fuels such as geothermal, wind, and hydro is limited.

However, if prices of oil remain high or even rise further in the next few years there will be fuel substitution across a wide range of

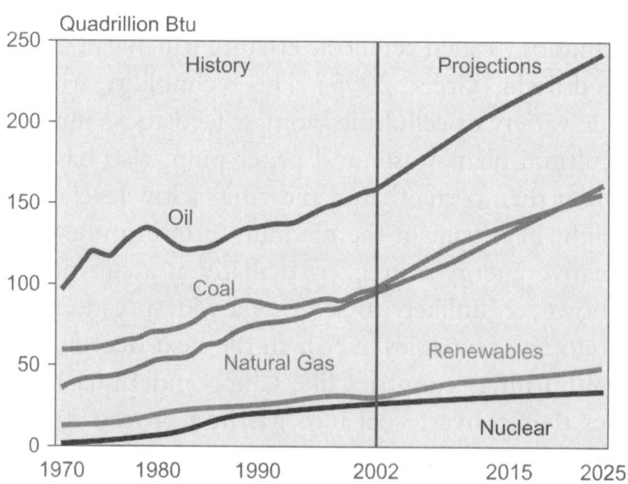

Sources: History: Energy Information Administration (EIA), *International Energy Annual 2002*. DOE/EIA-0219(2002) (Washingtion, DC, March 2004), website www.eia.doe.gov/iea/. Projections: EIA, System for the Analysis of Global Energy Markets (2005).

Figure 6.6 World Marketed Energy Use by Fuel Type, 1970–2025
Source: US Energy Information Administration (2005c).

activities. Ethanol and other crop-based fuel alternatives have become popular as the experience of Brazil has been publicized and venture capital has begun to flow to development of alternative fuels using corn waste in the United States. These developments have been moving forward in tandem with the wider spread of cars that can use hybrid fuels which mix gasoline with ethanol or other crop-based fuels. Further, the efforts of China to reduce automobile ownership by developing mass transit for large urban areas cannot be overlooked. Nevertheless, while laudable, these developments alone can do little to stem the rise in fossil fuel consumption over the next decade or so. About 4 million cars were sold in China in 2006, an increase of over 100% in 3 years. At the same time, China's highway network is expanding rapidly to accommodate more vehicles. By the end of 2004, China had 21,000 miles of motorways, more than double the 2000 figure. Seventeen years ago, it had no motorways. Only the United States has more limited access roads. By 2020, China plans to double again the length of its motorways (*Economist*, 2005a).

Government policies can be developed to support the shift away from transportation by automobiles and fossil fuels to alternative transportation modalities and fuels. Subsidies for mass transit, higher taxes on automobiles and on petroleum have to be seriously considered by all countries in the Asian region. In addition, subsidies for the development of alternative fuels must also be considered. Renewable energy now provides only about 1% of total energy requirements worldwide. This can be increased by further integration of renewables such as ethanol and fuel cells into the transportation network and also by increasing subsidies for attractive alternative fuels including solar and wind power. Of course, these will have only a marginal impact compared with the gains that will eventually result from shifting away from the use of the petroleum-driven automobile and the development of fusion technology to supplying clean energy for a variety of uses in the medium term. Development of new technologies will also be needed (see later section on Technology). According to the OECD (2007) in the medium term however, fossil fuels will still contribute about 77% of global primary energy demand by 2030 assuming a variety of shifts in production to other fuels and adoption of new technologies. This represents only a slight decrease from 81% reliance on fossil fuels with no changes in technology.

Air

Japan, South Korea and some European countries demonstrated that concerted efforts to reduce urban air pollution are possible and yield large gains in reduced respiratory infections and general levels of health. Simulations carried out by UNEP (1999) show that significant gains can be made in reducing the levels of SO_2 and NO_2 emissions by a combination of improvements in transportation efficiency, some fuel switching, and accelerated introduction of clean technology. These improvements are reflected in Figure 6.7.

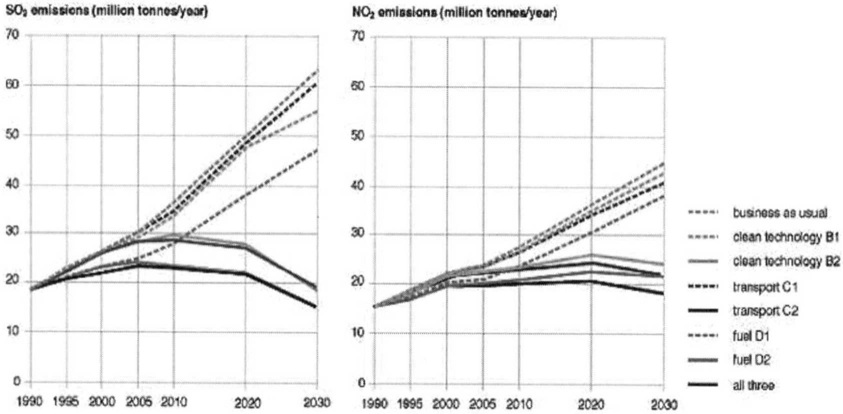

Figure 6.7 Options for SO_2 and NO_2 Emissions

Source: UNEP (1999).

Note: Business as usual shows the biggest increase in emissions; the two clean technologies alone (mostly clean technologies for industry introduced at different levels of per capita income) show a slight reduction, while adding transport policies in the form of switch to mass transport (C1) and combined with cleaner fuels (C2) add further to the benefits. Finally, fuel switching (D1 and D2) and a combination of all three result in the biggest reduction in sulfur and nitrous gases.

To enhance the introduction of clean technology, efforts would have to include replacement of existing vehicles and technologies with more efficient ones. Some examples are replacing two-stroke engines with four-stroke engines, the use of compressed natural gas in place of gasoline where feasible, phasing out leaded gasoline, and the adoption of strict emission standards for all vehicles. Fuel switching can be brought about by a carbon tax aimed at replacing coal where possible and adopting cleaner technology (scrubbers) if relative prices and availability suggest continued use of coal. Implementation of these sets of policies combined with promotion of public transport systems would result in a reduction of sulfur emission below their 1990 levels by 2030 and only a small increase in nitrogen oxides. Fuel switching and promotion of mass transit would also result in a reduction in carbon dioxide emissions which would help in reducing global warming (see below).

Biodiversity

Scientists are alarmed at the current rate of extinction of species. According to recent research one-third of amphibian species are threatened with extinction (Reichhardt, 2004). The 1992 Convention on Biological Diversity (CBD) at The Hague pledged to achieve a significant reduction of the current rate of extinction of species by 2010. Most of their efforts focus on protecting remaining habitats.

What further policy actions can be undertaken to stop this loss of biodiversity? Most conservation biologists believe that protecting habitat should take priority over biodiversity despite concerns over losing species. Trans-boundary protection — protected areas based on watershed or ecosystem boundaries rather than national borders — is also an important priority. Despite efforts to protect habitats and monitor endangered species the UNEP report projects that there will be significant losses in habitat and increasing pressure on biodiversity both on land and in the oceans (UNEP, 1999) due to further urbanization and declining forested areas. Furthermore, increasing economic activity will put additional pressures on all species. In all four scenarios that the report examined, South Asia and Southeast Asia are most vulnerable and will experience the most significant deterioration between 2002 and 2025. Even in the most optimistic and environmentally friendly scenario, South Asia's biodiversity index will fall by around 30% and even more in Southeast Asia where more habitats are under threat (UNEP, 1999). To slow this loss of biodiversity, greater efforts have to be made to protect existing habitats by increasing resources devoted to monitoring encroachment and enforcing existing regulations that prohibit logging and other practices that further encroach on protected areas. Fines and other punitive measures have to be strengthened in order for this to happen. This will require education of legislators and the public on the environmental costs of these encroachments.

Spending on environment

The environmental Kuznets analysis suggests that countries will spend a larger share of income on environmental protection as they reach a certain middle income level. Cross-section information on spending for environmental protection in Asia tends to support this hypothesis. The wealthiest country in the region, Japan, spends the most — around 2% of GDP, while the newly industrialized countries of Singapore (1.2%–1.5%), South Korea (1.3%–1.6%), and Taiwan (1%–1.2%) spend a bit less. The comparatively richer Southeast Asian countries of Thailand and Malaysia spend about 1%, China, Indonesia and Philippines between 0.5% and 0.7%, and Vietnam between 0.1% and 0.3%. The poorer countries in the region spend even less (Boyd, 2002). Given the relative scope of environmental difficulties and challenges faced by these countries these proportions should be reversed. Will they be in the future? Certainly environmental awareness is increasing and some policies are changing, however, a "pollute now and pay later" mentality is still strongly entrenched in the minds of policy makers in the Asian region. This has to change dramatically if spending for environmental protection is to increase.

Governance and institution building

Dealing with environmental issues is one of the most important challenges faced by developing countries in Asia and the survey by UNEP in Table 6.4 suggests that governance was high on the list of critical issues for the future. Adjusting attitudes, framing, implementing, and enforcing new policies that protect the environment are daunting challenges. Economic growth remains the primary goal of the economies of the Asian region. Lifting incomes and reducing poverty are the primary focal points of policy. The environment is usually a secondary priority. Problems in dealing with environmental issues are compounded by the fact that environmental issues are complex and interrelated with other aspects of the economy. Consequently, effective policies require coordination among many different branches of government. Furthermore, new initiatives and programs to protect the environment require shifts in priorities, introduction of taxes on polluters, and rewards to those who protect the environment. Powerful economic interests are often at stake and those who have been receiving free environmental services are not easily persuaded to start paying. The environment has its supporters who are often diffused and not well organized. The government has to act on its behalf in order to effectively deal with the issues at hand and also work to convince the public that environmental protection makes good economic sense (UNEP, 1999; World Bank, 2003). To do this, spending on environmental awareness has to be increased and environmental NGOs mobilized to develop priorities in each country that can be presented to policy makers and used to implement an environmental protection strategy that encompasses all of the critical areas that require attention from water and air pollution to maintaining biodiversity and reducing soil degradation.

Technology

Concerns about global warming, lack of secure supplies of fossil fuels, exploitation of non-renewable resources, and high prices for oil have created an environment where alternative fuel sources are being widely explored. What are the chances that new technologies will provide substantial alternatives to coal, oil and natural gas in the next few years? Currently, genetically modified crops offer higher yields and the possibility of saving on water and artificial fertilizers. Nuclear energy has

Table 6.5 Nuclear Generating Capacity, 2002–2025 (in gigawatts)

Country	2002	2010	2015	2020	2025
China	2.2	10.8	18.1	21.1	26.0
India	2.9	7.9	10.5	14.2	15.3
South Korea	13.7	22.0	23.3	24.1	25.5
Other Asian	5.6	8.3	8.2	11.3	13.0

Source: US Energy Information Administration (2005).

many advantages over fossil fuels, although there are also disadvantages which limits its widespread use around the world. Projections of various organizations reflect these concerns and the limited plans for the expansion of nuclear energy. For Asia, the reference case projects significant increases in nuclear energy starting from a low base (Table 6.5). China's nuclear power output will increase ten fold between 2002 and 2025, and India's will increase by five times. It will nearly double in South Korea where capacity is already large. However, given the low level of nuclear energy production in China and India (currently at the 0.4% and 0.9% respectively in 2001), an increase in nuclear generating capacity will make little difference in their overall energy balance.

Solar energy has been touted as a clean alternative source of energy for many years. However, it is mainly used to supply some residential water heating and cooling in regions which have good solar exposure. Worldwide, only 0.6% of total energy is supplied by solar power and in Asia (except Japan) solar power's contribution is nil. The chances of solar power making a dent in the energy balance in Asia over the next two decades are quite low. Solar cell technology is being developed and implemented in industrial countries, yet solar energy contributes very little to overall consumption. Biomass is widely used as a renewable energy source in South Asia and in parts of Southeast Asia (Table 6.6). However, biomass, primarily used in rural communities, is not a clean fuel. It contributes to air pollution and is particularly dangerous when used indoors. As an alternative to fossil fuels, biomass use will decline in the future as the migration to the cities continues.

Table 6.6 Biomass as Proportion of Total Energy Consumption (2005)

Country	Percent
China	18.8
India	38.5
Indonesia	31.6
Myanmar	77.4
Nepal	84.9
Pakistan	37.2
Philippines	23.1
Sri Lanka	52.9
Thailand	17.1
Vietnam	58.3

Source: US Energy Information Administration (2005).

For all other Asian countries, biomass consumption as a percent of total consumption is less than 10%.

Global warming

Scientific estimates suggest that the atmospheric concentrations of CO_2 have increased from around 275 ppm to 370 ppm over the past couple of hundred years (Hoffert *et al.*, 2002). Left unchecked, scientists estimate CO_2 concentrations will reach over 500 ppm by the end of this century. Climate models and paleontological evidence suggest that this will eventually produce global warming in line with past periods of warm average temperatures. The current average global temperature is between 13.8°C and 14.6°C (see Figures 6.3 and 6.8), while the average temperature during the last Ice Age was 12°C. During warm periods, temperatures averaged 22°C. (For further reference see table at www.scotese.com/climate.htm and Figure 6.3.) During periods of warm global temperatures, the ocean levels were also much higher than they were now and there were no polar ice caps and the amount of land mass was considerably smaller (see video at www.scotese.com/climate.htm). Even an average rise in the ocean's average tidal swell would cause extraordinary damage to coastal cities around the world.

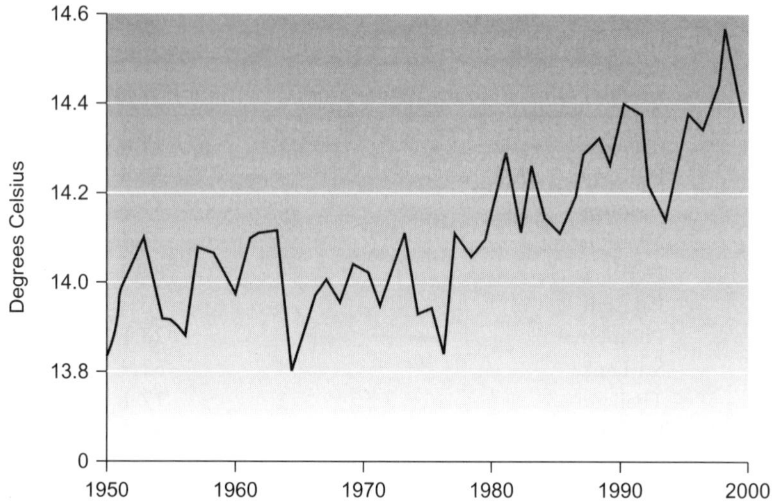

Figure 6.8 Average Temperatures at Earth's Surface — 1950–1999

Sources: Goddard Institute for Space Studies, Number 70 (GISS). Published in *GESAMP* (2001).

Most of the CO_2 is caused by emissions from three sources: Asia, Russia and the European Union and North America (Table 6.7).

To make rough forecasts of the pattern of emissions in the future, some scientists have used a simple equation — $I = PAT$, where I is the environmental impact, P is population, A is affluence as measured by per capital income level, and T is the impact of technology. Looking at the three drivers of climate change as reflected by CO_2 emissions in Table 6.7, it is evident that Asia is going to be the biggest source of emissions in the next few decades. Its rate of growth in income is probably going to be faster than the other two regions, its population growth more rapid, yet its ability to apply new technology is limited because of its previous low level of income. Estimates by several research institutes suggest that China will be the largest single contributor to higher levels of carbon emission between now and 2025 (US Energy Information Administration, 2005a). By 2025 China will be as large a producer of carbon emissions as the United States, compared to 60% in 1997 (Table 6.7; National Academy Press, 2005).

Table 6.7 Billion Tons of Co_2 Emissions in 1996

Country/region	Oil	Coal and others*	Total	Total by region
China	0.5	3.3	3.8	Asia = 8.3
Rest of Asia	1.9	2.6	4.5	
Russia	0.8	1.0	1.8	Europe including Russia = 6.7
Europe	1.4	3.5	4.9	
North America	2.5	3.8	6.3	6.3
Rest of World	2.0	1.5	3.5	3.5
Total	9.1	15.7	24.8	

*Coal is the predominant source. Other emissions come primarily from natural gas.
Source: Derived from World Resources Institute, quoted in *International Herald Tribune* (2002).

A discussion of alternative policies to reduce the pace of emissions requires some understanding of elementary science and an evaluation of alternatives using cost/benefit analysis. The problem is complicated by its global nature, the length of time leads and lags between introduction of policies and its impact on climate. Furthermore, there are natural forces beyond our control and, for the time being, beyond our understanding, that drive global climate changes over the long run. A cursory look at Figure 6.3 is enough to suggest that we are now in a period of relatively cool climate compared with the past history of the earth's temperature over the last few billion years. In addition, the impact and cost of global warming is difficult to ascertain and calculate. There are costs resulting from the rise in sea levels in low-lying areas such as Bangladesh, Shanghai, and other sea coasts around the Asian region (Eilperin, 2006). Inevitable climate change is also expected.

One troubling aspect of global warming is the nonlinear reaction of the global ecosystem as carbon dioxide emissions increase. The amount of carbon dioxide released by soil invertebrates and microbes is 10 times higher than the level of annual carbon emissions from fossil fuels. At the other end of the cycle the oceans and forests absorb carbon from the atmosphere. However, soils hold more carbon than the trees and the atmosphere combined. If the carbon cycle in the soil is disrupted, the amount of carbon could increase, contributing further to carbon dioxide emissions and to global warming. Research in Antarctica and elsewhere suggests that this might be happening (Heath *et al.*, 2005). Further research needs to be carried out to determine the long-term consequences of these findings.

So what can countries in the Asia region do in the next couple of decades to reduce the rate of release of carbon dioxide into the atmosphere? We have listed several policy options that can be easily adopted by Asian governments without additional scientific research while recognizing the anticipated rapid rate of economic growth in these economies within the next two decades. These options are listed in Table 6.8 by level of priority for Asia.

These suggested policies focus on improving efficiency in energy utilization and slowing the move toward more energy-intensive transportation modalities by raising the prices of fossil fuels and supplying alternatives for the automobile. These are supplemented by increasing forest cover, which is an effective and relatively simple method to slow the amount of CO_2 released into the atmosphere. The implementation of these policies would differ from country to country and would depend upon the willingness of governments to begin new initiatives, and introduce carbon taxes and a system of monitoring and taxing industrial pollution. There is a risk of a race to the bottom as countries pursue economic growth as an exclusive pre-occupation while failing to address environmental challenges.

To stimulate win/win policies that reduce emissions and still maintain growth, governments will have to design policies that result in the so-called "super externalities" which contribute to economic efficiency and growth while dealing with environmental degradation at the same time. Some possible policies are suggested in Table 6.9.

The final suggestion, capture of emissions, requires further research and development in partnership with industrial countries to find ways

Table 6.8 Policies for Reducing Emissions in Asia

Proposed policy	Impact of policy	Cost of policy
Increase automobile fuel efficiency and raise price of fuel to reflect environmental costs	Can easily double the miles per gallon and halve auto emissions as a result	Government and private sector commitment to build more fuel-efficient cars. Explore hybrids for Asia
Slow movement to private cars and motorcycles by better mass transit	Would make transport more efficient and would also reduce traffic congestion	Requires substantial infrastructure investment and perhaps subsidies
Improve efficiency of energy use in residential and commercial buildings	Reduce power, heating and cooling costs	Systematic efforts to monitor new building codes, review older buildings, and retrofit where needed
Improve efficiency and reduce level of pollution of coal-based power plants by installing scrubbers	Coal produces about 1.7 times as much per unit of energy when burned as does natural gas. Therefore, improving efficiency of coal-based plants would give greatest carbon reduction	Promote policy of building new coal plants that are efficient, close inefficient smaller plants, and monitor emissions of others
Plant trees where possible and stop deforestation	Deforestation is responsible for about a quarter of all carbon emissions entering atmosphere. Reversal will cut emissions dramatically, particularly in tropical countries	Enforcement of logging bans, start reforestation, and new plantations where possible. Tree planting programs in cities and countryside
Capture emissions from power plants fueled by coal and oil and pump into ocean or deep wells — also known as sequestering	Trap carbon emissions and inject into oceans or underground thus eliminating carbon emissions	New research that could be undertaken with the help of industrial countries

Table 6.9 Super Externality Policies

Policy	Impact on economy and environment
Women's education	Improves gender equality, reduces infant mortality, improves outlook for children's health and education, can promote environmental awareness, and reduces rate of population growth
Forestry policies	Reforestation and reduced deforestation both provide an additional sink for carbon, and help to reduce soil erosion, flooding, landslides. It can also provide livelihood in case or orchards or tree crops (e.g., coconuts, palm oil, rubber). Forestry policies are particularly powerful in tropical countries where carbon absorption is high
Mass transit	Cuts fuel emissions, reduces traffic congestion which reduces commuting time, and reduces urban air pollution
Vehicles	Increased efficiency by introducing hybrid technology and increased carbon taxes based on emissions reduces urban air pollution and carbon emissions at the same time

to store carbon dioxide produced by factories and power plants. This should be a priority within the next decade.

A number of other suggestions have been made for reducing emissions. Some of them can save some carbon emissions at the margin. These include building more nuclear power plants, reducing methane emissions from flooded rice paddies and bovine flatulence, introduction of solar panels for heating water and cooling of buildings, and adoption of wind and biomass fuels. These options may produce some savings at the margin but they are probably not worth the investment. Wind and solar power are subject to interruptions, do not easily fit into a grid, and are geographically limited in scope. They are unlikely to bring large additions to the amount of clean energy needed in the next two decades. Hydroelectric power is near the end of its possible expansion, while nuclear energy contributes less than 5% of the energy needs of India and China and is not projected to increase any time soon (see the US Energy Information Administration, 2004b).

In the longer run, scientists from industrialized countries are betting that sequestration and some new technology that can de-carbonize in conjunction with sequestering will become feasible. These include reducing emissions from fossil-fueled systems based on coal, oil or natural gas; biological engineering approaches to removal of carbon from the atmosphere; re-engineering of the entire fuel-cycle system; and geochemical and other approaches to storing carbon. Another focus would be to develop carbon-free energy supplies for automobiles using hydrogen and fuel cells. They are also hopeful that fusion technology will provide an unlimited source of clean fuel toward the end of this century (Hoffert *et al.*, 2002).

A more concrete scenario is presented by Gore (2006, p. 280/81) who adopts a table from Socolow and Pacala (2004) that shows the reduction in gigatons of carbon emission that can be saved by 2050 from a combination of six policies which can contribute to bringing emissions far below the current level of emissions. They are (1) more efficient use of electricity in heating, cooling, lighting, running of electrical appliances and equipment; (2) better building design; (3) more efficient automobiles using hybrid fuels and fuel cells; (4) better mass transit, more efficient trucks; (5) more wind and bio-fuels use; and (6) capture and storage of carbon. Each of these policies would require a sustained commitment of resources to achieve the projected reductions.

Item (6) is the most challenging of the Socolow and Pacala wedges[5]. Table 6.7 shows that over 60% of global carbon emissions comes from coal and there is no way to stop these emissions. Efficiency can be increased so that more energy is obtained from each ton of coal, but that ton of coal is still going to emit the same amount of CO_2. And the amount of coal being burned is increasing rapidly. It is cheap, abundant, and the fuel of choice in India and China. There are three possible technologies that have been suggested for carbon capture, that is, removing the carbon dioxide from the gases created during coal combustion. The first method extracts the carbon dioxide from the flu gases by passing it through a solution of chemicals called amines and absorbing the carbon dioxide. The carbon dioxide can be released by heating, and liquefied for storage. This process and

[5]The paragraph draws heavily on *Economist Technology Quarterly* (2006).

another (option 2) that separates the carbon dioxide from the nitrogen in the flu gases, are both expensive and require additional energy which reduces the thermal efficiency of the plant. The third approach is called integrated gasification combined cycle (IGCC). This process heats coal in oxygen, it reacts to create carbon dioxide and hydrogen. An amine (see solution 1) is more efficient now because the carbon dioxide is more concentrated. There are four IGCC demonstration plants operating in America and Europe, but they do not store carbon dioxide. IGCC, is advantageous since it produces hydrogen as a byproduct. Critics say it is too expensive and is prone to break. Unless stricter laws on carbon emission are adopted globally it is unlikely that IGCC or any other carbon capture technology will become widely adopted. And such legislation does not seem to be on any policy maker's agenda anywhere in the world right now.

Looking at the overall agenda of reforms including items (1)–(5) as well as (6), is such an agenda relevant for Asia? First, note that this is a very long run projection and many of these savings do not occur until 2015 or later. Second, many of these gains require both extensive implementation of existing technology, building of infrastructure, changes in people's habits, and retrofitting of existing infrastructure. In Asia all of these suggestions would definitely save energy and the proposals suggested in this chapter are not inconsistent with them.

There are some global agreements that propose limits on emissions for industrial countries and some developing countries as well. The most important of these, the Kyoto Protocol, is an agreement signed by 141 countries accounting for about 55% of greenhouse gas emissions. It became legally binding in 2005, seven years after it was signed. It requires signatories to cut greenhouse gas emissions by 5.2% by 2012 from 1990 levels. The United States, the top polluter, and Australia have signed but did not ratify, and China and India, two other larger polluters, are exempt, although they have signed. What will be the effect of the Kyoto Protocol on Asian countries? Countries can set their own targets according to their own levels of pollution. Korea will have some difficulties in complying with the treaty. There is only sketchy evidence of compliance in Asia since the Protocol has been binding for a short time. However, industrial countries can buy pollution rights from poor countries through a permit trading facility that allows industrial countries to buy pollution rights from poorer countries in order to meet their Protocol obligations. The World Bank's Carbon Finance Business uses the Protocol's flexible market mechanisms which allow rich countries to meet some of their commitments through the purchase of emission reductions from developing countries. The World Bank argues that these market mechanisms are absolutely essential to the long-term engagement of the global community and offer cost-effective solutions to combat climate change. World Bank research shows that if half the emission reductions are achieved by OECD countries domestically, the "compliance gap", to be met through trade with developing countries and transition economies through 2012, would be 2.5 billion tons — 10 times the current carbon purchase contracts. At a selling price of US$5–US$10 a ton, carbon payments to developing countries and economies in transition between 2005 and 2012, could be worth between US$12.5 billion and US$25 billion to developing countries (see World Bank, 2005). The longer term implications are less clear. These trades do not reduce the global level of pollution. They result in a transfer of resources to poor countries, which are not required to spend the money on pollution controls. Asian countries will have to comply by adopting controls such as the methods suggested in Tables 6.8 and 6.9.

One important point to note is that the system of pollution permits is being used more and more in industrial countries and is discussed further in Chapter 3. It is clear that the US economy is growing much more slowly that the economies of Asia and its appetite for energy is not growing nearly as fast. Furthermore, many of the suggestions for implementing existing technology on a much wider scale will entail costs and resource commitments that Asian economies may not be willing to consider even within the next decade. It is up to policy makers to convince them of the importance of considering these and other solutions that involve a long planning horizon and substantial budget allocations. Cooperation with industrial countries is also a possibility for lowering costs and increasing Asia's ability to implement new technologies (see Chapter 3 for more details).

Time delays

The nature of the response of the ecosystem to changes in policy and the pattern of pollution is difficult to summarize (WHO, 2005b). The time lags between environmental actions and the response of the ecosystem to these changes are long and variable. There is inertia and the amount of resilience to consider in the ecosystem. Inertia is the delay in response in the system after the cause of a change in the system has been removed, while resilience is the amount of stress the system can absorb and still remain capable of returning to its undisturbed state. As an example, consider the case of Lake Erie, one of the great lakes in the United States. For many years, Lake Erie was a dumping ground with industrial establishments lining its shores. It eventually became so polluted that all fish and vegetation died. Years later attempts were made to clean up Lake Erie. Pollution was stopped or dramatically cut back. Eventually, Lake Erie began to return to its state before pollution began (see http://en.wikipedia.org/wiki/Lake_Erie for further details on the Lake Erie experience).

Time delays complicate the task of environmental management as the impact of pollution build up over time and often go unnoticed for many years. For example, excessive phosphorous from fertilizer use accumulates in soils, and runoff carries the pollution to rivers, lakes, and coastal waters resulting in eutrophication.[6] However, it may take many years before this process becomes fully evident. If long-term impacts are not fully realized then policies that concentrate on short-term benefits will seriously underestimate environmental impacts. Climate change and loss of biodiversity and species extinction are two of the most highly publicized examples of environmental lags. There are many other examples including the impact of higher mercury levels on fish, birds and human populations. Mercury takes the form of methyl mercury which inflicts increasing levels of toxicity on species

[6]Eutrophication is the process whereby lakes and streams receive excess nutrients that stimulate excessive plant growth including algae that reduces dissolved oxygen in the water when dead plant material decomposes and can cause other organisms to die. Nutrients can come from sources, such as fertilizers, raw sewage, and soil erosion. Water with a low concentration of dissolved oxygen is hypoxic.

higher up the food chain through biomagnification, whereby persistent substances like methyl mercury will increase in concentration in microorganisms, to fish, to fish-eating predators like otters and loons, and to humans. Elevated methyl mercury levels may lead to the decline of affected wildlife populations and may affect human health when people consume contaminated foods. The mercury poisoning case at Minamata, Japan is a very famous example of environmental time lags. Local residents were poisoned by fish contaminated with toxic levels of methyl mercury originating from an industrial sewer discharge, which killed more than 1000 people (see http://www.ec.gc.ca/MERCURY/EN/bf.cfm).

Loss of habitats of endangered species is another example of environmental time lags. When species are driven close to extinction by changes in land use, it takes a long time to return to normal levels. In addition to time lags, many environmental impacts act in a nonlinear way. Changes may be gradual until a certain pollution threshold level is reached, after which changes become much more rapid. Some examples include the spread of infectious diseases. Infections that were not particularly harmful can mutate into a more virulent form as the Avian flu and SARs demonstrated. High population densities and global mobility increase the risk. The bubonic plague killed millions in Europe in the 14th century, the Spanish flu of 1918 killed between 20–40 million people, and the El Nino flooding in 1997 and 1998 caused cholera outbreaks in many parts of Africa. The collapse in fishery yields is another nonlinear example. The collapse of cod fish stocks in the Newfoundland area in 1992 forced the closing of the cod fishing industry. The strain on fish stocks by gradual overfishing resulted in too few adults remaining to produce enough fingerlings to support the harvest level. This dynamic is also affected by the volatility of the recharge of the fish stocks since the production and fertilization of eggs are influenced by ocean temperatures and currents. Introduction or removal of species can cause nonlinear changes in ecosystems while a change in the dominant species of an ecosystem can cause dramatic changes. The natural balance of the Great Barrier Reef in Australia has been disturbed by the rapid growth of crown of thorns starfish, a predator that eats other living animals. It moves across the coral reef leaving behind only white coral skeletons. A natural control is the Giant Briton shellfish *Charonia Tritonis*, one of the rare marine creatures that can feed on adult crown of thorns. However, the predator–prey balance has been disturbed along with a very strong growth in the crown of thorns population in the early 1990s. A female crown of thorns releases millions of eggs each year but usually only a very small number survive to become mature starfish. However, if conditions are favorable, it allows more larvae to settle on the coral reef and survive which can then lead to a population explosion of crown of thorns on the reef. The threat has subsided in recent years as survival rates declined through human intervention (see http://library.thinkquest.org/10009/cots.htm).

Some scientists believe that rapid growth in human populations and further stress on environmental systems are increasing the likelihood of nonlinear changes and the possibility of irreversible changes occurring. The hole on the ozone layer over the Antarctic is one example of the latter.

Aging and environment

As Asia's population ages, there could be some reduction in the overall impact of its population on the environment. Older people generally follow lifestyles that are less environmentally demanding. Changes in aging patterns will depend to a large extent on the level of per capital income now and in the future. Projections for richer countries such as Japan and Korea show a rapidly aging population through 2025 and onward, while populations in poorer countries will age more slowly. For example, India's population is aging slowly while China's population will age more rapidly (Figure 1.8). As people age, they tend to move around less, reducing demand on the transportation system. They may also live in smaller houses, apartments, or assisted dwellings where they may have a room with other facilities supplied jointly for the resident community. How would these developments impact on the environmental projections? Simulations run by the International Institute for Applied Systems Analysis (2005) for the United States suggest that carbon dioxide emissions could be reduced by as much as one-third by accounting for age over the next few decades. In the second-half of this century the percentage of people living in elderly households could rise from 10% to nearly 40%. Similar patterns may hold for future aging Asian populations. The pattern will depend upon when the demographic transition is reached. Populations in more prosperous countries which experienced the transition are already aging rapidly (Korea and Japan) and their ecological footprint is becoming smaller as they put less pressure on resources.

Environmental Kuznets Curve (EKC)

The relationship between environmental awareness, spending on environment, and environmental impact has been studied by looking at the environmental Kuznets curve. The EKC is the systematic relationship between income changes and environmental quality. It argues that rapidly growing economies in the industrialization phase of development where per capita incomes are low but growing rapidly tend to ignore environmental issues; however, as incomes increase they begin to spend more on environmental issues. Japan is a good example of this phenomenon. Environmental performance was weak in the 1960s and 1970s, and improved significantly in the past 25 years. Is there widespread evidence that this pattern exists in Asia and if so, what can be the expected impact on the environment in the next 25 years? Evidence suggests that the "pollute now, pay later" mentality in the United States, Japan and Europe has had a powerful impact on Asia's environmental protection perceptions and legislations in the past. However, this is changing rapidly, particularly where incomes are rising rapidly and a growing middle class is more aware of environmental issues. It is difficult to quantify these newly emerging trends on the projections for the future. An optimistic view is that while rising incomes put pressure on the environment, they also create public awareness, and a growing commitment to deal with environmental issues. Furthermore, the Asian experience as assessed by O'Connor (1994) found that countries that chose a relatively clean path to industrialization did not grow at a slower rate than those pursuing growth at the expense of the environment. The thrust of the policies suggested in this chapter suggests that there is a middle ground whereby win/win strategies and superexternalities can be adopted that clean the environment while achieving greater economic efficiency and higher productivity. The issues are discussed in more detail in Panayotou (2001) and Evans (1998).

Ecological footprint and international cooperation

The United Nations, the World Wildlife Fund (WWF), and the World Bank have worked to develop the concept of an ecological footprint. By comparing the amount of resources consumed with the biological/ecological capacity of the global environment, a measure of sustainability can be developed. The ecological footprint of a country is the total area required to produce the food and other resources required to service consumption demand plus resources and area required to absorb waste and construct infrastructure (Table 6.10). According to WWF International (2004), the world's demand for resources now exceeds the global regenerate capacity by 20%. This disequilibrium is sustained by drawing down natural resources, both renewable and non-renewable. The Asian region's use of the world's ecological resources has increased from around 15% in 1961 to 40% in the early years of the 21st century, and this trend is increasing. Moreover, the imbalance between the footprints of rich industrial countries and poorer developing countries will narrow and the environmental load will increase unless policies are taken to slow the trend toward greater resource use. This pressure can be reduced in four ways. Some of these have been touched on above. Others involve a possible shift in consumption patterns across countries to help reduce the global environmental footprint.

- Increase or at least maintain the existing global biocapacity. This can be done by reducing or eliminating the use of toxic chemicals, reducing soil erosion and degradation and preserving cropland for agriculture, protecting wetlands and watersheds, and maintaining (and expanding where possible) forests and fisheries.

Table 6.10 Ecological Footprints of Selected Countries

Country	Hectares per person			Change per capita (1991–2001)	
	Ecological footprint	Biocapacity	Ecological deficit	Ecological footprint*	Biocapacity**
Global	2.2	1.8	0.4	–2	–12
United States	9.5	4.9	4.7	7	–11
Asia Pacific	1.3	0.7	0.6	6	–11
Australia	7.7	19.2	–11.5	16	–6
New Zealand	5.5	14.5	–9	16	–13
Japan	4.3	0.8	3.6	6	–6
Korea	3.4	0.6	2.8	30	–12
Malaysia	3.0	1.9	1.1	10	–48
Mongolia	1.9	11.8	–9.9	–33	–11
Thailand	1.6	1.0	0.6	20	–1
China	1.5	0.8	0.8	14	–7
Papua New Guinea	1.3	2.6	–1.3	–8	–16
Indonesia	1.2	1.0	0.2	4	–14
Philippines	1.2	0.6	0.6	–6	–22
Sri Lanka	1.1	0.4	0.7	20	–12
Cambodia	1.1	1.0	0.1	9	–3
Lao PDR	1.0	1.4	–0.4	–4	–12
Myanmar	0.9	1.3	–0.4	10	1
Vietnam	0.8	0.8	0.0	14	6
India	0.8	0.4	0.4	1	–15
Pakistan	0.7	0.4	0.3	2	–18
Nepal	0.6	0.5	0.2	–4	–12
Bangladesh	0.6	0.3	0.3	0	–11

*Ecological footprints change with population size, average consumption per person, and resource efficiency.

**Biocapacity changes with the amount of biologically productive and its average productivity.

Source: WWF International (2004).

- Increase investment in technology to increase the efficiency of resource use. This would mean producing more (and environmentally friendly) goods and services from a given stock of ecological resources.

- Reduce consumption of luxury goods and services in countries where the ecological footprint is already large. This would allow for greater consumption in poorer countries, yet still maintain the overall global footprint while improving the global distribution of resources.
- Reduce the rate of population growth in Asia.

The balance between living standards and ecological footprints can be demonstrated by comparing the ecological footprint of each country with its Human Development Index (HDI) — a measure of how countries meet basic needs and minimum living standards through an index of educational attainment, life expectancy, and income per capital adjusted for purchasing power. These comparisons, displayed for some industrial and Asian economies in Table 6.11, suggest that the footprints of industrial countries can still be reduced, while Asian economies can reduce theirs through policy and technology measures as purchasing power and population growth continues, as mentioned earlier in the equation $I = PAT$. There are only slight differences in the HDI among industrialized countries, although Italy and Japan use far fewer resources than the United States and Sweden. By reducing resources used by the industrial countries the stress on the global environment can be reduced.

A win/win situation can be created if countries increase their HDI to levels of around 0.75 while keeping their footprints down to below 2 and preferably to 1.7 or 1.8. This can only be achieved if there is cooperation between industrialized and developing countries to share technology and work together to reduce global emissions.

There are a number of organizations already committed to this agenda including the United Nations, environmental NGOs, international financial agencies such as the World Bank and the ADB. A number of protocols and agreements have also been initiated including the Millennium Development Goals. Major accords reached on issues related to global environmental change include The Montreal Protocol on Substances That Deplete the Ozone Layer (1987), The Framework Convention on Climate Change (1992), The Convention on Biological Diversity (1992), the various agreements forged as part of the United Nations Conference on Environment and Development (UNCED) in

Table 6.11 Human Development Index and Ecological Footprint

Country	Human development index	Ecological footprint
Bangladesh	0.50	0.5
Cambodia	0.56	0.8
China	0.72	1.5
India	0.59	0.8
Indonesia	0.68	1.2
Korea	0.88	3.4
Lao PDR	0.53	0.9
Malaysia	0.79	3.0
Myanmar	0.55	0.9
Nepal	0.50	0.6
Pakistan	0.50	0.7
Papua New Guinea	0.55	1.2
Philippines	0.75	1.2
Sri Lanka	0.73	1.1
Thailand	0.77	1.6
Vietnam	0.69	0.7
Japan	0.93	4.3
USA	0.94	9.5
Sweden	0.94	7.1
United Kingdom	0.93	5.5
France	0.93	5.8
Italy	0.92	3.8

Source: WWF International (2005b).

1992, and the London Convention on the Prevention of Marine Pollution by Dumping of Wastes and Other Matters (1972 and 1996). Furthermore the CIESIN Thematic Guide essay on Formulating International Global Environmental Change Policy presents a general discussion of treaty formation and the role of treaties in environmental policy.

Final thoughts

The scope and complexity of the environmental challenge in Asia is daunting. To help fix ideas on the various issues, Table 6.12 presents

Table 6.12 Matrix of Environmental Developments, Causes, Impacts and Policy Recommendations

	Historical developments	Causes	Impacts	Future predictions/ recommendations
Population	• Dramatic demographic transition from high to low birth and death rates has resulted in a dramatic abatement in population pressure in the richest countries of East Asia (Japan and the newly industrialized economies (NIE))	• Economic growth and family planning policies are primary contributing factors	• South Asia and China will continue to be the major sources of population increase in the next few decades • Sustained pressure on a wide range of environmental resources from both urban cities and less developed rural areas	• Mid-range estimates project world population to peak at around 10 billion people toward the end of this century. Much will depend upon progress made in reducing population growth in South Asia, particularly in Bangladesh, India and Pakistan

(*Continued*)

Table 6.12 (*Continued*)

	Historical developments	Causes	Impacts	Future predictions/ recommendations
Water availability	• Fresh water withdrawal increased in Asia during the past century than in any other regions of the world, from 600 km^3 per year in 1900 to around 5000 km^3 in the mid-1980s (UNEP, 1999) • Less than half of the Asian population has access to proper sanitation, less than any other region in the world (ADB, 1997)	• Extensive use of irrigation has been responsible for the lion's share of increased water usage, with service to a larger population and industrial development responsible for the rest of the increase • World Resources Institute (1996) estimates that all Pacific fisheries are overfished and are at the risk of depletion	• Dirty water and poor sanitation are responsible for more than half a million infant deaths in the region as well as considerable illness and disability in the general population (WHO, 1992) • Fouling of waterways and surrounding river basins has even resulted in environmental refugees (WWC)	• By 2025 India is expected to be water-stressed with per capita water availability decreasing to around 800 m^3 per capita per year. China is also expected to reach the water-stress threshold of 1000 m^3 per capita per year by 2025 as water usage increases from a current estimated level of around 2300 m per capita per year

(*Continued*)

Table 6.12 (*Continued*)

	Historical developments	Causes	Impacts	Future predictions/ recommendations
Air quality	• Asia ranks as the world's most polluted region — more than five times as high as in industrial countries and Latin America (ADB, Asian Environmental Outlook, 2005) • Of the 41 cities ranked by WHO, 13 of the 15 dirtiest were in Asia	• China is the second largest emitter of carbon pollutants, right behind the United States with 1 billion tons annually. India is in fourth place (Worldwatch Institute, 2006) • At least 2/3 of acid rain is caused by coal-fired power plants with outdated pollution control equipment (Worldwatch Institute, 2006)	• Air pollution results in lost time at work, lower life expectancy, and resulting pain and suffering • Forest fires in Southeast Asia over the last decade or so, particularly in the 1997 fire in Indonesia which spread the haze to surrounding countries	• Nuclear energy is an alternative that is beginning to be considered by Asian economies. Only a few nuclear reactors are currently operating although they only contribute a small fraction of the power needs of the region and have not resulted in any reduction in carbon dioxide emissions • Fuel switching and promotion of mass transit would also result in a reduction in carbon dioxide emissions which would help in reducing global warming

(*Continued*)

Table 6.12 (*Continued*)

	Historical developments	Causes	Impacts	Future predictions/ recommendations
Land & forest resources	• WRI (1999) estimates that 60% of the remaining frontier forests in Asia are under threat from economic development, and Wilson (1998) estimates that forest cover is declining at the rate of 1% per year	• Agriculture and the expansion of farmland as well as shifting cultivation have also contributed to a decline in forest cover and decreasing soil fertility • Collection of wood for fuel mainly by poor farmers, and lack of clear property rights have also contributed to forest degradation and depletion	• Land degradation problems are widespread in northern India, Bangladesh, Cambodia, Malaysia, Thailand and Vietnam • The worst affected by desertification is Central Asia although there are significant areas of desertification in South and Northeast Asia (UNEP, 1998)	• Bringing this trend to a halt will require concerted efforts to rebuff the pressures from loggers, increased incentives for reforestation, and development of new plantations, like rubber, palm oil, fruits, and other tree crops • Checks and balances to be put in place to ensure that these subsidies are not misused by logging interests for other purposes

(*Continued*)

Table 6.12 (*Continued*)

	Historical developments	Causes	Impacts	Future predictions/ recommendations
Solid & waste management	• Beede and Bloom (1995) estimate that on average a 1% rise in income results in about a third of a percent rise in municipal solid waste • 4 million personal computers are discarded each year in China. India alone generated electronic waste worth $1.5 billion in 2003 (UNEP, 2005)	• Waste collection and management are costly. Municipal governments in Asia spend between 50% and 70% of their budgets on waste management, yet collections remain low. Between 50% and 70% of residents receive service (ADB, 1997; UNEP, 1993)	• The result is an increase in pollution of the land and waterways in and around urban areas as well as further degradation of coastal estuaries and the ocean • In India, an outbreak of the bubonic plague was linked to inadequate waste disposal (Tysmans, 1996)	• Efforts to systematically deal with electronic waste have been lagging because it has only recently been recognized as an environmental problem

(*Continued*)

Table 6.12 (*Continued*)

	Historical developments	Causes	Impacts	Future predictions/ recommendations
Energy	• Commercial energy use is also growing rapidly at around 6% per year, much faster than in the rest of the world where the average growth rate is less than 2% and less than 1% in OECD countries (ADB, 2001)	• Income growth and industrialization further heighten the need for energy • Inefficient energy management evident in many Asian countries. For example, Indonesia and Bhutan require more than 30,000 BTUs per dollar of GDP compared with 5000 BTUs for Japan and UK (EIA, 2003)	• As a result the Asian region as a whole now produces more carbon emissions than any other region • Between 1990 and 2004 carbon emissions from China and India rose by 67% and 88%, respectively (WWI, 2006)	• Half of total growth in oil consumption could come from Asia in the next 15 years (Yergin, 2006) • However, current projections of energy demand take very little notice of downward shifts in demand as a result of lifestyle changes that refocus on mass transit, or the use of environmentally friendly products such as ethanol and hybrid cars

(*Continued*)

Table 6.12 (*Continued*)

	Historical developments	Causes	Impacts	Future predictions/ recommendations
Biodiversity	• The Asian region's use of the world's ecological resources has increased from around 15% in 1961 to 40% in the early years of the 21st century, and this trend is increasing • One-third of amphibian species are threatened with extinction (Reichhardt, 2004)	• The resources of the region have been increasingly exploited both for international trade and to sustain the growing population (UNEP, 1999) • Poaching and the illegal harvesting and trade of medicinal plants and animals have increased both in Mongolia and the Republic of Korea where items are exported to the lucrative black markets of Pacific neighbors (JEC, 1997)	• In Indonesia, for example, some 1500 varieties of rice disappeared during 1975–1990 (WRI, UNEP, and IUCN, 1992) and similar trends have been observed throughout the region • By 2005, India is expected to produce 75% of its rice from just 10 varieties compared with the 30,000 varieties traditionally cultivated (Ryan, 1992)	• In all four scenarios examined, South Asia and Southeast Asia are most vulnerable and will experience the most significant deterioration between 2002 and 2025 (UNEP, 1999) • To slow this loss of biodiversity greater efforts have to be made to protect existing habitats by increasing resources devoted to monitoring encroachment and enforcing existing regulations

(*Continued*)

Table 6.12 (*Continued*)

	Historical developments	Causes	Impacts	Future predictions/ recommendations
Global warming hazards	• Scientific estimates suggest that the atmospheric concentrations of CO_2 have increased from around 275 ppm to 370 ppm over the past couple of hundred years (Hoffert *et al.*, 2002). Left unchecked, scientists estimate CO_2 concentrations will reach over 500 ppm by the end of this century	• Most of the CO_2 is caused by emissions from three sources: Asia, Russia, and the European Union and North America • Looking at the three drivers of climate change as reflected by CO_2 emissions in Table 6.7, it is evident that Asia is going to be the biggest source of emissions in the next few decades. Its rate of growth in income is	• Impact and cost of global warming is difficult to ascertain and calculate • There are costs resulting from the rise in sea levels in low-lying areas such as Bangladesh Shanghai, and other sea coasts around the Asian region. Climate change is expected (Eilperin, 2006)	• By 2025 China will be as large a producer of carbon emissions as the United States, compared to 60% in 1997 • Suggested policies focus on improving efficiency in energy utilization and slowing the move toward more energy-intensive transportation modalities by raising the prices of fossil fuels and supplying alternatives for the automobile

(*Continued*)

Table 6.12 (*Continued*)

	Historical developments	Causes	Impacts	Future predictions/ recommendations
		probably going to be faster than the other two regions, its population growth more rapid, yet its ability to apply new technology is limited because of its previous low level of income		• There is an increasing risk of a race to the bottom as countries pursue economic growth as an exclusive preoccupation while failing to address environmental challenges, if we were to "allow" human nature to take its course

Source: Author.

a matrix relating pertinent issues to historical developments, causes, impacts, and pertinent policy issues and recommendations. Population control is one important way to reduce future environmental damage. Although the size and age distribution of the workforce in 20 years is already more or less determined, additional efforts can be made to reduce the impact of population pressure particularly in densely populated areas where environmental impacts are significant. Efforts to conserve and effectively utilize dwindling air, water, and land resources are critical if Asia is to avert further environmental degradation. Scarce water resources are expected to be a particularly challenging issue in the coming decades. More effective pricing of water as well as provision of potable water for urban areas are keys to meeting these challenges. Development of cleaner alternative fuels and more effective measures to reduce air pollution are necessary to reverse the upward trend in particulate emissions and global warming. Implementation and development of new alternative sources of power, fuels and technologies along with cooperation with industrial countries will be necessary in order to make progress in reducing air pollution and effectively reducing global warming. Better energy management is also necessary to slow the increase in carbon emissions. Lifestyle changes as well as development of alternative fuels are critical components of such a strategy. Biodiversity is threatened by population pressure, industrialization, and various forms of pollution. A strategy that protects existing habitats and establishes new habitats is necessary to slow this loss of biodiversity. More resources also have to be devoted to monitoring encroachment and enforcing existing regulations.

Bibliography

Adeel, Z. and N. Nakamoto (2002). The Future of Environmental Governance in East Asia. In Z. Adeel (ed.), *East Asia Experience in Environment Governance: Response to a Rapidly Developing Region*. United Nations University Press.

Adhikari, B. (2001). Literature Review on the Economics of Common Property Resources: Review of Common Pool Resource Management in Tanzania. Report prepared for NRSP Project R7857. Natural Resources Systems Programme (NRSP).

American Association for the Advancement of Science (AAAS) (2005). Atlas of Population and the Environment. http://atlas.aaas.org.

ASEAN (1997). *First ASEAN State of the Environment Report.* ASEAN Secretariat, Jakarta, Indonesia.

Asian Development Bank (ADB) (1997). *Emerging Asia: Changes and Challenges.* Asian Development Bank.

Asian Development Bank (ADB) (2001). *Asian Environmental Outlook 2001.* Asian Development Bank.

Asian Development Bank (ADB) (2005). *Asian Environment Outlook 2005.* Asian Development Bank.

Australia Uranium Association (2003). Global Warming: Nuclear Issues Briefing paper 24. http://www.uic.com.au/nip24.htm [Retrieved 1 January 2003].

BBC News (2006). Booming Nations 'Threaten' Earth. *BBC News*, 12 January. http://news.bbc.co.uk/2/hi/science/nature/4604556.stm.

Beede, D. and D. Bloom (1995). The Economics of Municipal Solid Waste. *The World Bank Observer*, 10(2), 113–150.

Boyd, A. (2002). Environmental Cost of Asia's Development. *Asia Times*, 26 November.

Braatz, S., G. Davis, S. Shen and C. Rees (1992). *Conserving Biological Diversity: A Strategy for Protected Areas in Asia-Pacific Region.* Asia Technical Department Series No. 1993. Washington, DC: World Bank.

Bradsher, K. and D. Barboza (2006). Pollution from Chinese Coal Casts a Global Shadow. *New York Times*, 11 June.

Brandon, C. and R. Ramankutty (1993). Toward an Environmental Strategy for Asia. World Bank Discussion Paper No. 224.

Bumb, B. L. and C. A. Baanante (1996). Policies to Promote Environmentally Sustainable Fertilizer Use and Supply to 2020. IFPRI 2020 Vision Brief 40. International Food Policy Research Institute (IFPRI). http://www.ifpri.org/2020/briefs/number40.htm.

Butler, Rhett A. (ed.) (2006). Tropical Rainforests of the World. http://rainforests.mongabay.com/0101.htm [Retrieved 28 February 2006].

CAI News (2006). Pollution: A High Price for Growth. *CAI News*, 17 January. http://www.cleanairnet.org/caiasia/1412/article-59873.html.

Canada Dept of the Environment, Transport and the Regions (2000). The cost of climate change-UK report is a world first. *News release No. 354 from Dept of the Environment, Transport and the Regions.* http://www.solcomhouse.com/CostofGlobalWarming.htm [Retrieved 25 May 2000].

Clements, B., R. Hugouneng and G. Schwartz (1995). Government Subsidies: Concepts, International Trends, and Reform Options. IMF Working Paper No. 95/91, September.

Cropper, M. and C. Griffiths (1994). The Interaction of Population Growth and Environmental Quality. *American Economic Review*, 84(2), 250–254.

Cruz, W. and R. Repetto (1992). *Environmental Effects of Stabilization and Structural Adjustment Programs: The Philippines Case.* World Resources Institute.

Culverco.com (2004). Alternative Fuel Vehicles. *Energy Efficiency World website.* http://www.midamericanenergy.com/eew/more/alt.html [Retrieved 16 January 2006].

da Cunha, L. V. (1989). Sustainable Development of Water Resources. International Symposium on Integrated Approach to Water Pollution, Lisbon, Portugal.

Dauvergne, P. (1993). The Politics of Deforestation in Indonesia. *Pacific Affairs*, 66(4), 497–518.

Degrading Environment Threatens Health in East Asia (2005). China Internet Information Center website. 21 April 2005. http://www.china.org.cn/english/index.htm.

DeShazo, J. R. (1996). The Level of and Demand for Environmental Quality in Asia. Background paper for *Emerging Asia: Changes and Challenges*. Manila: Asian Development Bank.

Djuhari, L. T. (2001). Asian Ministers Pledge to Crackdown on Illegal Logging, Trading. *Environmental Network News*, 14 September.

Douglass, M. and G. L. Ooi (2000). Industrializing Cities and the Environment in Pacific Asia: Towards a Policy Framework and Agenda for Action. In D. Angel and M. Rock (eds.), *A Clean Revolution in Asia*. Sheffield: Greenleaf.

Dowd, M. (2006). Ozone Man. *International Herald Tribune*, 6 February.

Dowling, J. M. and R. Valenzuela (2004). *Economic Development in Asia*. Thomson Learning.

Dua, A. and D. C. Esty (1997). *Sustaining the Asia Pacific Miracle: Environmental Protection and Economics Integration*. Institute for International Economics.

Dutta, M. (2005). *China's Industrial Revolution and Economic Presence*. World Scientific Publishing Co. Inc.

Economist (1999). Why Greens Should Love Trade. 19 October.

Economist (2005a). 2 June.

Economist (2005b). Climatology: Changing Science. 10 December.

Economist Technology Quarterly (2006). 2 December.

Eilperin, J. (2006). Debate on Climate Shifts to Issue of Irreparable Change. *Washington Post*, 29 January.

Eiras, A. J. and B. D. Schaefer (2001). Trade: The Best Way to Protect the Environment. The Heritage Foundation. 27 September 2001. http://www.heritage.org/Research/TradeandForeignAid/BG1480.cfm.

Ekman, I. (2006). Plugging Up Potential for Oil Spills in Europe. *International Herald Tribune*, 9 February.

Elert, G. (ed.) (2001). Power Consumption of the United States. In *The Physics Factbook*. http://hypertextbook.com/facts/2001/KatyHo.shtml [Retrieved 11 March 2006].

Energy Information Administration (EIA), US Department of Energy website (2006). Energy Kid's Page: Ethanol Timeline. http://www.eia.doc.gov/kids/history/timeline/ethanol.html [Retrieved 28 February 2006].

Engelman, R. (1994). *Stabilizing the Atmosphere: Population, Consumption and Greenhouse Gases.* Population Action International.

Engelman, R. and P. LeRoy (1993). *Sustaining Water: Population and the Future of Renewable Water Supplies.* Population Action International.

Environment Canada website (2005). Climate Change Overview: The Science of Climate Change. *The Green Lane.* http://www.ec.gc.ca/climate/overview_science-e.html [Retrieved 10 May 2005].

Environment News Service (1999). 2 August.

Evans, P. J. (1998). Industry and the Environment in Asia. *TDRI Quarterly Review*, 13(3).

Falkenmark, M. and C. Widstrand (1992). Population and Water Resources: A Delicate Balance. *Population Bulletin.* Population Reference Bureau.

Financial Times (2006). 12 April.

FAO, UNDP and UNEP (1994). Land Degradation in South Asia: Its Severity, Causes and Effects Upon the People. World Soil Resources Report No. 78. Rome, Italy: FAO.

Food and Agriculture Organization of the United Nations (FAO) (2004). Earth Trends Data Tables: Coastal and Marine Ecosystems, Freshwater Resources. *Water Resources and Fisheries.* FAO.

Friedman, T. L. (2006). A Declaration of Energy Independence. *International Herald Tribune*, 27 January.

GEMS (1996). Annotated Digital Atlas of Global Water Quality. GEMS Water Collaborating Center, Ontario, Canada. Available on diskette and at http://www.cciw.ca/gems/intro.html.

GESAMP Report No. 70 (2001). A Sea of Troubles. Global programme of Action for the Protection of the Marine Environment from Land-Based Activities (GPA), United Nations Environment Programme (UNEP), The Hague, The Netherlands.

Giridharadas, A. (2006). Asian Giants Have Just Started to Leave Their Marks. *International Herald Tribune*, 25 January.

Gleick, P. (ed.) (1993). *Water in Crisis: A Guide to the World's Freshwater Resources.* Oxford University Press.

Global Environment Monitoring System (GEMS) (1996). Airs Executive International Programme. http://www.epagov/airs/aeint. World Health Organization.

Gore, A. (2006). *An Inconvenient Truth.* Rodale Books.

Greer, D. (2005). Creating Cellulosic Ethanol: Spinning Straw into Fuel. *BioCycle eNews Bulletin*, May.

Hammer, J. and S. Shetty (1995). East Asia's Environment: Principles and Priorities for Action. World Bank Discussion Paper No. 287.

Hansen, J. (1999). *The Global Warming Debate.* NASA Goddard Institute for Space Studies. http://www.giss.nasa.gov/edu/gwdebate.

Hare, B. and M. Meinshausen (2004). How Much Warming are We Committed to and How Much Can be Avoided? PIK Report No. 93. Potsdam Institute for Climate Impact Research (PIK).

Heath, J., E. Ayres, M. Possel, R. D. Bardgett, H. I. J. Black, H. Grant, P. Ineson and G. Kerstiens (2005). Rising Atmospheric CO_2 Reduces Sequestration of Root-Derived Soil Carbon. *Science*, 309(5741), 1711–1713.

Hettige, H., M. Marri and D. Wheeler (1998). Industrial Pollution in Economic Development: Kuznets Revisited. World Bank Development Research Group Working Paper No. 1876.

Hillstrom, K. and L. C. Hillstrom (2003). *Asia, A Continental Overview of Environmental Issues.* ABC-CLIO.

Hoffert, M., K. Caldeira, G. Benford, D. Criswell, C. Green, H. Herzog, A. Jain, H. Kheshgi, K. Lachner, J. Lewis, H. D. Lightfoot, W. Manheimer, J. Mankins, M. Mouel, L. J. Perkins, M. Schlesinger, T. Volk and T. Wigley (2002). Advanced Technology Paths to Global Climate Stability: Energy for a Greenhouse Planet. *Science*, 298(5595), 981–987.

Intergovernmental Panel on Climate Change (IPCC) (2001a). Climate Change 2001: Mitigation. *IPCC Third Assessment Report: Climate Change 2001.* Intergovernmental Panel on Climate Change.

Intergovernmental Panel on Climate Change (IPCC) (2001b). Climate Change 2001: Impacts, Adaptation and Vulnerability. *IPCC Third Assessment Report: Climate Change 2001.* Intergovernmental Panel on Climate Change.

Intergovernmental Panel on Climate Change (IPCC) (2005). *IPCC/TEAP Special Report: Safeguarding the Ozone Layer and the Global Climate System: Issues Related to Hydroflourocarbons and Perflourocarbons.* Intergovernmental Panel on Climate Change.

Intergovernmental Panel on Climate Change (IPCC) (2007). Climate Change 2007: The Physical Science Basis, Summary for Policy Makers. IPCC Secretariat, Paris and Geneva.

International Atomic Energy Agency (1999). Annex II: Nuclear Power Case Studies. *International Atomic Agency 2nd Scientific Forum: Sustainable Development and Nuclear Power.* http://www.iaea.org/Publications/Booklets/ScienticForum2/scientificforum2.pdf [Retrieved 28 September 1999].

International Herald Tribune (2002). Coal is the Predominant Source. 2/3 November.

International Herald Tribune (2006a). Chemical Plants Called 'Risk' for China's Rivers. 25 January.

International Herald Tribune (2006b). Array of New Species Discovered in Indonesia. 9 February.

International Herald Tribune (2006c). Business: How US Policy Undermines Hybrid Cars. 9 February.

JEC (1997). *Asian Environmental Report 1997/98.* Toyoshinsya, Japan: Japan Environmental Council.

Jha, R. (2004). Alleviating Environmental Degradation in the Asia-Pacific Region: International Cooperation and the Role of Issue-Linkage. Departmental Working Papers 2005–2001, Australian National University.

Joyce, C. (2006). Study Backs Ethanol as Gasoline Substitute. National Public Radio, 1 February.

Kataoka, Y. (2002). Overview paper on Water for Sustainable Development in Asia and the Pacific. Asia Pacific Forum for Environment and Development (APFED) First Substantive Meeting, 12 January.

Khor, M. (2006). *Land Degradation Causes $10 Billion Loss to South Asia Annually.* Third World Network.

Landler, M. (2006). Europe's Image Clashes with Reliance on Coal. *New York Times*, 20 June.

Larsen, J. (2002). Illegal Logging Threatens Ecological and Economic Stability. Earth Policy Institute. 20 May 2002.

Litvin, D. (1998). Development and the Environment: Yellow Skies, Smarting Eyes. *Economist*, 21 March.

Lynch, O. and K. Talbot (1995). *Balancing Acts: Community-Based Forest Management and National Law in Asia and the Pacific.* Washington, DC: World Resources Institute.

Macan-Markar, M. (2001). Logging Bans in Asia Prove Ineffective. *Malaysiakini.com CP-List.* Mkini Dotcom Sdn Bhd. 18 November 2001.

MacKinnon, J. (1994). Analytical Status Report of Biological Conservation in Asia-Pacific Region. In *Biodiversity Conservation in Asia-Pacific: Constraints and Opportunities, Proceedings of a Regional Conference.*

MacKinnon, J. and K. MacKinnon (1988). *Biodiversity.* National Academies Press.

Magrath, W. and P. Arens (1989). The Costs of Soil Erosion on Java: A Natural Resource Accounting Approach. Environment Department Working Paper No. 18, World Bank.

Maixner, L. Challenges and Solutions: Water Issues Facing Southeast Asia. Unpublished paper from Waverly-Shell Rock High School.

Mason, B. A. (2006). Kentucky's Lost Mountains. *International Herald Tribune*, 9 February.

McNeely, J., K. Miller, W. Reid *et al.* (1990). *Conserving the World's Biological Diversity.* Washington: World Bank, WRI, IUCN WWF.

Montana State University — Bozeman website (1999). Milankovitch Cycles and Glaciation. *Glaciers and Glacial Geology: A hypertext for the Appreciation of Glaciers, and How They Work by the Students in Geology 445 — Glacial Geology - Spring, 1999.* http://www.homepage.montana.edu/~geol445/hyperglac/index.htm [Retrieved 20 February 2006].

Morgan, G. and T. Smuts (1994). More on 'What Can be Done About Climate Change?' *Global Warming and Climate Change.* Carnegie Mellon University. http://www.gcrio.org/gwcc/booklet3.html.

MRC/UNEP (1997). Mekong River Basin Diagnostic Study: Final Report. Mekong River Commission, Bangkok, Thailand.

Murray, C. J. L. and A. D. Lopez (1996a). *The Global Burden of Disease: A Comprehensive Assessment of Mortality and Disability from Diseases, Injuries, and Risk Factors in 1990 and Projected to 2020.* Cambridge: Harvard University Press.

Murray, C. J. L. and A. D. Lopez (1996b). *Global Health Statistics: A Compendium of Incidence and Prevalence Estimates for Over 200 Conditions.* Cambridge: Harvard University Press.

Murray, G. and I. G. Cook (2002). *Green China Seeking Ecological Alternatives.* Routledge-Curzon.

National Academy Press (1991). *National Academy of Sciences 1991: Policy Implications of Greenhouse Warming.* National Academy Press.

National Academy Press (2005). *Understanding and Responding to Climate Change: Highlights of National Academies Reports.* National Academy Press.

New York Times (2006). The State of Energy. 1 February 2006.

Nordhaus, W. D. (1991). Economic Approaches to Greenhouse Warming. In R. D. Dornbush and J. M. Poterba (eds.), *Global Warming: Economic Policy Approaches.* MIT Press.

Nordhaus, W. D. (2006). The Stern Review on the Economics of Climate Change. Mimeo, November.

Nordhaus, W. D. and J. Boyer (2000). *Warming the World: Economic Models of Global Warming.* MIT Press.

Nuclear Power Corporation of India, Ltd (2005). 7N-plants in advanced stages of completion. http://www.npcil.nic.in [Retrieved 11 March 2005].

O'Connor, D. (1994). *Managing Environment with Rapid Industralization: Lessons from East Asian Experience.* Paris: Organisation for Economic Co-operation and Development.

OECD (2007). Energy: Finding a New Gear. *OECD Observer*, January.

Oldman, L. (1994). The Global Extent of Soil Degradation. In D. Greenland and T. Szaboles (eds.), *Soil Resistance and Sustainable Land Use.* Wallingford: Commonwealth Agricultural Bureau International. http://www.isric.nl/GLASOD.htm.

Ostro, B. (1994). *Estimating the Health Effects of Air Pollution: A Method with an Application to Jakarta.* Policy Research Paper WPS 1301. Washington, DC: World Bank.

Pan, P. P. (2005). Chinese Officials Sought to Hide Toxic Spill. *Washington Post Foreign Service*, 26 November.

Panayotou, T. (1990). The Economics of Environmental Degradation: Problems, Causes and Responses. Development Discussion Paper No. 335, Harvard Institute for International Development (HIID).

Panayotou, T. (2001). Economic Growth and the Environment. Working Paper No. 56, Center for International Development (CID) at Harvard University.

Paoletto, G. and C. Termorshuizen (2002). Chemical Governance in East Asia. In Z. Adeel (ed.), *East Asia Experience in Environment Governance: Response to a Rapidly Developing Region*. United Nations University Press.

Pergamon Press (1991). *World Development*, 19(1), 1 January.

Perlez, J. and R. Bonner (2005). Below a Mountain of Wealth, a River of Waste. *New York Times*, 27 December.

Pesek, Jr, W. (2005). Factoring in China's Gross Domestic Pollution. *International Herald Tribune*, 5 December.

Pierce, F. (2001). Death in a *Glass of Water. The Independent (UK)*, 19 January. http://www.arsenicfilters.com/glass.htm.

Population Bulletin, Population Reference Bureau; Sustainable Water: Population and the Future of Renewable Water Supplies, Population Action Council International website http://www.cnie.org/pop/pai/water-12/htm.

Potera, C. (2004). Air Pollution: Asia's Two-Stroke Engine Dilemma. *Environmental Health Perspectives (EHP)*, 1 August. http://www.pubmedcentral.nih.gov/articlerender.fcgi?artid=1247506.

Reichhardt, T. (2004). Can We Halt the Loss of Biodiversity? Science Developments. http://www.scidev.net/Features/index.cfm?fuseaction=readFeatures&itemid=344&language=1.

Renner, M. and Z. Chafe (2006). Turning Disasters into Peacemaking Opportunities. *State of the World 2006 Special Focus: China and India*. Worldwatch Institute.

Repetto, R., W. Magrath, M. Wells, C. Beer and F. Rossini (1989). *Wasting Assets in the National Income Accounts*. World Resources Institute.

Revkin, A. C. (2002). Dire Warning on Fossil Fuel. *International Herald Tribune*, 2 November.

Romero, S. (2006). Baby Steps Not Enough on Oil Use. *International Herald Tribune*, 4 February.

Root, T., J. Price, K. Hall, S. Schneider, C. Rosenzweig and A. Pounds (2003). Fingerprints of Global Warming on Wild Animals and Plants. *Nature*, 421, 57–60.

Ryan, J. C. (1992). Conserving Biological Diversity. In L. Brown *et al.* (eds.), *State of the World 1992*. New York, United States: W. W. Norton.

Sanders, R. (2006). Ethanol Can Replace Gasoline with Significant Energy Savings, Comparable Impact on Greenhouse Gases. UC Berkeley Press Release, 26 January. http://www.berkeley.edu/news/media/releases/2006/01/26-ethanol.

Schlamadinger, B., L. Ciccarese, M. Dutschke, P. M. Fearnside, G. Brown and D. Murdiyarso (in press). *Should We Include Avoidance of Deforestation in the International Response to Climate Change?*

Scotese.com (2002a). Climate History. *PALEOMAP Project website*. http://www.scotese.com/paleocli.htm [Retrieved 20 April 2002].

Scotese.com (2002b). Paleoclimate Animation: Paleomap Project. *PALEOMAP Project website*. http://www.scotese.com/paleocli.htm [Retrieved 20 April 2002].

Smil, V. (1996). *Environmental Problems in China: Estimates of Economic Cost*. East-West Center Special Reports No. 5. Honolulu, Hawaii: East-West Center.

Smith, K., C. Corvalan and T. Kjellstrom (1999). How Much Global Ill Health is Attributable to Environmental Factors? *Journal of Epidemiology*, 10(5), 573–584.

Socolow, R. and S. Pacala (2004). Stabilization Wedges: Solving the Climate Problem for the Next 50 Years with Current Technologies. *Science*, 305(5686), 968–972.

Spotts, P. N. (2004). Stabilizing the Global Greenhouse May Not Be So Hard. *Christian Science Monitor*, 13 August.

Srinivas, A. (2005). All That Gas. *Outlook India*, 25 April.

Stern, N. (2006). *Review on the Economics of Climate Change*. http://www.hmtreasury.gov.uk/independent_reviews/stern_review_economics_climate_change/sternreview_index.cfm [Retrieved various dates November 2006].

Stevenson, G. G. (1991). *Common Property Economics: A General Theory and Land Use Application*. Cambridge University Press.

Stockholm Environmental Institute (2003). Regional Air Pollution in Asia. Regional Air Pollution in Developing Countries (RAPIDC) International Co-Operation. Stockholm Environmental Institute. http://www.york.ac.uk/inst/sei/rapidc2/asia/aspoltex.

Tickell, C. (2002). Cold Reality of Global Warming. *Financial Times*, 2 November.

Tisdell, C. (1998). Sustainable Development: Differing Perspectives of Ecologists and Economists and Relevance to LDCs. *World Development*, 6(3), 373–384.

UNCCD (1998). *The Social and Economic Impact of Desertification in Several Asian Countries: Inventory Studies*. Geneva: Interim Secretariat of the Convention to Combat Desertification.

Union of Concerned Scientists website (2005). Each Country's Share of Global CO_2 Emissions. *Climate Science: The Science of Global Warming*. 10 August 2005. http://www.ucsusa.org/global_warming/science/each-countrys-share-of-co2-emissions.html [Retrieved 26 January 2006].

United Nations Development Programme (UNDP) (2005). *Human Development Report 2005*. United Nations Development Programme.

United Nations Economic and Social Council (2005). Review of the State of the Environment in Asia and the Pacific. Ministerial Conference on Environment and Development in Asia and the Pacific 2005. United Nations Economic and Social Council.

United Nations Environment Programme (UNEP) (1993). *Environmental Data Report 1993–1994*. Oxford, UK: Blackwell.

United Nations Environment Programme (UNEP) (1998). The Social and Economic Impact of Desertification on Several Asian Countries: Inventory Study. *United Nations Convention to Combat Desertification*. United Nations Environment Programme.

United Nations Environment Programme (UNEP) (1999). *GEO-2000 Global Environmental Outlook 2000*. United Nations Environment Programme.

United Nations Environment Programme (UNEP) (2001). Protecting the Oceans from Land-Based Activities: Land-Based Sources and Activities Affecting the Quality and Uses of the Marine, Coastal and Associated Freshwater Environment.

GESAMP Reports and Studies, 71: IMO/FAO/UNESCO-IOC/WMO/WHO/IAEA/UN/UNEP Joint Group of Experts on the Scientific Aspects of the Marine Environmental Protection. United Nations Environment Programme.

United Nations Environment Programme (UNEP) (2002). *GEO-3 Global Environment Outlook 3: Past, Present and Future Perspectives.* United Nations Environment Programme.

United Nations Environment Programme (UNEP) (2005). *GEO Yearbook 2004/2005: An Overview of Our Changing Environment.* United Nations Environment Programme.

UNESCAP/ADB (1995). *State of the Environment in Asia and the Pacific, 1995.* United Nations Economic and Social Commission for Asia and the Pacific, and Asian Development Bank. New York, United States.

Uranium Information Center Limited website. http://www.uic.com.au/nip24/htm.

US Department of Energy (2005). DOE Outlines Research Needed to Improve Solar Energy Technologies. US Dept of Energy Press Release, 12 August.

US Department of Energy (2006). Carbon Sequestration Focus Areas. Office of Science. http://cdiac2.esd.ornl.gov/scienceman.

US Department of the Interior, US Geological Survey (1992). *The Great Ice Age.* Interior. US Geological Survey.

US Energy Information Administration website. http://www.eia.doe.gov/ooiaf/ieo/enduse/html.

US Energy Information Administration (2003). *International Energy Outlook 2003.* US Department of Energy.

US Energy Information Administration (2004a). *India's Environmental Issues.* US Department of Energy.

US Energy Information Administration (2004b). *Energy Information Administration Country Analysis Briefs.* US Department of Energy.

US Energy Information Administration (2005a). *Energy Information Administration China Country Analysis Brief*, China. US Department of Energy.

US Energy Information Administration (2005b). http://www.eia.doe.gov/emmmeu/cabs/china.html.

US Energy Information Administration (2005c). *International Energy Outlook 2005.* US Department of Energy.

US Environmental Protection Agency (EPA) (2004). Global Warming: Climate's come a long way. *Kid's site US Environmental Protection Agency website.* http://www.epa.gov/globalwarming/kids/history.html [Retrieved 15 March 2006].

US Environmental Protection Agency (EPA) (2005). Global Warming: So what's the big deal? *Kid's site US Environmental Protection Agency website.* http://www.epa.gov/globalwarming/kids/bigdeal.html [Retrieved 15 March 2006].

Wald, M. L. and E. L. Andrews (2006). Call to Cut Foreign Oil is a Refrain 35 Years Old. *New York Times*, 1 February.

WCMC (1995). Protected Areas Data Unit of the World Conservation Monitoring Centre. Unpublished data. World Conservation Monitoring Centre (WCMC), Cambridge.

Wheeler, D. (2001). Racing to the Bottom? Foreign Investment and Air Quality in Developing Countries. Development Research Group, World Bank.

Whittington, D. (2003). Municipal Water Pricing and Tariff Design: A Reform Agenda for South Asia. *Water Policy*, 5(1), 61–76.

Wilson, E. (2001). Is Agriculture Raiding South Asia's Water Supplies? Water Scarcity and Water Reform in South Asia. In P. Pinstrup-Andersen and R. Pandya-Lorch (eds.), *The Unfinished Agenda: Perspectives on Overcoming Hunger, Poverty and Environmental Degradation*. International Food Policy Research Institute (IFPRI).

World Bank (1992). *World Bank Development Report 1992*. Oxford University Press.

World Bank (1999). *World Development Report 1999/2000: Entering the 21st Century: The Changing Development Landscape*. World Bank.

World Bank (2001a). *Globalization, Growth and Poverty. A World Bank Policy Research Report*. World Bank.

World Bank (2001b). Development, Poverty and the Environment — Tracing the Connections. In *Making Sustainable Commitments: An Environment Strategy for the World Bank*. World Bank.

World Bank (2001c). The Strategic Framework. In *Making Sustainable Commitments: An Environment Strategy for the World Bank*. World Bank.

World Bank (2003). *World Development Report 2003: Sustainable Development in a Dynamic World*. World Bank.

World Bank (2005). Kyoto Protocol Enters into Force, press release, 15 February.

World Commission on Forests and Sustainable Development (1999). *Our Forests, Our Future*.

World Health Organization (WHO) (2005a). Tuberculosis. *Fact Sheet No. 104*. World Health Organization.

World Health Organization (WHO) (2005b). Ecosystems and Human Well-Being: Health Synthesis. *Millennium Ecosystem Assessment*. World Health Organization.

World Health Organization (WHO) and UNICEF (2000). *Global Water Supply and Sanitation Assessment 2000 Report*. Geneva and New York. http://www.who.int/water_sanitation_health?Globassessment/Glassessment7pdf.

World Resources Institute (WRI) (1996). *World Resources 1996–97: A Guide to Global Environment*. Electronic Database. New York: Oxford University Press.

World Resources Institute (WRI) (2005a). Ecosystems and Human Well-Being: Wetlands and Water Synthesis. *Millennium Ecosystem Assessment 2005*. World Resources Institute.

World Resources Institute (WRI) (2005b). Ecosystems and Human Well-Being: Opportunities and Challenges for Business and Industry. *Millennium Ecosystem Assessment 2005*. World Resources Institute.

World Resources Institute (WRI) (2005c). Ecosystems and Human Well-Being: Desertification Synthesis. *Millennium Ecosystem Assessment 2005.* World Resources Institute.

World Resources Institute (WRI) (2005d). Living Beyond our Means: Natural Assets and Human Well-Being Statement from the Board. *Millennium Ecosystem Assessment 2005.* World Resources Institute.

World Resources Institute (WRI) (2005e). Ecosystems and Human Well-Being: Biodiversity Synthesis. *Millennium Ecosystem Assessment 2005.* World Resources Institute.

World Water Commission (2000). *World Water Vision: Commission Report. A Water Secure World: Vision for Water, Life and the Environment.* Paris.

Worldwatch Institute (2006). *State of the World 2006 Special Focus: China and India.* Worldwatch Institute.

WRI, UNEP and UNDP (1992). *World Resources 1992–93: A Guide to the World Environment.* New York, United States, and Oxford, United Kingdom: Oxford University Press.

WWF International (2004). *Report on the Living Planet.* Washington, DC.

WWF International (2005a). Fishing Subsidies. *Global Challenges and World Trade.* WWF International.

WWF International (2005b). *The Ecological Footprint and Natural Wealth.* Washington, DC.

WWF-World Wide Fund for Nature (2005). *Asia-Pacific 2005, The Ecological Footprint and Natural Wealth.* WWF-World Wide Fund for Nature.

Yergin, D. (2006). Ensuring Energy Security. *Foreign Affairs*, 85(2).

Young People's Trust for the Environment website (2006). Factsheet: Global Warming. http://www.yptenc.org.uk/docs/factsheets/env_facts/glob_warm.html.

Zimov, S., E. A. G. Schuur and Chapin, III, F. S. (2006). Climate Change: Permafrost and the Global Carbon Budget. *Science*, 312, 1612–1613.

Chapter 7

China and India

Study the past if you would divine the future.

— *Confucius (551–479 BC)*

If you want the present to be different from the past, study the past.

— *Spinoza (1632–1677)*

Introduction

We devote a separate chapter on the two giant economies of the Asia region for two reasons. First, these economies alone have over 2 billion people, nearly half the world's population. And many are poor. For this reason alone they deserve special attention. Secondly, they are among the most dynamic and rapidly growing economies in the global economy and are exerting growing influence in commerce, international trade, and environmental issues. Better understanding of these forces underlying economic growth and social change in China and India are important for understanding and projecting the future course of economic development in Asia.

China

History

There are several institutional developments that paved the way for rapid economic growth in China beginning in the late 1970s, even in the continuing presence of a highly structured communist party, central planning, and the absence of private property. These institutional developments created zones of free enterprise and competitiveness that were allowed to exist alongside the centrally planned apparatus. Resources flowed into less regulated activities where market prices prevailed and high profit opportunities were opening up. Naughton *et al.* (2002) describes it as follows:

> High market prices and low taxes were a recipe for high profits so people and resources continued to be attracted to the market sector. The result was a kind of Swiss cheese economy, in which the old government-run economy provided the matrix for all economic activity, while holes were opened up to provide space for market activity. The state-run economy was gradually hollowed out, but the economy as a whole continued to grow. Agriculture, township and village enterprises, foreign trade sparked by the growth of the SEZs — all these growth areas provided power arguments for the continuation and strengthening of reforms (p. 340).

In this section we review briefly, the major sources of competition and growth.

Agriculture and the household responsibility system

The first source of competition and growth in China was the household responsibility system (HRS). This was a contracting system that revolutionized agriculture in the early years after Deng Hsiao Ping had gained control of the leadership reins. Historically, agricultural collectives had certain output quotas determined by the state that were to be sold at pre-determined prices. Incentives for increasing productivity, undertaking innovation, technological improvement, or increasing output were minimal. Output growth in agriculture stagnated and rural poverty levels were high. To bring more dynamism to the sector and increase profits, some local leaders began dividing up the land among work units or families and then sub-contracting production quotas to these smaller units rather than contracting to large agricultural collectives. One particularly popular method was called "contracting everything to family", wherein families had to sell a set amount of produced goods to state officials at stipulated prices, but anything produced above these quotas could be sold at higher prices to other buyers. They buyers were usually entrepreneurs that sold this excess production to the free market where prices were free to fluctuate.

One tenet of the HRS as envisioned by the central government was that land tenure within the village was secure and also that those with tenure could legally transfer these rights to others. It is also clear from several detailed village level studies that local village leaders have extensive latitude in establishing land tenure rights as well as the transfer of these rights. This village power tends to supersede the general dictates handed down from the central government.

Growth in agricultural production for the first few years following the reform was spectacular (see Table 7.1). Agricultural sector growth averaged 7.1% from 1979 to 1984 compared with 2.7% from 1970 to 1978. This agricultural sector growth provided growth impetus to the rural economy while providing income for rural residents to buy industrial goods. Output growth and industrial growth also accelerate but not as fast as agriculture. After 1985, agricultural growth slowed, although still a strong 4% growth in the subsequent decade. In the meantime, industrial growth strengthened further.

Table 7.1 Production Growth in Agriculture Industry and TVEs

	1970–1978	1979–1984	1985–1995	1996–2000	2001–2005
GDP growth	4.9	8.5	9.7	8.2	9.5
Agriculture	2.7	7.1	4.0	3.4	3.9
Industry	6.8	8.2	12.8	9.6	10.7
Non-farm rural enterprises (TVEs)	n.a.	12.3	24.1	14.0	

Source: Anderson (2003); ADB (2006).

Industry and growth of town and village enterprises

Town and village enterprises (TVEs) are a special feature of the Chinese economy found nowhere else in the world. They are capitalist profit making enterprises even though there is no private ownership of the means of production in the Chinese communist system[1]. As is evident from Table 7.1, TVEs have grown at a spectacular rate since the 1980s, and they are largely responsible for the rapid growth of industrial production in the last 25 years. Employment has grown from 28 million in 1978 (7% of the workforce) to 140 million (25% of the workforce) in 2005. They contribute more than 55% of industrial output, more than 30% of GDP, and more than 50% of exports. What is even more remarkable is that there are 22 million TVEs spread over China and they are generally classified as small and medium enterprises, with less than 100 employees per establishment.

Town and village enterprises are a cooperative venture between the town and village authorities and local rural businessmen. The land and improvements come from the local towns and villages, and the technology and labor force from the local villages and towns themselves. Legal arrangements vary although they have characteristics in common. In the beginning TVEs produced simple consumer goods to satisfy demand from farmers and others in the rural communities that benefited from the rapid growth of agricultural production that followed the partial privatization of the agricultural production system as household responsibility system was implemented. The growth of TVEs was also the result of pricing decisions and price controls that made agricultural enterprises much less profitable than TVEs, and TVEs also fill holes in the structure of production by large-scale State Owned Enterprises (SOEs) that created profit potential for the TVEs (Putterman, 1997). TVEs absorbed surplus rural labor as a result of rising productivity in agriculture and they helped to smooth out the seasonal work cycle in rural areas by providing off-season work. They also served to increase the supply of consumer goods and helped reduce inflationary pressure when price liberalization was phased in.

---

[1]Housing in rural areas, and to a lesser extent in cities, can be privately owned.

TVEs also facilitated the development of local markets and contributed to the expansion of exports. In addition the development of TVEs helped pave the way for the emergence of private firms as some TVEs were transformed into privately owned ventures. To support the growth of TVEs the government provided some incentives for the development of TVEs to local authorities.

The interaction between rapid agricultural growth and income of the output supplied to rural residents by the TVEs created an upward spiraling cycle of economic growth and income generation that fueled overall growth in the Chinese economy of close to 10% per annum for more than a decade. Before reforms in agriculture were adopted in 1978 consumption was constrained to low levels and many commodities were available only through a ticket rationing system. As this system was dismantled, agricultural incomes mushroomed. In the early years following reform TVEs began to supply a wide variety of labor-intensive consumer goods including textiles and garments, bicycles, processed food products, and light consumer manufacturers such as sewing machines. Demand shifted and was augmented by growing demand for washing machines, TV sets and electrical appliances. Whether TVEs were as efficient as the private sector and whether they created as much employment as they might have if the ownership set up had been different has been the subject of considerable discussion (Putterman, 1997). As more private enterprises and greater emphasis on foreign trade developed with the formation of the special economic zones, the importance of the TVEs began to wane in the late 1990s.

Foreign trade and SEZs

Foreign trade has grown dramatically over the past two decades and has been a major factor in the rapid growth of the Chinese economy. As a share of GNP international trade has increased rapidly. Imports are now 30% of GDP, up from 5% in 1978. In recent years the trade surplus with the United States has increased. This surplus is part of an overall pattern such that the Chinese trade surplus with European Union, Japan and the United States has grown to about $350 billion by 2005. The deficit with the rest of the world — primarily other countries in greater China such as Taiwan and Hong Kong and also the members of ASEAN — has also grown to around $250 billion leaving an overall surplus of $100 billion. China imports raw materials and semi-finished parts from the rest of the world, including from other countries in greater China and ASEAN, and exports finished goods to the EU, the United States and Japan. This pattern of Chinese trade can be thought of as a trade bridge where surpluses with industrial countries are used to liquidate deficits to other Asian economies. When viewed in this light the large Chinese surpluses with the United States and Europe can be viewed as a part of the global pattern of trade where China plays a key role as the production platform for a wide array of goods. It may also be part of a wider international trade and development strategy to raise incomes by attracting foreign direct investment and maintaining a favorable exchange rate. (See Chapter 3 for more details on this relationship with industrial countries.)

As part of its strategy to undertake reforms in 1978, the Chinese government included an initiative to open up its international trade regime. As a component of this policy it established special economic zones in several southern provinces in 1980 including Shenzhen, Zhuhai, and Shantou in Guangdong Province and Xiamen in Fujian Province. It also designated the entire province of Hainan as a special economic zone. In 1984, China further opened 14 coastal cities to overseas investment, and in 1985 extended the open economic zones of the Yangtze River Delta, Pearl River Delta, Xiamen–Zhangzhou–Quanzhou Triangle in south Fujina province, the Shandong Peninsula and the Liaodong Peninsula, Hebei, and Guanzxi into an open coastal belt. In 1990 China further opened the Pudong New Zone in Shanghai to overseas investment and opened up more cities in the Yangtze River valley. Beginning in 1992 a number of other cities were opened up to foreign investment and also all the capital cities of inland provinces and autonomous regions. In addition, 15 free trade zones, 32 state-level economic, and development zones were opened as well as 53 industrial development zones.

In one sense SEZs have served as a policy experiment where a variety of policy instruments and institutional arrangements have been tested and where effective arrangements have been extended to the rest of the country. As a policy tool, the SEZ's role has been more significant than the amount of foreign trade generated or the amount of foreign direct investment attracted.

In the next phase of expansion the development of inland cities is designed to link the coastal areas with inland areas via river transport. The government has invested significant resources into developing and/or improving the infrastructure located both within the SEZ and in support of the SEZ. Locating businesses in the SEZs affords tax breaks and waiver of import and export taxes and other restrictions. The SEZs have been a spectacular success in achieving their stated purpose, namely to attract foreign direct investment and to develop export platforms for a variety of industries. They have also attracted new technology and managerial expertise as foreign firms have brought these into the special economic zones. In the case of the Pudong New Zone, the state also permits the zone to open financial institutions and run tertiary industries. The Pudong area has also spearheaded capital and skill-intensive industries in automobiles, microelectronics, household electrical appliances, biomedicine and optical, mechanical and electrical products.

The SEZs and the foreign investment they have attracted have resulted in sustained rapid growth in production and exports for the past 25 years. About half of Chinese exports are from foreign funded enterprises, primarily in the SEZs in the southern provinces. Foreign direct investment (FDI) has grown from almost nothing in 1980 to over $50 billion annually in 2003. The upturn in FDI began in earnest in the early 1990s and continued at an unrelenting pace until the Asian financial crisis when it dipped a bit only to return to its upward trend in 2000 (see Figure 7.1).

These official estimates of FDI have to be discounted somewhat because of "round tripping" of investment to take advantage of tax breaks. Round tripping occurs when money from China is sent overseas (to Hong Kong, Caribbean Islands and many other locations) and then returned as FDI to take advantage of fiscal incentives and improved investor protection offered by China to foreign investors. World Bank economists estimate round tripping at around 25% of

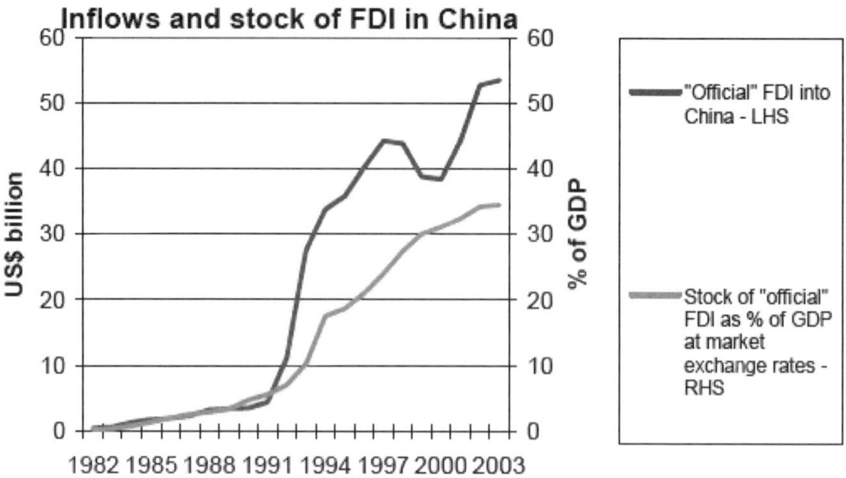

Figure 7.1 FDI in China

Source: Erskine (2004).

Table 7.2 China's Utilized FDI Investment Inflows by Country for 2004

Country	FDI in billions
Hong Kong	18.9
Virgin Islands	6.7
South Korea	6.2
Japan	5.4
United States	3.9
Taiwan	3.1
Singapore	2.0
Germany	1.0

Source: US Department of State (2005).

official FDI in recent years and estimates by the Asian Development Bank (ADB) are even higher at around 40% (Gunter, 2004; Xiao, 2004; see also Table 7.2 which shows that a large share of FDI came from Greater China and Virgin Islands, a popular round trip site).

Compared with other developing countries this adjusted stock of FDI may not be that high. Erskine (2004) estimates it could be as little as 7% of GDP, which is well below the global average of 22% of GDP. Nevertheless, the inflow of FDI has been remarkable given the short time horizon. Even though FDI has been responsible for around 10% of fixed capital formation since 1979 it has been well utilized and has combined with domestic investment as the main driver of growth in the reform era. Since 2001, investment has contributed an average of over 6% to overall growth of 9% per annum, about two-thirds of the overall increase in GDP.

Rapid growth in exports has led to persistent trade surpluses and increased tensions with the United States and the European Union. Trade surpluses and current account surpluses, which reached 7% of GDP in 2005, have also resulted in a build up in foreign reserves, which exceeded $1 trillion by the end of 2006 up from $212 billion in 2001 and much lower levels in earlier years. China's international reserve position puts it three times higher than its closest rival in Asia (outside of Japan), Taipei, China, which has a reserve position of $250 billion. Trade and current account surpluses have resulted in increasing pressure for China to revalue its currency, which it did earlier in 2005. However, the revaluation was small compared to expectations of its trading partners, particularly the United States and the European Union.

World Trade Organization

China's accession to the World Trade Organization (WTO) in 2001 has created a number of challenges and opportunities for China. From the trade point of view, China will have to lower tariffs in accordance with WTO rules. There has been speculation that this is exerting hardships for Chinese farmers who would be subject to more competition from foreign products. While agricultural surpluses have been high for several years, a trend toward increased foreign imports has not appeared yet. In general, import tariffs are already relatively low at around 10% and effective tariff rates are even lower because of the large number of exemptions (Bergsten *et al.*, 2006, pp. 81, 82).

Trade in agriculture is a particularly interesting area since accession to the WTO requires dismantling a number of restrictions in agriculture and it also presents a number of opportunities depending on the outcome of the ongoing trade negotiations in the DOHA round. The adaptation of biotechnology in agriculture is also an important area for study.

Rather than explore these areas, this section focuses on trade in textiles and clothing. Quantitative restrictions existing under the multifiber agreement were phased out over a 10-year period which ended at the end of 2004. When China's exports of textiles and clothing shot up in the first quarter of 2005, the European Union and the United States took steps to invoke a safeguard clause which is allowed in the WTO commitment by China and which again allows the United States and the European Union to control the increase in textile and apparel exports. These agreements limit growth to the United States and the European Union in 21 categories of textile products to ranges varying from 8%–12% to the European Union from 2005–2007 and to the United States of 10%–15% in 2006 and slightly higher levels in subsequent years. Clothing manufacturers will feel the impact of new restrictions as about 70% of the restricted imports are apparel products. Asian Development Bank (ADB) (2006, p. 121) suggests that these restrictions on Chinese exports will continue until 2016 as China can be treated as a non market economy under WTO rules and as a result China could become the target of trade safeguards in other countries. Since China has had a longstanding internationally competitive position in textiles and apparel account for 18% of total exports in 2001 to 2005 and employing 10 million people, these actions will have deleterious effects on the sector.

Intellectual property protection has been a major source of contention between China and foreign investors in the United States and the European Union. China has pirated videos, CDs, and computer software, and these infringements have not abated but rather continue to increase. China is an especially big violator of software copyrights. Software piracy resulted in an estimated $527 million in losses to the copyright owners. At least 30 CD-ROM plants pirating software in China are reported, according to the Business Software Alliance's report (http://www.american.edu/TED/hpages/ipr/cheng.htm).

By not enforcing intellectual property rights the Chinese authorities are reducing the incentives of domestic inventors, researchers and scientists. Furthermore these actions also hurt the bilateral relations with many countries, especially the United States. China has taken some steps to deal with intellectual property rights including making copyright infringement a criminal offence. However, China continues to top the list of countries with copyright violations. Recent developments with internet services such as Google and Yahoo also point to the importance not only of copyright but also of freedom of speech and the rights of the Chinese government to censor the flow of internet material. There have also been recent lawsuits by several US companies against the leading Chinese search engine Baidu for providing free music downloads. The tension between western internet providers and the Chinese authorities as well as the tensions within China between potential dissidents and the regime can only be expected to intensify as the internet becomes more well established and accessible.

Issues and Challenges Facing China

Demographic transition and aging

China has the largest population in the world and the amount of arable land per capita is limited by comparison with many other countries. It has only 7% of the global supply of arable land. As a result, there have been concerted efforts to reduce the rate of population growth for many years. From 1970 to 1979 concerted efforts were made by urging women to adopt a "late, long, few" policy. This called for later childbearing, more space between births, and fewer children. The late, long and few policy was successful or at least the policy coincided with a dramatic reduction in fertility. During the decade of the 1970s there was a significant decline in the total fertility rate from 5.9 to 2.9. Such a decline in total fertility, although perhaps not quite as dramatic, would be expected as part of the demographic transition that has been experienced in many developing countries even without a government campaign. The power of such a demographic transition depends primarily on a few key variables, most importantly the educational status of women and their labor force participation rates.

Evidence from a variety of countries suggests that where women are educated and are also participating in the labor force, the pressure to have fewer children is very strong. Both of these factors were present in China. Literacy rates for women were high (around 65% in 1980 and climbing to 86.6% by 2000 (UNESCO, various issues) and so were labor force participation rates (75% in 1980 according to ILO, 2003). In addition a large proportion of women in the child bearing years were using birth control (87% — IUD for the most part — according to Hesketh *et al.*, 2005).

Therefore, it is not surprising that the total fertility rate declined even though the rate of urbanization was not particularly high because of restrictions on migration from rural areas to the city. After 1979, government policy on child bearing became more restrictive. The one-child policy consisted of a set of regulations governing the appropriate size of Chinese families. Without going into detail, the one-child policy applied to urban residents and government officials, and the policy was generally enforced with few exceptions. In rural areas, a second child was allowed after 5 years although the provision was sometimes applied only if the first child was a girl. The one-child policy was enforced by local authorities. As a result there was some variation in compliance. Abortion rates were low because birth control methods were long-term, monitored by health workers and effective.

After 1979, the total fertility rate declined further, although at a more modest rate to the range of 1.7 in 2004, 1.3 in urban areas, and just under 2 in rural areas (Hesketh *et al.*, 2005). The strength of the impact that the one-child policy had on total fertility is debatable. Further declines in fertility could have been expected as a result of the underlying forces that led to the decline from 1970 to 1979 compounded by the continuing pace of urbanization, high labor force participation rates of women, and continued high literacy and further upward trends in years of schooling completed. In highly urban and well-educated countries with high per capita income such as Singapore and Japan, fertility rates are between 1.5 and 1. With increasing income and urbanization there is no reason why China could not match these figures in a few decades.

We do know that the sex ratio, defined as the proportion of male live births to female live births has been changing rather dramatically since the one-child policy was established. As a guide we know that in industrialized countries this ratio ranges from 1.03 to 1.07, that is, a few more males are born than females. In China there has been a steady increase in the sex ratio, from 1.06 in 1979 (well within the industrialized country boundaries) to 1.17 in 2001 (Kang and Wang, 2003, pp. 88–98).

What is happening in total to change this ratio is unclear. Though they are technically illegal there have been some abortions, the pattern of which is related to sex selection with the first pregnancy in urban areas. In rural areas, second children could be aborted if they are females. Non-registration of female births also contributes to the sex-ratio gap. The more ominous outcome of infanticide is reportedly low although infant mortality of girls is somewhat higher than for boys as a result of less aggressive health maintenance for girls. Incidentally sex ratios are also high (in the range of 1.12 to 1.20) in many other Asian countries, where male children are also preferred, probably as a result of abortion as ultrasound tests for sex during pregnancy have become more widely available.

The longer term consequences of the sex-ratio imbalance in China are open to speculation. Men unable to marry and have children may become more aggressive or antisocial. There may also be more mail order brides as the bride price goes up to attract women from other Asian countries, many of which are having similar difficulties. Furthermore, attitudes toward women may be changing as more women become educated and their relative scarcity makes them more desirable (Hesketh *et al.*, 2005).

Finally, the one-child policy has been modified somewhat in recent years as its importance and popularity have waned. Couples no longer need to obtain permission to have their first child, so that local birth quotas will be dispensed with. However, pressure to scrap the policy completely has been resisted.

How will China's population policy affect the population age distribution and the necessity for providing social security for the elderly? Given the pattern of population growth, the United Nations (2004)

Table 7.3 Changes in Working Population in China

Decade	Changes in working population (aged 15–59) by millions of people
1975–1985	163
1985–1995	124
1995–2005	107
2005–2015	33
2015–2025	–29
2025–2035	–77

Source: United Nations Population Division (2004).

quoted in Bergsten *et al.* (2006) shows that China's working age population (15–59) is expected to grow only slightly between 2005 and 2015 and with decline thereafter (refer also to Table 7.3) as the total fertility rate continues to fall and longevity increases. By 2030 the projected share of the elderly in total population will exceed that of the United States. Currently, without a well-funded pension system this suggests that the majority of the elderly will have to be cared for by those in the workforce. Of course, as the population ages the ratio of working age to those over 60 will also decline from 8 in the 1950s to around 2 by 2020. This will present a dramatic challenge to the society over the next few decades in terms of how to care for and support the elderly. The rising dependency ratio may also undermine household savings and constrain future growth.

Urban/Rural divide and income inequality

Migration from rural to urban areas is a prominent feature of most models of economic development. The classic article by Nobel Prize winner Arthur Lewis (1954) and expanded upon by Fei and Ranis (1961), features the movement of low productivity labor from the rural sector to more productive employment in the urban sector. Such movement raises both the productivity of the rural sector as population movements raise the productivity of remaining workers and the urban sector where the new migrants earn more money than they did previously working on the farm. Such rural to urban migration tends to bring wages in the two sectors into equilibrium, although a disparity remains because urban workers typically have more capital to work with and are more highly skilled and educated. In industrial countries and also in developing countries the rural/urban standards of living measured by per capita income differentials would be a factor of between 1.5 and 2 times that of rural residents. In China these differentials are substantially higher, as much as 6 to 1, once the higher volatility of rural incomes and the differences in social benefits are taken into account (see Whalley and Yui, 2006; Zhao, 1997).

In China a set of historical circumstances has created a situation whereby this natural flow of labor which contributes to overall economic efficiency has been interrupted. The household registration system, *Hukou*, which was installed as a regulation in 1958, requires that all changes of residence have to be registered with both origin and destination governments, and that approval of said movement from the place of birth origin must be approved by both origin and destination government offices, usually the police station.

The result of this policy was to effectively restrict rural migration to those who were willing to risk breaking the law. As the pressure to migrate increased, particularly following the growth in agricultural output between 1978 and 1983 which brought an end to food rationing and allowed migrants to buy their own food, there was some relaxation of migration restrictions. Peasants no longer had to report to collectives for daily work schedules since the collectives had been dismantled. With the rise of TVEs rural workers were allowed to move to small towns to work. In fact, during the mid-1980s rural enterprises provided an average of 10.8 million jobs per year (Zhao, 1997). Nevertheless, over the next two decades from 1985 to 2005 the rate of rural to urban migration did not increase beyond 0.25% per annum. Failure to obtain *Hukou* certification prevented rural migrants from obtaining any of the perquisites available to legitimate urban residents, including children's education, housing and other social benefits. They were often employed in the construction industry and provided temporary housing by their employers.

The *Hukou* system and the restrictions on rural to urban migration have been significantly relaxed in recent years and this has resulted in a rapid increase in the size of what became known as the "floating population". Because their status was so tenuous, most migrants never established permanent residence and they became known as the "floating population", moving back and forth between rural and urban locations depending on the availability of work and other personal factors.

Because of all of the hardships and uncertainties that potential migrants face, many potential migrants simply elect not to migrate, despite the higher wages they could earn in the city. The barriers to migration have imposed a substantial dead weight loss on the economy as well as creating a social situation where urban residents earn more, have more benefits, and are blessed with more social spending for health and education. They feel superior to those living in rural areas and the disparity in living standards and income creates further divisions in the social fabric of the communist system. Social tension, crime, protests by farmers and other rural residents have increased. Nevertheless, despite these problems and relaxation of some restrictions, the basic features of the *Hukou* system remain intact.

The slow pace of migration combined with rapid economic growth in the coastal provinces as a result of increased foreign and local investment has resulted in growing income inequality. Five provinces along China's Eastern Seaboard account for one-fifth of the country's population, two-fifths of its GDP, three-quarters of its exports, and its entire trade surplus. Even more surprising, Guangdong and Zhejiang account for 80% of China's trade surplus and these two provinces dominate China's international presence and provide its competitiveness on the global stage.

These provinces are now experiencing faster increase in costs (e.g., wage, land, etc.) as a result of such rapid growth. However their dominance in the country's international trade is still increasing. Furthermore, rich provinces are more likely to understate their GDP and poor provinces to overstate, and as a result the real gap could be larger. However, living costs are also lower in the interior provinces and that could offset some of the apparent real income differences.

There are other possible explanations for regional disparity besides coastal location. For example, export value per capita for the five provinces along the Eastern Seaboard is about four times larger than that of the four coastal provinces directly to their north. Such a large gap would suggest that being coastal may not even be the most important factor in explaining China's regional disparity. Rather, it could be the cluster effect of FDI in these southern coastal provinces and the economies of scale that result from this clustering which has been reinforced by brain drain away from the inland provinces. This has been reinforced by government policy that has supported growth in these provinces. To offset the growing divergences between coastal and interior provinces the government is trying to channel more investment to the interior.

The differences in natural endowment could also help to explain the origin of the regional disparity. The Yangtze Delta has benefited from favorable climate conditions that made the region's agriculture more productive than elsewhere. The advantage in agriculture evolved into advantages in commerce, industry and technology as well (see discussion in Morgan Stanley, 2006).

The regional disparity severely limits China's ability to use macro-instruments like interest rates and exchange rates to manage its business cycle. This is why China has a tendency to use administrative measures to manage its business cycle. But, political decentralization has made administrative measures much less effective than before.

Another measure of growing income inequality is reflected by rising measures of overall income inequality such as the Gini coefficient. The Gini coefficient for the country as a whole has been rising since 1952. Between 1952 and 1978, the Gini rose from 22.4 to 29.3. For most of the 1970s it hovered between 28 and 30. In the 1990s it rose further to 33 in 1995 and further to 37.2 by 2000 (Gajwani and Kanbur, 2006). Morgan Stanley (2006) estimates Gini to be even higher, from 40 to 50 which would put China among the economies with the greatest income inequality globally. These estimates by Stanley, in line with anecdotal evidence suggest that the Gini coefficient has increased even further since 2000. Other evidence supports the deterioration in income distribution. For example, Barro type regression (Barro *et al.*, 1995) carried out by Raiser (1998) suggests that income per capita has been diverging across the provinces of China.

Consistently growing income inequality and income divergence have not been observed in other developing countries in Asia (Kim, 2001), nor are these trends consistent with economic theory. Very simply, without restrictions on mobility, labor resources are expected to move to locations where they receive higher remuneration and their marginal productivity is higher. This would normally be in locations where the amount of capital per capita is higher and that would be in urban areas where industrialization is taking place. The systematic discrimination against rural workers in China both by restricting the natural flow of internal labor migration and the systematic allocation of government resources to urban residents has created a dual labor market that can only create further difficulties in the future.

Environment

We have touched on many of the environmental problems faced by China in Chapter 6. This section crystallizes the main policy issues facing the country and the governmental authorities. One major issue is land use and protection of the environment. Local authorities, as part of the town and village development programs lease land (use rights) to industries that want to establish new factories, expand existing facilities or to relocate from other locations. Re-allocation inevitably means obtaining the land from peasants now occupying and tilling the land. Local governments are primarily interested in making income from these new industries and are not particularly interested in environmental pollution. Local government officials sometimes permit the operation of highly polluting industries because such industries provide significant sources of income for the local government, although local residents suffer.

Energy

Another issue is energy. The main source of energy for electricity for industrial use and commercial consumption is coal. Coal supplies about 75% of China's energy needs. Coal is abundant and cheap and China leads the world in both the production and consumption of coal. China uses more coal than the United States, European Union and Japan combined, and consumption continues to increase at double digit rates. Most of the coal mined in China has high sulfur content and also is high in fly ash. Many of China's coal-fired plants are old, not particularly energy efficient and have not been retrofitted to remove sulfur and fly ash from smoke stack emissions. Coal-fired plants also produce carbon emissions that contribute to global warming.

How do China's emissions stack up with the rest of the world? Focusing first on carbon dioxide, currently China is the second largest emitter of energy-related carbon dioxide emissions after the United States at nearly 13% of the world's total and is expected to increase further to nearly 18% by 2025 (US Department of Energy, 2003a). While carbon dioxide emissions are expected to continue to rise, China has been undertaking various policies to conserve energy including reduction of coal subsidies and the amount of energy consumed and carbon emissions per dollar of GDP have fallen. There are some initiatives to adopt new technologies such as coal gasification (IGCC) which allows for sulfur and mercury to be filtered from the gas and disposed. It may also be possible to separate the carbon dioxide from the emissions and pumped underground although this technology is still in the experimental stage. Another option is coal heated steam generators that are more efficient in converting coal into energy which would reduce carbon emission per kilowatt and also lower sulfur emissions.

However, neither of these new technologies is likely to be widely used in the next decade or so. Bradsjer and Barboza (2006) present an interesting account of the attitudes of Chinese government officials toward new coal technology. Chinese authorities are reluctant to adopt expensive new technology particularly when they do not have the rights to that technology and have to buy it from overseas firms. Therefore, the outlook is for continued build up in carbon emissions rather than a sharp reduction even though the per capita emissions will remain small compared with industrial countries. The size of the population and the voracious appetite of the Chinese industrial engine that continues to demand more energy is enough to guarantee that China will be the biggest emitter of carbon emissions within a few years.

> The increase in global-warming gases from China's coal use will probably exceed that for all industrialized countries combined over the next 25 years, surpassing by five times the reduction in such emissions that the Kyoto protocol seeks (Bradsjer and Barboza, 2006, p. 1).

Turning to sulfur the outlook is more optimistic. Acid rain resulting from high sulfur content of smoke stack emissions is a continuing problem for northern Chinese provinces and also for Japan and South Korea. Acid rain pollutants of sulfur and mercury have also been recorded as far away as the west coast of the United States. Acid rain causes damage to crops and soil as well as presenting a health hazard for people. The Chinese government has made a commitment to install sulfur filters (scrubbers) on every power plant by 2010. The Japanese are so concerned that they have agreed to lend \$125 million to Shanxi province to pay for coal fired steel plants in the provincial capital of Taiyuan. Sulfur emission trading permits are also being tried out. The question that remains is whether the filtering equipment, which is costly to maintain and which uses a lot of electricity, will be kept in good running order. If it is, and new sulfur-free technologies (coal-heated steam and coal gasification) are implemented, sulfur emissions could be reduced significantly. If not, emissions could increase further with devastating health and environmental consequences.

The final issue for air pollution and global warming is the potential use of alternative fuels. All of these alternatives lumped together as renewable energy source (hydroelectricity, solar, wind, nuclear and geothermal) now supply around 20% of China's electricity generation (US Department of Energy, 2003a).

In order for these to make a dent in the coal emission story they would have to increase faster than overall energy use. Is that possible? The answer is unclear but the outlook is not optimistic. Wind power is being utilized in the northern and western regions and along the coast. Solar energy is concentrated with households and as a substitute for coal for cooking in rural households. However, both solar and wind together are unlikely to provide a big offset to coal. Prospects are better for hydropower from the Three Gorges Dam, although this is a one-shot addition to the power grid, and nuclear power, which is projected to rise to 4% from the current 1% by 2025. The development of renewable energy sources such as nuclear power is also constrained by the availability of new technologies, which, in turn, is affected by the export restrictions of source countries.

Better pricing policies, greater consumer and industrial awareness, implementation of a variety of different approaches to offset the use of polluting coal will all help to reduce the growth of carbon and sulfur emissions. In the section "future growth of income and its components" we will discuss another alternative, which is to slow down the growth rate of industrial production and the overall use of energy.

Petroleum

We turn now to petroleum and the contribution that the internal combustion engine is making to air pollution and to global warming. Currently the contribution of automobiles and other transportation equipments is small. Car ownership is quite low, less than one vehicle per 100 people. Remember, however, that there are 1.3 billion Chinese and that vehicle demand has been accelerating rapidly. New policies are being implemented to get a handle on auto-emissions. Emission standards for new cars in China are already higher than they are in California, where emission requirements are the most demanding of all the US states. Fuel economy is also quite high and Chinese cars are relatively small.

Nevertheless, the force of numbers is likely to swamp these changes in regulations. China is already one of the largest automobile manufacturers in the world. In 2005 China produced nearly 5 million vehicles, making it the third leading producer of automobiles after the United States and Japan. This is up from just over 1 million units in 2001 and virtually no production just a few years earlier. Car ownership is still much lower in China than it is in other developing countries and only about 1% of the ownership rate in the United States where there is nearly 1 vehicle for every person. However, the demand for automobiles is strong and growing in double digit rates every year. If this trend continues for long the amount of carbon pollution from automobiles will begin to rival that of carbon pollution from factories burning coal.

The challenge to planners is to slow the increase in automobile vehicular traffic by developing a comprehensive plan that includes more mass transit and expansion of the rail system. The plan should keep the bicycle as a viable form of transportation in cities for short trips and would levy tolls for traveling in the central business districts during rush hours, increase taxes on petroleum and on cars while increasing fuel economy and developing alternative fuels such as cellulosic ethanol.

From the current perspective and recent trends in vehicle production the prospects for bringing about a dramatic slowdown in the increase in automobile production in China are not good. There are many companies, both domestic and foreign, that care little about the global and local consequences of carbon emissions. They are more focused on the bottom line and prospects for selling cars to hundreds of millions of new customers. However, some cities are already restricting the growth in automobile registrations. Shanghai, for example, is restricting the number of new cars registered per year.

If China is really interested in slowing the rate of increase in automobile traffic in urban areas they can look to Singapore and Hong Kong, where resources have been poured into mass transit (buses, light rail and taxis) while making parking expensive and levying high taxes on cars, gasoline, and annual car registration. As a result car ownership per capita is much lower than in other cities of comparable size and income. Hong Kong has only 43 private cars per 1000 residents and Singapore 110, compared with over 200 per 1000 residents in Bangkok, a city with a much lower per capita income than either Singapore or Hong Kong. However, mass transit development in China is only relevant to inner cities and more urbanized regions along the coast.

There are a host of other environmental problems facing China including air and water pollution, deforestation and soil erosion. These issues were taken up separately in Chapter 6 and analyzed within the context of the challenges facing Asia.

Industrial policy

Growth accounting can be used to decompose the underlying growth in GDP. The usual methods add the contribution of labor and capital and total factor productivity (TFP) (Solow, 1956; Heytens and Zebregs, 2003). Since the adoption of reforms in 1978 the industrial sector has been the growth engine of the Chinese economy. Between 1978 and 1993 average GDP growth of over 9% was composed primarily of increased labor productivity. The bulk of this increase (about 7%) was the result of an increase in labor productivity. This increase in labor productivity came from TFP of 3.7% due to better technology and 3.2% increase in the capital to labor ratio. The balance came from an increase in employment in the labor force. The increments to the capital stock resulted in an investment rate average of just over 30% annually. Between 1993 and 2004 the contribution of TFP diminished somewhat to 2.7% and the increase in the capital–labor ratio was responsible for an even greater contribution to growth of 5.1%. The investment rate rose further to 36.8% of GDP. Employment growth was responsible for only 1.1% of GDP growth. Consequently, over 90% of the growth in industry in the last decade was in the form of labor productivity growth rather than employment growth. This growth in productivity flowed from a continued increase in the capital–labor ratio and a high investment rate. Urban employment growth slowed from 5.4% per annum in 1978–1993 to 2.7% in 1993–2004 and much of the agricultural surplus labor was also absorbed in the service sector.

The inability to absorb growth in the labor force and the labor shedding of state-owned enterprises has been partially a result of the slow growth in labor-intensive employment both in the service sector and in industry. Contrary to the early years of industrialization, growth in manufacturing and industry has become more capital-intensive. This is the result of a mammoth increase in investment that took place across the board in all components of industry — in large state-owned enterprises, in small-scale industries, and in foreign-owned conglomerates located in the various EPZs.

Subsidies played a significant role in this build up of investment. Over time the build up in investment has resulted in overcapacity, a profit squeeze, and rising inventories. Despite growing overcapacity, further expansion is under way. The steel industry is operating below capacity and the aluminum industry is operating at 75% of capacity yet further investment is ongoing (see ADB, 2006, China chapter). As a result of such a buildup in the capita stock, the industrial sector has been able to implement many new capital-intensive technologies to go with its existing labor-intensive industrial base. By 2004, China was producing a wide array of capital-and skill-intensive consumer goods and exporting many of these products to industrial countries and the rest of Asia.

The focus on industrial development and on raising the investment rate raised the share of overall value-added in industry to nearly 60% of GDP by the early years of the new century as the share of agriculture continued to fall. Contrary to patterns found in other rapidly growing economies the share of services in total income has not changed much in China, remaining around 24% for the past few years. This is much lower than the service share in other countries in the region (Table 7.4). However, recent revisions of national income statistics suggest that service sector value-added have been underestimated. It is likely that such revisions will raise official estimates of GDP (World Bank, 2006a).

Consumption share of GDP has fallen as the share of investment has risen. From 60% in the 1960s, consumption fell to around 45% of

Table 7.4 Sector Share of Services in GDP (%) — 2004

China	24.2
Vietnam	40.3
Indonesia	40.5
Thailand	44.0
Philippines	47.2
Bangladesh	49.2
Malaysia	49.6
Pakistan	52.6
India	53.2
Sri Lanka	55.8

Source: ADB (2006).

GDP while investment has risen from 20% in the early 1960s to 45% of GDP in recent years.

This relatively low rate of domestic consumption is also partially responsible for the widening trade surplus and the build up of foreign reserves. As investment has risen so has the rate of savings, driven by lack of medical insurance, lack of unemployment compensation, limited credit facilities for consumers to purchase durable goods, housing, and rising costs of education. Furthermore, bank-lending has been skewed toward investment and away from consumption.

How long will the pattern continue and, indeed, should it continue? In the future the Chinese authorities must give considerable attention to whether policies that favor industrial development should be redirected toward promoting growth in labor-intensive services and toward boosting the level of consumption expenditures while moderating investment. Factories continue to complain about excess capacity even as more capacity is added. By redirecting some of this investment toward important service sectors such as education, health and rural development, as well as the development of financial and insurance services, the pressure on the balance of payments could be relieved along with the growing tendency for the distribution of income to deteriorate. By emphasizing the quality and efficiency of investment total factor productivity could be raised along with GDP with the same level of gross investment. The World Bank has done a simulation that takes into account such changes and the ADB in its China chapter in ADB (2006) also suggested such a shift in strategy.

This strategy may have already begun as reflected by the substantial shift in employment that has begun in the past few years. McGuckin and Spiegelman (2004) note that labor productivity has been increasing and those foreign-owned firms are largely responsible for these gains. The productivity gains have occurred as job losses in the industrial sector have taken place in steel processing (557,000), non-metal mineral products (429,000), industries dominated by SOEs. On the other hand, employment gains were made in electronics and telecommunications (374,000) and garments (160,000). Most of these employment gains were made in foreign-owned firms.

McGuckin and Spiegelman go on to note that over 12 million jobs have been lost in SOEs between 1995 and 2002 and this was not totally offset by gains in the private sector as the industrial sector lost over 4 million jobs.

Some of these job losses have been offset by increases in service sector output and employment, but not enough to overcome the combination of downsizing and productivity increases that have characterized the changes in the industrial sector. Furthermore, revisions in the national accounts completed in 2004 suggest that the service sector has been playing a larger role in the economy than early data led observers to believe. The recent revisions, which also suggest that the overall Chinese GDP is also greater after the revision, show that the services sector share of GDP in 2004 was just over 40% compared with 32% from previous estimates. This reinforces the belief that the service sector is playing a larger role in the Chinese economy although output is still weighted heavily toward industry and job creation is still not strong enough to compensate for the loss of industrial sector jobs.

Social policy

As free market forces have gathered strength over the past two decades, the social services network traditionally provided by SOEs for many workers has broken down. In the 1970s and 1980s, about two-thirds of all hospital beds were in hospitals run as part of the government health system. And pensions were paid directly by SOEs from current earnings rather than as a retirement fund financed by contributions from workers and employers. This is one reason why the government has been reluctant to let SOEs fail, since it would have implications for these retirement benefits as well as increase unemployment. The rural health care delivery system also had difficulties and a rural cooperative medical scheme was initiated in 2002. Still many people are outside the social safety net. Health care spending has fallen as a proportion of the national budget and individual expenditures on health care have increased. From 1980 to 2004, those covered directly from the government budget fell from 36% of total employment to 17% while the share financed through enterprise run health schemes also fell, from 43% to 27%. The remainder was paid directly by patients, whose share went from 21% to 56% of total costs.

Between 1980 and 2004 private individual health expenditures accounted for two-thirds of the increase in health care expenditures. Health care has become more and more unequal. Those rich enough to pay, benefit, those less able suffer (Bergsten *et al.*, 2006).

However, a turnaround is still possible. China's health expenditures account for less than 5% of budgetary outlays while private outlays are nearly twice as much (UNDP, 2005). Furthermore, the country's medical insurance program covers less than half the urban residents and around 10% of rural residents. This has reinforced the rural/urban divide and has resulted in a further deterioration in the government's attention to the rural sector. The delivery of health services could be raised and the mix could be redirected toward the rural poor and others without insurance. This would be consistent with a shift of resources toward services and lifting consumption and a reduced emphasis on investment and industrial growth.

Education has been a strong point in China since the 1950s and since 1985 Chinese law has required a mandatory 9 years of education. Illiteracy rates have declined as well (see Table 4.13 in Chapter 4 for a comparison with other Asian economies). At less than 3% of GDP in 2004 overall government spending on education is somewhat lower than the average of just under 4% for Asian economies (Bergsten *et al.*, 2006; Gupta *et al.*, 1998). However, education spending has been skewed toward the coastal provinces and away from the poorer rural regions where poor local villages charge students fees to help cover operating costs. Furthermore, good teachers often move out of rural school districts and girls often lose out when families have limited resources to send more than one child to school in rural areas. In recent years there has been a big push for higher education and there are currently 20 million students registered. Again, they are predominantly from the richer urban areas in the coastal provinces. To redress this imbalance more resources need to be directed to rural and interior provinces.

The public pension system, which has many holes in it that need to be addressed, is also skewed toward the wealthy urban residents.

55% of urban workers are covered compared with 11% of rural workers (Jackson and Howe, 2004).

International trade and exchange rates

As its economy has opened up and it has lowered its trade barriers, China has built up its foreign trade. Imports are now 30% of GDP, up from 5% in 1978. Using this ratio as a measure of openness, China is now twice as open as the United States and three times that of Japan. China has built up the largest holdings of foreign exchange in the world in a matter of a decade, rising from less than $50 billion before 1994 to $286 billion in 2002 and nearly $900 billion in 2006 (Table 7.5). The current account balance has also increased rapidly, from $17 billion in 2001 to a projected $180 billion in 2006 and from 1.3% of GDP to around 7% in 2005 and 2006.

Many reasons have been suggested for these rapid build-ups in external imbalances, which have created tension between China and industrialized countries, particularly the United States. These include China's low wages, which have to be weighed against its low productivity. Also, unlike India which also can provide low cost labor, China has provided an attractive environment for foreign investors including physical infrastructure to support such a rapid build-up in international

Table 7.5 China's Foreign Reserves Minus Gold

Year	Foreign exchange reserves minus gold (in $billion)
1977	2.3
1982	11.3
1987	16.3
1992	20.6
1997	142.8
2002	286.4
2006 (March)	875.1

Source: State Administration of Foreign Exchange, PRC.
Notes: In 1992, reserves were redefined to exclude foreign-exchange deposits of state-owned enterprises with the Bank of China.

trade. China has also consolidated its position as the focal point for final assembly of a wide variety of industrial goods. Given its large pool of unskilled and semi-skilled labor and strong physical infrastructure, China is now the final assembly point for many goods including electronics and information technology products. Much of this is funded by foreign-owned enterprises that have production platforms in many different Asian countries. Goods assembled from imported parts and components now comprise more than 50% of China's total exports and an even larger fraction of goods exported to the United States (Bergsten *et al.*, 2006, p. 89).

The rapid build-up of foreign exchange reserves and current account surpluses has resulted in increased pressure for China to allow its currency to appreciate. In 2005 the authorities allowed a modest revaluation of the Yuan, although not nearly as much as the industrial countries would have liked and so pressure for further revaluation continues. As the coastal provinces continue to thrive, labor migration continues to be restricted, income distribution deteriorates, and the Chinese low wage advantage is beginning to be eroded. However, because of the other attractions of China, including the historical build up of expertise and FDI and the continued development of the infrastructure network China continues to be a highly competitive exporter and as long as trade and current account balances remain strongly positive there will be continued pressure for currency appreciation (see also the discussion of international trade and the interaction of Asian economies with the United States and Europe in Chapter 3).

State-owned enterprises

Before 1978 enterprises were all state-owned, organized according to traditional central planning where prices and output were determined by the central planning authority. Efficiency was not considered an important feature of industrial performance and managers were paid to deliver a certain amount of output without much regard for minimizing costs. After 1978, efforts were made to reform the SOEs as part of the program to introduce more market orientation into the economy. There were many challenges as the SOEs continued to operate as more private sector and market reforms were introduced and they were put into competition against domestic firms operating in the private sector as well as foreign-funded firms operating in the EPZs. There has been extensive analysis of the evolution of SOEs over the last 25 years which we will not go into here (Naughton, 1995). It is sufficient to point out a few salient features to give some background for the issues and challenges facing SOEs. The first point is that plan targets and the responsibility of the SOEs to provide a variety of social services for its workers tended to dominate the SOE agenda in the early stages of industrialization. As market forces were introduced, a dual track system evolved where planned output was sold to the government at one price and any excess production was sold at another. Market forces and good management practices were introduced as part of the reform process, particularly in light manufacturing. Resources began to be allocated by market forces, productivity increased, and economic efficiency improved. Total factor productivity increased, particularly in machinery and electronics during the 1980s (Table 7.6).

While non-state output dominated these markets some output was also traded in the market. In heavy manufacturing and the minerals sector, the SOEs still predominated in the early years of the reform where shortages of energy and raw materials required state control over the production, distribution, and sales of these commodities.

Table 7.6 Growth in Total Factor Productivity by Sector

Sector	Annual growth in total factor productivity, 1980–1989 (%)
Food	2.3
Textiles	2.5
Chemicals	2.7
Building materials	3.4
Machinery	6.1
Electronics	7.9

Source: Groves *et al.* (1995).

The dual track system created many opportunities for corruption which became better established as time went on (see Chapter 5 for more details). Secondly, the existence of the side-by-side operation of the state sector and the private sector created many distortions and allocation anomalies as well as tax issues. In the steel industry local governments began to invest in steel mills outside their own territory in exchange for a claim on the output that resulted from their investment (Naughton, 1995, p. 223) and similar developments occurred in iron and coal mines. In fact, in the coal industry most of the additional output came from non-state production. Taxation reform and creation of a level playing field for SOEs and the non-state sector continues to challenge reformers and the government in general.

Despite these difficulties with the dual track system, by the second half of the 1990s many features of a competitive market economy became evident. Monopoly profits were falling and converging, productivity improved, and the rate of bankruptcy increased as loss-making firms were taken over or went out of business. In fact profit rates in SOEs fell dramatically in the late 1980s and early 1990s and SOEs began to show losses, particularly in the minerals sectors of coal and petroleum mining. These losses have continued as the SOEs have become more of a drag on the economy, with the dynamic private and foreign-funded sectors becoming the driving force in the industrial sector.

Jumping to the present era of the early part of the 21st century we can have a quick look at the issues facing the SOEs and how the government is facing these challenges. Drawing on the work of the World Bank (Mako and Zhang, 2003) several observations are relevant. First, the share of SOEs in overall industrial output has fallen dramatically since the beginning of the reform era (Table 7.7) being displaced by collective private sector and foreign-owned firms.

Nevertheless, SOEs remain an important component of the industrial sector and one where significant reforms remain necessary. Secondly, the reform of SOEs has many aspects including market-based ownership changes, resolution of financially distressed SOEs, governance of remaining SOEs, funding of SOEs and how financial shake outs will impact the state-owned commercial banks that hold

Table 7.7 Industrial Output Share by Form of Ownership

Industrial output share	1980	1984	1988	1992	1996	2002
State-owned	80.2	75.8	68.2	61.6	43.2	30.0
Collective	19.8	24.2	28.9	29.2	31.3	29.0
Foreign			1.0	7.1	18.5	34.0

Source: Chinese Central Bureau of Statistics.

the IOUs of the SOEs. Third, the decision to establish a central commission to deal with SOE ownership, rights of shareholders, and other aspects of SOE management (SASAC, State-owned Assets Supervision and Administration Commission) could improve state asset management if it exercises prudent and effective policies to deal with a variety of these issues. Fourth, the World Bank assesses the current portfolio, highlights key issues in management of State capital, and looks at international experience as a guide to possible lessons for China. SOE restructuring is to be the subject of a future paper by the World Bank. Briefly, the record of SOEs has improved in the past few years up to 2001 as a result of privatization of SOEs, mergers and bankruptcies. Nevertheless, SOEs seem to have developed a liquidity problem as (uncollectible) accounts receivable and inventory are tying up capital. The most distressed sectors are building materials, chemicals, forestry, food processing, textiles, machinery urban utilities, construction, transportation and storage, and commerce. In many cases SOEs have been "decapitalized" in the face of ongoing losses. Only a few SOEs have acceptable liability/equity ratios.

One of the conclusions from this analysis by the World Bank is that small and medium SOEs should be culled from the SOE portfolio. About half of these (150,000 according to World Bank) could be sold as viable businesses and the rest could be liquidated or assets sold.

To restructure SOEs, two questions have to be addressed — how to allocate losses from SOEs and how this will be done institutionally. Claims of employees, suppliers, creditors, social insurance programs, and tax authorities will likely be higher than the value of SOE assets. So losses will have to be shared in a way that balances social needs with the growth of a market economy (Mako and Zhang, 2003, p. 13). Asset management companies will bear much of the responsibility as the process of restructuring continues and staff requirements will be beefed up as this process goes forward. Those SOEs that are not in trouble will need assistance in improving management and corporate governance as well as development of boards of directors and appointment of board members.

Overall SOE management will require the upgrading of a variety of skills and capabilities including auditing and accounting, portfolio management, dividend policy and capital reinvestment, sharing of gains and losses between central and local governments, and budget and risk management. Finally, ownership issues will have to be addressed including possible sale to employees, contracting to private management, and corporatization. The latter could include possible IPOs and/or sales to domestic or foreign investors. There are many issues here which become more pertinent as China continues to more toward more market-based systems. The experience of other countries in Asia will be particularly useful. Transparency in the process of transformation from state to private interests is essential. As part of this process, corporate governance is essential, particularly because of the variety of interests in a society without private property and the establishment of a legal system that recognizes these rights.

Financial sector[2]

China's financial system is dominated by four state-owned commercial banks that account for the vast majority of financial activity and bank lending. These banks were created in the 1980s as part of the economic reform program when the central bank function was separated from its commercial bank function. Previously, under the central planning framework, banks acted as an instrument of the government to finance its economic programs. In addition to the four commercial banks of the financial system there is the central bank (People's Bank of China (PBOC)), three policy banks created to provide finance to

[2]This section draws heavily on Huang *et al.* (2005).

the central government and to finance grain procurement, and a series of small regional banks and rural credit cooperatives. The four commercial banks are wholly owned by the state and the other banks are also owned by state entities such as SOEs or local governments. Until 1998 the four commercial banks, which control the bulk of bank assets, served as the financial arm of the SOEs. Lending to non-state firms was and continues to be a small proportion of their loan portfolio. Credit plans were issued by the central bank to the commercial banks and their regional offices in cooperation with enterprises and the government planning agencies. After negotiation they were approved by the State Council. Because most private firms were not part of the planning apparatus they did not submit investment plans and as a result many non-state firms were excluded from access to credit from the four state banks. Interest rates were also rigged to favor SOEs. Working capital loans to non-state firms were 20% higher than similar type loans to SOEs. In recent years interest rates have become more flexible, although SOEs still have an advantage over private firms.

This system of control and credit allocation to SOEs has created a credit crunch for domestic private firms including TVEs and the squeeze on lending tightened in the late 1980s and early 1990s. According to Sehrt (1998, p. 83), loans to collectives, TVEs and private firms accounted for 5% of new loans by the four state banks in 1995. Private firms have received little support from the rapidly growing equity market. Even in the late 1990s, after supposed liberalization of the Shanghai and Shenzhen stock exchanges, only about 1% were non-state firms.

As a result of these systematic distortions in lending favoring SOEs it is not surprising that surveys of private firms listed saving from labor compensation and loans from friends and relatives as the most important sources of funds. There was also probably substantial borrowing from the informal curb market which charged much higher rates than the formal sector.

There are some indications that the system is now responding more to market signals. There has been some de-control of interest rates for large foreign currency loans and deposits and the interest rate structure has been simplified. However, liberalization of other interest rates including checking and saving accounts has not yet begun (see http:www.buyusa.gov/china/en/bank.html. for further details).

How has this central planning type of credit allocation system without market signals and the usual screening of credit applications that characterize western banking impacted on the quality of lending? It is not surprising that many loans were non-performing. The official estimate of NPLs in the four state banks for 1998 was 25%, of which about 20% was considered non-recoverable. Outside estimates put the figure even higher. In addition to the fact that many loans to SOEs were not commercially viable, the size of the NPLs also reflects the social obligations that the SOEs were expected to fulfill including education, housing and health expenditures.

While progress in dealing with NPLs is shrouded in some mystery because of changes in definition and opaqueness in the way NPLs have been transferred to Asset Management Companies (AMCs) that were formed in 1999, there are some signs that banks are beginning to respond to market signals despite the continued strong hand of the central government. Some banks in provinces where the private sector is strong are more willing to lend to larger-scale private firms and there has also been some growth in the financing of home mortgages. And the AMCs are beginning to repack NPLs and are going forward with plans to sell them to investors.

In conclusion, we have to say that the health and viability of the country's financial system is complicated and depends on a number of factors. Even in well-developed financial markets difficulties such as the saving and loan crisis in United States can occur. In developing countries such problems are compounded by a lack of institutional development, government agendas that conflict with market development, and poor planning. In China this is compounded by the lack of private property and the insistence on government ownership of the main financial institutions.

This short review shows that the financial system is highly controlled and politicized and the stock market has been used as a vehicle to extend the exercise of this power and commitment to supporting SOEs. In addition the government has not been able to provide a transparent regulatory apparatus that deals with potential agency problems that arise between borrowers and lenders. These problems are not particular to China — they exist in industrial countries as well — but there is still a need to establish an efficient regulatory system. Otherwise the financial system will be plagued by agency, moral hazard, and information asymmetries. It is incumbent upon the state to allow financial markets to develop and function as they have not in the past by relaxing control over the banking system and allowing private ownership to flourish. At the same time it is also incumbent on the state to develop a transparent and comprehensive regulatory system that can deal with the myriad problems that are bound to arise in a complicated market-driven financial system.

A key feature of this transformation is the recognition of the legitimacy of private firms and their access to bank loans as well as the ability of the banking system to evaluate lending options based on market criteria. There is a clear consensus among most economists that financial sector reforms have lagged seriously behind reforms in the real sector. This is primarily because of the refusal of the state to relinquish control and ownership of the banking system.

One nagging question in all of these is how China has been able to grow so fast with such an inefficient financial system. There are several possible explanations. First, perhaps China could have grown even faster had it not experienced the distortions that favored SOEs and resulted in credit rationing to the rest of the industrial sector, the stagnation of the rural economy, and the increase in income inequality between rural and urban sectors. Secondly, foreign financing of investment and joint ventures in EPZs through FDI have created an additional growth driver even as the TVEs have languished. The size of this impetus is obvious from a glance at Table 7.7. And this shift in ownership is likely to continue as SOEs are sold to the private sector or continue to languish as the private sector prospers. A final reason could be that western views of the importance of market-based financial systems may not be nearly as critical in socialist countries.

Reform of the financial sector goes beyond bank ownership and relaxation of regulations that have created financial repression and resource misallocation. China's weak fiscal capacity is also a result of the inability of the government to find a viable solution to provision of social services formerly supplied by SOEs. By relinquishing the need to supply credit to SOEs that are no longer viable and by focusing on developing an alternative system to finance these expenditures the government could begin to effectively deal with pressing social sector issues.

Will changes in bank regulations following WTO accession open the door for more foreign banks and international competition into the banking sector? http://www.buyusa.gov/china/en/bank.html contains a review of the potential for US bank involvement in China. The conclusion from this analysis is that credit card business is quite viable at present and there has been some foreign equity taken in local credit card companies. Currently opening of foreign banks outside the major cities is prohibited. Nevertheless, HSBC, Standard Chartered and Citigroup have all applied to incorporate locally to meet government regulations for doing local currency retail banking.

Macroeconomic management

Many investment decisions are made by local governments that are not effectively controlled by the central government. While the central government will ultimately bear the responsibility for the quality of loans and whatever bad debts that arise, local governments are making the loans. As a result, fiscal policy is not particularly effective. Monetary policy bears the brunt of macroeconomic stabilization. This also explains why savings and investment rates are so high. Local governments compete against each other to attract funds from savers and also foreign investors. As a result, investment flows are determined more by external demand than by domestic factors. As discussed in Chapter 3, China is willing to build up unproductive foreign exchange reserves, allowing foreign firms to make the bulk of investment decisions in cooperation with local governments. Local governments' revenues depend upon more rapid growth and these revenues are used to fund more growth. Instead of government budgets being countercyclical in the traditional western sense, Chinese budgets are procyclical as revenues feed into further foreign funded investment.

Overheating means more overcapacity. Since the investment decisions are made at the provincial level to attract investment away from others the usual indicators used by the federal government to gauge whether the economy is overheated such as inflation are not relevant.

Regional and global issues

As China becomes a more important player on the global stage, its role as a pivotal player in Asia will become more important. We have noted above that accession to the WTO will bring several long-term challenges to China including agriculture policy, intellectual property and general trade issues. In addition, as telecommunications, banking, insurance and professional services have been opened up for foreign investment, competitive pressures on domestic companies will accelerate in the service sector. Because there are few SOEs in the service sector, it is likely that greater competition will arise as new foreign-owned enterprises vie with local firms. However, the state is not likely to relinquish control over the banking sector as quickly as it might allow further competition in other professional services. It is still too early to tell how this will unfold. In telecommunications we have seen the control struggles that have already begun between the central government, the internet and other information technology providers including Google and Yahoo. It may be difficult for the government to control the access of information as the Chinese economy continues to grow and competitive forces exert their influence in markets for products and services. The state sector will continue to shrink and foreign-owned and domestically operated private sector companies gain more strength.

Aside from the internal issues relating to WTO and the growth of the service sector, China can be expected to play a bigger role on the global and regional trade scene. In the chapter dealing with these issues, we speak about the importance of the growth of preferential free trade agreements as well as the growing importance of trading blocs such as ASEAN and ASEAN + 3. In this context the relative role of the WTO and the ongoing Doha Round of Trade Negotiations compared with the rapid growth of free trade agreements will play a significant role. China will be an important player in this dialogue.

An Asian trading system, similar to although not modeled after the European Union, and an Asian Monetary Area have also been discussed. As one of the largest economies in the Asian region, China will play a critical role in determining the future of these initiatives.

China has become a truly global economic power in a few decades. FDI and international trade have certainly played a critical role. Yet the system-wide reforms and institutional developments that have taken place are more fundamental. These include the changing behavior of government to become more supportive of the transition to a market economy. If current trends continue, it will not be long until the majority of industrial firms have foreign ownership (see last row of Table 7.7). Will the economy become more and more dominated by foreign firms and will there be a backlash against such foreign influence? How will the attitudes of the Party leadership react to the continued erosion of state control? Will the private sector gain a foothold in banking, insurance, and other services sectors as quickly as it did in manufacturing and industry? There are many foreign banks ready to bring their expertise and technology and many are hopeful that the Chinese market will open up. The thought of 1.3 billion new consumers and customers boggles the mind.

India

History

Until 1990 India followed a pattern of economic development based on government control and ownership, inward looking development policies and limited emphasis on international trade. The so-called "license raj" stressed government controls and supervision of many aspects of commerce, extensive monopoly power, and stable but comparatively slow overall growth. Combined with relatively rapid population, these forces combined to bring about relatively modest growth in per capita income and resulted in India falling behind many of the other developing countries in Southeast Asia and East Asia.

As a response to a fiscal crisis that put India on the brink of international bankruptcy, the government undertook a series of reforms in the early 1990s that resulted in a significant shift in policy toward international trade, competition, regulation and foreign investment that led to an acceleration in economic growth. In the remainder of this section we briefly review these and other developments in the Indian economy in years since these reforms were enacted, drawing where needed on past performance and policy. We begin with a review of agricultural sector background, performance and policy.

Agriculture

Agriculture is the backbone of the rural economy where nearly 60% of the population resides and where poverty incidence is high. The contribution of agriculture to GDP has been falling over the years as has the number of people depending on the sector for livelihood. It now contributes between 20% and 25% of GDP as compared with a much larger share a few decades ago.

There have been many success stories in Indian agriculture. Many factors have contributed to an increase in overall production and in labor productivity. Among the most important is technology transfer, adoption of new high-yielding varieties (of traditional crops) along with extension of the irrigation and farm extension networks as well as better rural education (see Bhattarai and Narayanamoorthy, 2002). Off-farm employment was also an important factor in raising rural incomes (Foster and Rosenzweig, 2003). Total factor productivity resulting from an increase in production as a result of better farm management, technological transfer, and other technological improvements accounted for about half of the increases in agricultural output during the "green revolution" of the 1960s and 1970s and into the 1980s (Evanson *et al.*, 2002 and Table 7.8).

During the Green Revolution period, agricultural output grew more rapidly than in any other period, yet still never reached an average of 3% per annum even when the drought and poor weather years are dropped.

These data also show another important aspect of agriculture, the importance of the monsoon and the general weather factor. Dropping the two drought years 1965/1966 and 1986/1987 out of a sample

Table 7.8 Annual Growth Rates in Crop Output, Inputs and Total Factor Productivity

Growth rate	1956–1965	1966–1976	1977–1987	1956–1987
Based on 3-year moving average				
Crop output	2.18	2.68	2.07	2.25
Crop inputs	1.08	1.28	1.00	1.11
Total factor productivity	1.10	1.39	1.05	1.13
Based on "normal years"				
Crop output	2.35	2.77	2.15	2.40
Crop inputs	1.1	1.31	1.02	1.14
Total factor productivity	1.27	1.49	1.14	1.31

Source: Evanson *et al.* (2002).
Notes: "Normal years" excludes years of drought and poor weather — 1965, 1966, 1979, 1980.

of 30 years still results in an increase in growth of 0.15% per year (comparing normal year crop output and total crop output shown in the final column of Table 7.8) which can be directly linked to loss of output from monsoons. The impact of weather on the rest of the economy is evident from the sharp contraction in GDP output during these three bad weather episodes. In 1965/1966 there were a reported 1.5 million deaths as a result of drought and related illness, starvation and malnutrition. In 1979/1980 agricultural output fell by 13% and in 2002/2003 agricultural output fell by 5.2%. There was also a drought in 1956 which accounts for negative growth in that year as well. These perturbations are reflected in the historical record of growth in India as displayed in Figure 7.2. Variability in agricultural output and incomes has slowed the growth in the sector and had a negative impact on the rest of the economy. Partly this is because 60% of India's farmland is still rain-fed and has decreased slowly from about 90% in the 1950s. During a drought, production in this component of agriculture is adversely affected.

Due to the weather's impact on yields, irrigation has played a key role in increasing the productivity of Indian agriculture. During the height of the green revolution, irrigated land accounted for about 20% of the gain in rice productivity, while about 75% came from

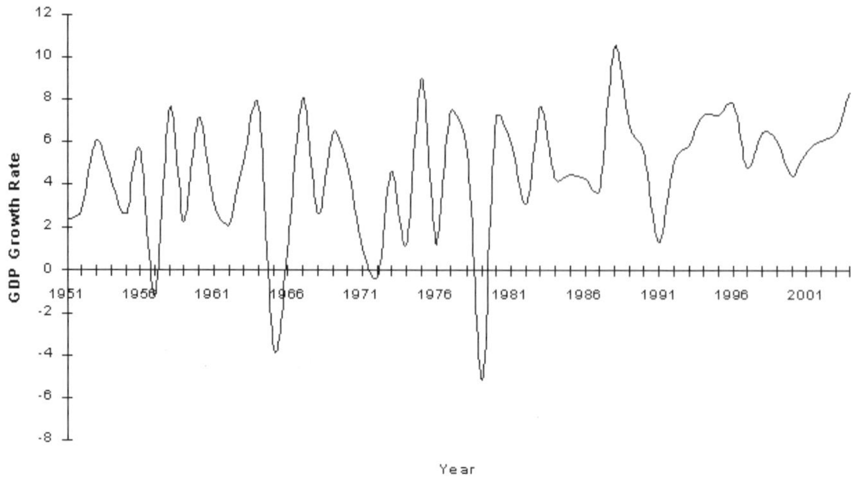

Figure 7.2 Growth in GDP in India

Source: www.economywatch.com/indianeconomy/indian-economy-overview.htm.

fertilizer and other inputs including new technology (Dowling and Valenzuela, 2004, p. 74; Mellor and Mudahar, 1992). Nevertheless, despite these gains in productivity and in agricultural output arising from a variety of sources, the rural sector remains poor and poverty levels are still high. As we suggested in Chapter 4, poverty is a persistent problem and has much to do with population growth, provision of rural infrastructure and education, and the performance of the rest of the economy. It also has to do with the failure of state governments to supply needed support because of lack of funds and the need to accelerate diversification and post-harvest farm management such as storage.

Growth in industry

The industrial sector has contributed to economic growth in India over the past few decades. However, its share in GDP remains relatively small compared with other competing economies, particularly in East Asia (see Table 7.9). Furthermore, the sector share of industry

Table 7.9 Sector Share of Industry, 2004 (in percent of value-added)

China	66.3
Korea	43.2
India	26.0
Indonesia	44.5
Philippines	33.2
Thailand	46.7
Vietnam	39.4

Source: ADB (2006, Table A4).

fell between 1990 and 2000 from 27.6% to 26.9% and further to 26% in 2004. The sector is not growing as fast as GDP! While growth in the industrial sector has picked up some in the past few years, it has been unable to generate enough employment to absorb the growing workforce. As a result, underemployment and unemployment have been a persistent challenge for the government. Compared with China, growth in Indian industry has been anemic.

What are the reasons for this? On the surface India has a large pool of unskilled and semi-skilled workers that can offer competitive wage rates. Yet, growth in industrial output of products requiring this kind of labor has been slow. There are several explanations for this pattern of industrial development.

Investment climate

Investment in an industry or a country will be a function of a variety of factors. To simplify matter we can think a simple risk and return payoff. If expected returns are high and the risks low, investment will take place. Conversely, a high-risk and low-return environment will result in a low level of investment. In the case of India we can identify several factors which have contributed to high risks and low expected returns. We can characterize some of these factors as components of the investment climate. As a result of reforms undertaken in the early 1990s, investment grew by an average of 18% between

1991 and 1995 only to fall back of growth of 8% in the second half of the decade. The failure to carry through with second-generation reforms is probably responsible for part of this decline in investment.

What exactly are the components of a "poor" versus a "healthy" investment climate? There are several and they all point to a healthy environment where firms can thrive and be competitive at the international level, which is by developing competitive exports that can be sold in global markets. Research has identified: (i) entry and exit policy, (ii) infrastructure bottlenecks and reliability of service, and (iii) regulations and bureaucratic procedures.

Entry and exit

A competitive industrial environment and a level playing field introduce dynamism into industry that encourages innovation, rewards efficiency, and penalizes firms that cannot make the grade. It also means that the dispersion in productivity among firms in a particular industry should be low — less efficient producers either increase their productivity or leave the playing field. In India it is neither easy to enter nor easy to leave the market. There are many requirements to start up a firm including various permits and other bureaucratic hurdles. Larger projects, such as in the power sector, require even more clearances. According to the Confederation of Indian Industry, typical foreign power project requires 43 clearances at the federal level and 57 at the provincial level (World Bank, 2002).

As a result, many approved foreign direct investment projects are abandoned. In the 1990s less than 20% of approved projects were implemented. On the other hand, bankruptcies are laborious. India has a miniscule proportion of bankruptcies as a share of total firms, much lower than the United States, where the ratio is between 3% and 4% and even lower than the rate in Thailand which has a low rate but still considerably higher than India. This lack of dynamism in the industrial sector suggests a lack of competition and also a lack of entrepreneurship and entrepreneurial initiative. In the past this industrial lethargy has been a direct consequence of the "license raj" where the government had its fingers in all the pies.

Labor mobility

Hiring and firing are heavily regulated and controlled by the government. Companies employing more than 100 people have to ask state governments for permission to fire or retrench workers and this permission is seldom given. To get around this regulation, firms resort to hiring contract workers and keep the number of employees less than 100. This is not good for firms seeking to take advantage of economies of scale and it also implies that a number of products cannot be produced by large-scale firms. Recently some export process-sion zones have been established that waive the rule, but they are small in number and these labor restrictions continue to constrain the business community. There are also proposals that the labor limit be raised to 1000 employees but they have not yet been adopted.

Land ownership

It is very difficult to buy and sell land in India. The World Bank reports that 90% of land parcels are subject to dispute over ownership which can take decades to settle. McKinsey and Company (2000) estimates that land market distortions account for 1.3% of lost growth in income per year. Zoning laws, rent controls and protected tenants freeze land in key city centers and restrict competition. Without the ability to assign land to its best use through the operation of markets, inefficiencies are created and land prices cannot reflect opportunity costs.

Lack of openness to foreign competition

India has been a relatively closed economy for most of the post-war period. Trade as a share of GDP is quite low compared with the export-oriented economies of Southeast Asia and East Asia. Tariff rates were high in the 1960s, 1970s and 1980s. While they came down somewhat following the liberalization measures of the early 1990s, they are still relatively high — 30% compared with 15% in China and 10% in Thailand. Furthermore, non-tariff barriers cover about a third of all goods, also a much higher rate than competitors

Table 7.10 FDI and International Trade as Share of GDP

Country	FDI as share of GDP in percent (1998)	Trade as share of GDP in percent (2004)
India	0.5	42
China	5.0	65
Malaysia	6.1	221
Thailand	6.2	136

Source: World Bank (2002) and World Development Indicators online.

in the Asian region. Foreign ownership has also been discouraged as India continued to follow import-substitution policies in the 1970s and 1980s while its neighbors in East and Southeast Asia were successfully following export promotion growth strategies. We need not go into the details of these restrictive policies. It is enough to note that the level of FDI in India is much lower than it is in China as is the share of exports and imports in GDP (Table 7.10). While some of the lack of FDI may be due to other factors such as low labor mobility, exit policies and poor infrastructure (see next section), the regulatory environment for foreign investment is also partially responsible. When protection has been removed and foreign competition permitted, efficiency has increased and many older local firms have had difficulty competing although new local entrants are often able to compete (see the machine tool example cited in World Bank, 2002).

Infrastructure bottlenecks and high costs

In terms of physical infrastructure needed to support industrial enterprises such as transportation, power, telecommunications and water supply, India is a high cost economy compared with competitors such as China. In the transportation sector, all Indian ports taken together are smaller than Manila and Bangkok and far behind Hong Kong and Singapore. There are no Indian ports in the top 20 world ports (see Table 2.10, Chapter 2). Hong Kong and Singapore ports handle 16,000 containers per month while all Indian ports handle around 2000 and India is at a shipping cost disadvantage *viz-a-viz* other Asian competitors. For textile exports to the United States it is at a 10% cost disadvantage compared with Indonesia, over 20% with Thailand, and over 35% with China and S. Korea. India has fewer paved roads than other countries in the region and its railroads carry mostly passengers rather than freight. China's railroads carry nearly 80% of traffic units as freights, while for India the ratio is 5%. However, the new Golden Quadrilateral highway — 3625 miles of national highway that connects India's four major cities, New Delhi, Calcutta, Chennai and Mumbai — and a number of small ones — has been modernized and is facilitating trade and labor mobility like never before. (See Waldman (2005) for an entertaining account of this movement.)

Energy is also more expensive for two reasons. Mismanagement of the power sector by a government monopoly until the 1990s resulted in underinvestment in maintenance and installation of new capacity. Power was not delivered reliably. Secondly, the unreliable provision of power by the government meant many firms invested in their own small generators which are much more expensive than an efficient nationwide power supply. Energy costs as a percent of total sales in garments and electronics on average are estimated to be between 3% and 3.5% of total sales in India, closer to 2.5% in China and less than 2% in Indonesia and Thailand (World Bank, 2002). Telecommunications is in better shape in India, probably because new developments in the internet and telecommunications in general did not fall under the regulatory umbrella that was in place in the 1980s and earlier. Water is another area where India has fallen behind, particularly in big cities. However, this has not been cited as a general problem for industry.

Regulations and corruption

We have touched on the problem of corruption in Chapter 5. Here it is important to note that in terms of general governance, which includes a variety of measures other than but also including corruption such as rule of law, number of regulations and political stability, India does fairly well on rule of law, political stability and regulatory framework. On the other hand, entrepreneurs face an inordinate amount of time spent with regulators and other officials. Compared with China, Thailand and Indonesia, about 16% of management time in India is spent with regulators, nearly twice as much as in China and

OECD countries where the ratio was around 9% (World Business Environment Survey, 2000).

Customs clearance is another area where India has developed a reputation for long delays. Not only are average delays longer in India than in China, S. Korea and Thailand (nearly 11 days versus between 7 and 8), but the variance in delivery time is also high. Garments and pharmaceuticals can take up the 27 and 24 days respectively, at great cost and inconvenience to customers. Complaints about taxes and tax rates come just behind customs as the biggest regulatory obstacle facing businesses in India. The complexity of the tax system gives the tax authorities even greater power in assessing taxes and in being open to bribery.

Taken together, all of these impediments to investment have resulted in slow if not anemic growth in the industrial sector. Those investments that have been made were often in industries that were thriving behind a web of protection from competition. There was little scope for rewarding innovation and improving efficiency. The automobile industry is a good example. Until foreign imports and some joint ventures with foreign firms were permitted in the 1990s, India continued to produce cars that were introduced by the British in the 1950s.

Growth in services

The service has grown quite strongly in the 1990s. The sector grew from a 41% share of GDP in 1990 to 54% in 2005, picking up the slack from a declining agricultural sector, which declined to 20% by 2005. The headline grabber has been IT, first the labor-intensive call centers that popped up in Bangalore and elsewhere in the early to mid-1990s and more recently, other high-tech services. However, analysis of the entire decade of the 1990s suggests that a number of other service sectors have also grown rapidly (refer to Table 7.11). As we would suspect, business services grew rapidly, at nearly 20%. However its share in GDP is miniscule. Larger service sub-sectors such as banking, communications, community services and distribution services (associated with trade) also grew rapidly. Tourist services also expanded although it is still a small sub-sector of the economy.

Table 7.11 Growth of Service Sector in the 1990s

Sector	Share in GDP (1990)	Sector growth (1990–1999) in percent per annum
Distribution services	11.9	7.3
Hotels and restaurants	0.7	9.3
Railway services	1.4	3.6
Other transport	3.8	6.9
Storage	0.1	2.0
Communications	1.0	13.6
Banking	3.4	12.7
Insurance	0.8	6.7
Real estate	4.8	5.0
Business services	0.3	19.8
Legal services	0.08	5.8
Public administration and defense	6.0	6.0
Personal services	1.1	5.0
Community services	4.3	8.4
Other services	1.0	7.1

Source: Banga and Goldar (2004).

Financial sector

The financial sector consists of the Reserve Bank of India, 274 commercial banks, 223 of which are publicly owned and 24 foreign banks. Foreign banks have only 7% of total commercial bank assets but a 40% share of foreign exchange transactions and 25% of the government securities market. The Reserve Bank exercises a good deal of control over commercial banks including directed lending to key sectors and interest rate controls. Foreign banks are not allowed to have local customers and their ownership of local banks is restricted to 5%. 40% of loans by commercial banks (both private and publicly owned) are earmarked for priority sectors including agriculture, small business and exporters. Foreign banks also are subject to earmarking of loans but only up to 32% of loans. There are caps on rural lending and deposit rates and this has made banking business unattractive in rural areas because of the high transactions costs and collateral requirements. There are also ceilings on earnings from deposits, controls on interest rates charged to borrowers and also for non-resident Indian (NRI) accounts. The reserve requirement is now close to the statutory minimum of 3%; having fallen from over 10% a decade ago. Lowering the reserve requirement reduces the tax on the banking system facilitating greater economic efficiency.

Restrictions imposed by the Reserve Bank of India such as the statutory liquidity ratio (SLR) and the cash reserve ration (CCR) as well as interest rate controls and directed credit, have contributed to the financial repression of the banking system. There has been some liberalization since the early 1990s. The government and the Reserve Bank of India have carried out a number of reforms that have resulted in improved prudential regulation, greater efficiency in bank operations, and greater scrutiny of loan applications, and the CCR and the SLR have been lowered, reducing the implicit tax on the banking system and improving bank efficiency as a result. It also lowers the rates of interest that have to be paid on government debt and indirectly provides an incentive to accumulate public debt.

The banking system does not have a large amount of bad debt. The level of net non-performing loans is estimated at less than 3% of outstanding net loans in 2004/2005 (IMF, 2005; Table 1.1) and about double that for gross loans. Nevertheless, the banking system in India is still highly risk-averse and lending to the industrial sector has been weak. Currently nearly 40% of its assets are held in government bonds far above the statutory requirement of 25%. The corporate sector also holds a large volume of government bonds and together with bank holdings, this continues to restrict the borrowing of business firms and investment in the industrial sector.

Continued financial repression, while lower than before reforms, suggests that capital account liberalization should be carefully reviewed before taking further steps (Kletzer, 2003).

Capital formation and overall economic growth

Changes in the regulatory environment in the early 1990s, the growth of the service sector and decline of agriculture have all contributed to a better climate for investment and capital formation in the economy as a whole. Private investment as a share of total investment has increased and now approaches 20% of GDP, double the rate in the mid-1980s. This has also resulted in an increase in the ratio of GDP to the capital stock. The strength of private sector investment reflects a variety of efforts to improve the investment climate, shortcomings reviewed earlier to the contrary notwithstanding, including shortening the list of industries reserved for small enterprises, simplification of rules for international investment and lowering of trade barriers and simplification of licensing requirements. Higher levels of investment and increases in productivity have been a significant factor in raising overall economic growth to between 6% and 7% and to even higher rates during years of strong agricultural growth.

External environment

The review of the industrial sector foreshadows developments on the trade scene. Exports and imports have grown as a share of GDP in recent years as reflected in Table 7.3 although still much lower than its neighbors in Southeast and East Asia. This has been the result of liberalization measures in the early 1990s that have begun to take hold. However, historically trade shares of GDP were quite small as India adopted a development strategy that stressed domestic factors and little reliance on foreign assistance or foreign direct investment. In the 1990s the trade share was only 31% of GDP and in the 1980s only 13.4%. So there has been substantial progress even though India's share of world exports is still only 1% compared with 5.8% for China. India's main exports are manufactured goods, mainly gems and jewelry (17% of total exports in the late 1990s) along with ready-made garments and handicrafts (16%) and cotton yarn fabrics (8.5%) (see http://www.indiaonestop.com/compofexports.htm). India has benefited from the ending of the Textiles and Clothing Agreement. Unlike China, whose textile and apparel exports have been placed under quota restrictions by the United States, India's exports of clothing and textiles to the European Union increased by 16% in 2005 while exports of clothing and textiles to the United States increased by 34% and 22%, respectively.

The pattern of Indian exports is very much like the composition of Southeast Asian exports before the boom in electronics. India exported almost no electrical goods in the 1990s although that pattern has been changing. The conclusion that flows from this very brief look at the external environment is that India has not made the transition out of narrowly based simple labor-intensive textile exports. Even leather and leather products comprise only about 5% of total exports.

Export processing zones

Although India was one of the first countries to set up export processing zones in 1965 to send goods to Russia, it has lagged behind other countries since then. However, in 2000 India passed a law setting up special economic zones (SEZs). Since then, SEZs or EPZs, as they are called in other countries, have been set up in a number of locations. These zones range in size from as little as 10 hectares to 1000 hectares for multi-sector zones. Another scheme which would involve even larger sites approaching the size of Shenzhen in China is being suggested.

According to the Indian SEZ website — http://www.sezindia.nic.in/sez/asp — there is approval for 42 SEZs in various parts of the country by the private sector with the State Government taking a major role in the set-up. Various schemes are being put in place to attract industry including infrastructure (power, water and telecommunications) and freedom from the usual bureaucratic regulations. However, the restrictive labor law that has been an impediment to investment in the past is still in force.

These zones have only begun to operate in the past few years and they do not yet have an established track record nor have their prospects been well established. Preliminary results reported by the Indian SEZ website show exports from the 11 SEZs now in operation close to \$4 billion in 2004–2005. This is about 3% of total exports in 2005. If the Indian SEZs do take off they could provide a needed boost to industrial production and exports. A list of some of the existing EPZs and a short description of their features are shown in Table 7.12 with exports

Table 7.12 Location and Exports of Some SEZs

Location of SEZ	Some features of zone	Exports in 2004–2005 — crores of rupees
Kandla	Provides an internationally competitive and hassle-free environment for export production and for trade operations and duties and tariffs.	1060
SEEPZ		8298
Noida	An export processing zone in Noida for the purpose of trade operations, duties and tariffs to provide adequate infrastructural and other necessary support services.	4266
Madras	A multi-product zone under the administrative control of the Ministry of Commerce and Industries housing 86 functional units and facilitating exports through reduction of transaction costs.	1376
Indore	A special economic zone for purposes of trade and customs duties, with special rules for facilitating foreign direct investment.	55
Visakhapatnam	This zone provides for cost-effective solution, for operational hurdles and it facilitates the enterprise by providing land, standard design factory, water, power, telecommunications, and support services.	579
Surat	An economic zone for purposes of trade and customs duties, with special rules for facilitating foreign direct investment.	1539
Cochin	A special economic zone for purposes of trade and customs duties, with special rules for facilitating foreign direct investment.	462
Santacruz	Special economic zone for manufacture and export of electronic items, gems and jewelry items and provides basic infrastructure such as constructed factory premises in well-built factory.	na

(*Continued*)

Table 7.12 (*Continued*)

Location of SEZ	Some features of zone	Exports in 2004–2005 — crores of rupees
Falta	An export processing zone in Kolkata for the purpose of trade operations, duties and tariffs to provide adequate infrastructural and other necessary support services to enable them to be competitive, quality-wise and cost-wise.	569

Source: http://www.sezindia.nic.in/sez/asp — website of SEZ India.
Note: 4.87 crores of rupees equals $1 million.

in crores of rupees in 2004–2005. Notice that the descriptions are very general and only mention possible products to be produced.

In addition to these SEZs, there are export sites specializing in information technology that are of recent origin.. They go under the generic name ITPL, or information Technical Park limited. There is one in Bangalore — International Tech Park, Bangalore (ITPB) and others in Hyderabad and Karnataka. The private sector and foreign investors appear to be interested, led by Singapore.

Population and human resources

After China, India has the largest population in the world, over 1 billion people. Because its rate of population growth has been more rapid than China's, it is expected to surpass China to become the most populous country in the world within the next decade. India is now in the middle of a demographic transition from high to low birth and death rates. This transition takes place as incomes and living standards rise in developing countries. There are a number of factors that influence the speed and intensity of this transition. Industrialization and urbanization are accompanied by migration from the countryside to cities, where living spaces are more constrained and the costs of raising more children and maintaining larger families are substantially higher than in rural areas. Income growth allows countries to spend more on education and health. Women's health and education are important. Birth control methods are also more available and more widely used as incomes grow and literacy increases.

In countries and regions where there has been greater spending on women's education and health, women's literacy, life expectancy, labor force participation, and years of schooling completed have also increased. These improvements in women's education and health have had powerful beneficial impacts on family formation and population growth. As women become healthier and more educated and as they enter the labor force in greater numbers, infant mortality and total fertility fall, and children's educational attainment and health increases. Women feel greater empowerment and take greater responsibility for the health, education, and general well being of their families. Population growth rates also fall as women wait longer to have children and also have fewer children over their lifetime.

These secondary effects of women's education on population and children's health and education suggest that raising the average level of women's education may be one of the most important policy objectives of a developing country. In India the demographic transition is taking place and women's education and literacy have increased. Chapter 4 provided some details of the Indian experience as well as the experience of other countries. In Chapter 4 it was noted that the overall poverty level in India remains high and that India is unlikely to reach the Millennium development goal of halving poverty between 2000 and 2015. It was also noted that literacy is still low and infant mortality levels high. These observations are consistent with a slow demographic transition characterized by limited progress in raising health and educational standards for women.

In this section we concentrate on highlighting the experience of poorer regions and states, pointing out that the states with the highest population growth are those with the poorest record of raising women's literacy and reducing infant mortality. Of the larger states in India, six of the poorest (Assam, Bihar, Madhya Pradesh, Orissa, Rajastan and Uttar Pradesh) with total population of around 400 million people are also at the bottom of the scale in terms of living standards and human resource development for women. Women's life expectancy (a proxy for women's health) is low, infant mortality and female illiteracy high. Use of contraceptives is also low and total fertility is high (see Table 7.13).

Sterilization is the most popular method of female contraception, accounting for about 75% of all contraception. Temporary contraceptive methods are used by a much smaller fraction of the population. Demographers suggest that sterilization is not a viable long run solution to population control and that India needs to adopt more temporary measures such as condoms, IUDs, and other measures widely used in other countries in order to accelerate the demographic transition, increase the rate of decline in fertility and reduce the rate of population growth (Pathak *et al.*, 1998). Poor education and health outcomes have interacted to keep women from taking measures to reduce family size, provide better education and health care for their children and participate more fully in the labor force. This, in turn, has resulted in more rapid population growth in the poorer provinces and reinforced the already slow growth in per capita income.

Living conditions in the rural areas of the poorest states remain grim. In Bihar, the poorest state, the population census of 2001 showed that 89% of households had a mud floor in their house, only 1.4% had access to tap water (80% had access from hand pump), 95% used kerosene for lighting (5% had electricity), 86% had no latrine or bathroom in the house and 65% of those with outside latrine had no drainage for waste water. Virtually all households used firewood (29%), crop residue (37%), or cow dung cakes (30%) as cooking fuel. Only 3% of households had both toilet and electricity. More than half the households had no substantial household assets, i.e., no car or other motorized vehicle, no bicycle, no telephone, television or radio. In Orissa, households were somewhat better off, 50% had a bicycle and 21% had a radio. Still 44% had none of these amenities and 80% had mud floors.

The divergence in income between rich and poor states has been reinforced in recent years by fiscal developments. In the 1990s there

Table 7.13 Population, Income and Demographic Indicators by State

State	Population in millions (1991)	Female life expectancy (1992–1994) — rank (low to high out of 15 states) in parenthesis	Infant mortality per 1,000 births (1992–1994) — rank (high to low out of 15 states) in parenthesis	Total fertility rate (per woman) (1993) — rank (high to low out of 15 states) in parenthesis	Percent of married women using any contraceptives (1992–1993) — rank (lowest to highest out of 15 states) in parenthesis	Percent of illiterate females (6 years of age and over) (1992–1993) — rank (high to low out of 15 states) in parenthesis	Rank of per capita income (2001) (low to high out of 15 states)
Poorer states							
Assam	22	54.4 (2)	78 (5)	3.3 (6)	42.8 (6)	61.5 (5)	4
Bihar	86	56.4 (5)	69 (6)	4.6 (2)	23.1 (2)	71.4 (2)	1
Madhya Pradesh	66	53.2 (1)	102 (2)	4.2 (4)	36.5 (5)	65.7 (4)	5
Orissa	32	55.1 (4)	109 (1)	3.1 (8)	36.3 (4)	58.6 (6)	2
Rajasthan	44	56.7 (6)	85 (4)	4.5 (3)	31.8 (3)	74.6 (1)	6
Uttar Pradesh	139	54.5 (3)	93 (3)	5.2 (1)	19.8 (1)	68.5 (3)	3
Richer states							
Andhra Pradesh	66	61.5	66	2.7	47.0	61.5	8
Gujarat	58	60.5	63	3.2	49.3	48.7	12
Haryana	62	63.2	69	3.7	49.7	54.1	13
Karnataka	61	63.6	68	2.9	49.1	53.5	10
Kerala	68	73.4	15	1.7	63.3	17.6	9
Punjab	65	67.2	55	3.0	58.7	48.0	15
Tamil Nadu	61	62.5	58	2.1	49.8	43.9	11
West Bengal	61	62.3	61	3.0	57.4	44.8	7

Source: Population Census and Registrar General of India.

was a tendency toward greater centralization of fiscal spending in India. As a result, transfers to the state governments were reduced. With fewer resources from the federal government, states were not able to undertake many new infrastructure projects or to increase social spending. State spending has also been constrained by slow growth in tax revenues, particularly in states that were growing slowly or had a low income base. Poor states such as Bihar, Orissa and Uttar Pradesh were hard hit. As a result new physical infrastructure spending fell and maintenance of existing facilities deteriorated. Similar spending problems constrained social spending for health and education.

Rapid population growth in poor states has contributed to this widening of the gap in income per capita between the richest and poorest states. In the 1970s and 1980s, Orissa and Bihar, the two states with the lowest per capita income, had per capita income of between 40% and 50% of that of Punjab, the richest state. By 2000 and 2001 the figure had fallen to 34% for Orissa and only 22% for Bihar. These stark contrasts between the richest and poorest states are often covered up by more aggregative measures like the Gini coefficient, which has remained more or less constant for the past few decades (see Macroscan, 2003).

While poverty in the poor states arises in part from a lack of human and physical resources and high population growth, it is also caused by systemic discrimination (lower caste Hindus and Muslims) (see also Chapter 4 on poverty and income distribution and Box 7.1 for discussion of treatment of dalits). A recently released report issued by a committee headed by Rajinder Sachar (2006) documents the latter. Among its findings is the general observation that Muslims have lagged behind the country's average in many respects, including income per household, average level of education, and literacy and health indicators. The report said that in urban areas the Muslim communities mostly live in slums with poor infrastructure. It also noted the need to compile a better database so that Muslim minorities can be identified and targeted for programs to raise incomes.

Box 7.1 Dalits — The Untouchables

There are an estimated 160 million dalits in India, nearly as many dalits as there are people living in Italy, France and England. The term "untouchable" was outlawed years ago but it is still being used. And this is because many dalits work in dirty and deplorable conditions. Most dalits are uneducated and many work in filthy conditions in lowly occupations where no one else will work. Some, mostly women, clean toilets. They use a broom and a pan to clear the dried excrement from dry latrines. They pile it into baskets and carry to a dumping place that can be far away. They work without gloves, boots or masks and are susceptible to a variety of viral and bacterial infections. Loads are heavy and the dalit women often suffer from back problems. Dalit men often work in sewers, diving through manholes into raw sewage to clean out blockages in the pipes. They wear no protective gear and suffer from a variety of diseases ranging from skin disorders and eczema, to eye infections and gastrointestinal problems.

Because they work in such an unclean environment and because of their former name "untouchables", dalits are systematically discriminated against by the rest of society. They are forced to live in parts of towns and villages away from others, are generally not permitted to own land, are discriminated by shopkeepers and employers and their children are forced to sit at the back of school classrooms and receive little attention from their teachers. Dalits live in a vicious cycle of poverty from generation to generation from which there is little chance of escape. Statistics numb the senses. Female literacy rates are below 20%, poverty rates are near 50%, school dropout rates for girls is close to 80%, life expectancy is less than 50, and infant mortality (from birth to 5 years) is over 170 per thousand births on average and 1.5 times higher for girls.

Beginning with Mahatma Gandhi, there have been social reformers in India pressing for uplifting the status of Dalits. There are educational programs for them and jobs reserved for them. Progress is slow. They have few champions in the political arena and there are no systematic efforts to apply modern technology to phase out these dirty jobs. The dalits, and other minorities including Muslims and scheduled tribes, are a continuing reminder of the depth of discrimination that continues in India as a grim reminder of past and current neglect.

Source: Pandita (2006); Dhesi (1998).

Issues and Challenges Facing India

Economic growth

The trend rate of growth in the Indian economy has increased gradually until it is now between 6% and 7% for the past decade and projections of growth close to 8% for the next few years are typical (ADB, 2006). Can it move even higher? Research at the Asian Development Bank suggests that moving beyond a trend rate of 8% would require substantially higher levels of investment. However, even higher levels of growth could be attained. How could this happen? First, FDI in industrial sector projects in the SEZs would have to increase substantially. Along with further liberalization of the sector this would serve to further revitalize the industrial sector, which is likely to continue to experience strong export growth in textiles and clothing exports. By adding this projected rapid growth in industry to the already strong service sector, and with appropriate investments in agriculture, overall economic growth could easily average more than 8% over the next decade. This assumes that the overall investment climate remains strong and overall capital formation remains high or even increases further to the range of 30%. This would also require reinvesting of profits from the successful software industry and also of using expertise gained in software to other aspects of industrial development. The ability to attract more FDI would require a focused policy of infrastructure enhancement, particularly in power and transportation, perhaps in partnership with the private sector. The Indian Committee on Infrastructure has estimated these needs to be something over 1% of GDP per annum, a large amount but not an unreasonable goal (ADB, 2006, p. 164).

Income inequality and regional disparities

More rapid growth and an increase in profits and revenue from a revitalized industrial sector should bring more tax revenue and a potential for addressing the growing regional disparities and income distribution issues. Higher oil prices will put pressure on these resources to some extent, although the recently introduced VAT will also bring additional

Table 7.14 Annual Percentage Change in Public, Private and Total Employment in the Organized Sector of the Indian Economy, 1970/1971 to 2001/2002

	Public	Private	Total
Poorer states			
Bihar	1.9	−1.8	0.9
Uttar Pradesh	0.7	−0.5	0.4
Orissa	2.5	0.4	2.2
Madhya Pradesh	0.8	−0.8	0.5
Rajasthan	2.3	2.7	2.4
Rich states			
Tamil Nadu	1.9	1.1	1.6
Gujarat	1.6	1.5	1.5
Haryana	2.8	2.9	2.8
Maharashtra	1.3	0.6	1.0
Punjab	2.1	2.5	2.2

Source: Ministry of Labor, Government of India (2006); Purfield (2006).

revenue to the states. What is needed is a commitment to address these issues directly. State governments in India have substantial power and the judicious exercise of anti-poverty and human resource development initiatives have been highly successful in states like Kerala. Kerala is in the middle of the states in terms of per capita income (Table 7.14), yet it ranks number one in terms of having low infant mortality, low total fertility and low levels of illiteracy. Other richer states could, if they had the political will, easily emulate Kerala. For the poor states some form of federal subsidy is needed.

How much of the regional disparity in income is due to neglect of scheduled classes and scheduled tribes? As noted in Chapter 4 the scheduled tribes and scheduled classes are widely subject to social exclusion despite efforts of the government to promote a classless society. Together, they comprise up to 25% of the population of India or over 250 million people, almost as many people as there are living in Western Europe. They have higher poverty rates and lower levels of social development including higher illiteracy and infant mortality and lower life expectancy than the national average.

Where do these two classes live? There are large numbers of scheduled classes (SC) and scheduled tribes (ST) living in Orissa, Bihar and Madhya Pradesh, three of the five poorest states. Together, the population of these three states is over 160 million and significant percentages are ethnic minorities. While data from the latest population census have not been released, older data from the 1981 census shows Orissa with 37.1% (14.7% SC and 22.4% ST) while Bihar had 32.8% (14.5% SC and 18.3% ST) and Madhya Pradesh with 37.1% (14.1% SC and 23% ST). These ratios are probably somewhat higher now because population growth is higher among these minorities.

We can only speculate that much of the poverty in these three states is the result of persistent discrimination against these ethnic minorities for a long time. It is worthwhile to note that significant proportions of SC in populations of Uttar Pradesh (21%), West Bengal (22%), Himachal Pradesh (25%), Punjab (27%) and Haryana (19%) did not keep some of these states from growing strongly. It is probably the case of a combination of high proportions of both minorities combined with poor state government and a vicious cycle of poverty that has resulted in slow growth for Bihar, Orissa and Madhya Pradesh.

It is more difficult to suggest how this pattern can be reversed unless there is a change in state governance — the attitudes toward these minorities backed up with resources from the central government and public action. Work by the IMF (Purfield, 2006) suggests that poor states need to diversify out of agriculture and industry or else adopt policies that make these sectors more productive. This research also suggests that better allocation of resources to provide better infrastructure and to improve its investment efficiency could improve performance. In particular, improved transmission efficiency and reduction of distribution losses in the electricity sector has a positive impact on growth. The work by Purfield also suggests that in general, poor states generate fewer jobs and this pattern will have to be reversed if these states are to grow more rapidly (Table 7.14).

Furthermore, rich states attract more capital (half of FDI goes to 5 richest states) and have a better record of reducing poverty. This is not a surprise since FDI is probably utilized more effectively in the richer states at the current level of development in India. Much of this FDI is in the service sector. The pattern of FDI in services and not industry is only a surprise if one is not aware of the slow growth of the industrial sector over the past few decades. Generally, growth strategies suggest more investment in industry and this strategy has paid off for many countries in Asia. Why not India? That brings us to the next sections on agriculture and industry.

Agricultural policies

After controlling for monsoon effects, trend in agricultural growth has leveled off in the past decade. Past experience tells us that irrigation has been responsible for a significant contribution to growth. Extending the irrigation network will require more expenditure at the grassroots level to develop a wider extension of the tube well network and also better conservation of existing water flows through a more rational pricing of water and more efficient maintenance of the existing irrigation canal networks and better efforts for flood control. Crop diversification, farm extension, and rural education along with the further development of small-scale rural enterprises that can absorb labor and bring in cash income are also important components of a growth strategy for agriculture (Foster and Rosenzweig, 2003; Bhattarai and Narayanamoorthy, 2002). The rural sector still employs the majority of the country's population. By raising incomes in agriculture, poverty levels can also be lowered. With appropriate policies and a more efficient distribution of resources for technological transfer, more rural education and rural employment generation outside of agriculture there is no reason why the sector cannot accelerate growth to levels of 3% per annum attained in the 1970s.

Industrial policies

There is a clear agenda for action for industrial policy that has been foreshadowed by the historical review of industrial sector performance in the previous section. This includes infrastructure investment, greater openness, development of SEZs, attracting foreign direct investment, more flexible labor market policies, better land use, and more flexible entry and exit policies for industry.

Infrastructure

We noted in the review of historical experience above that infrastructure bottlenecks were cited as the most troublesome impediment to investment by foreign companies. Poor infrastructure raises costs across the board for domestic and foreign firms alike. Transportation probably raises costs more than other infrastructure components simply because transport is a component of the production cost of any good. We noted how far India is behind China. What needs to be done to bring India more into alignment with its neighbors and to bring it to a more competitive footing? First, port facilities, particularly container ports, need to be expanded and upgraded as foreign trade and foreign direct investment become more important. The major ports of Mumbai and Chenai should receive immediate attention followed by smaller ports in the south, east and west coasts. Particular attention should also be made to making India competitive with other textile and apparel exporters. The Quadrilateral highway is a major first step toward upgrading the transportation network. Emphasis now has to be placed on other roads connecting smaller cities and rural centers where agricultural processing could expand. The rail network needs to revamp its pricing and incentive structure to encourage rail transport while continuing to address the needs to upgrade the rolling stock to handle more passengers in greater comfort. Airline freight service to handle fragile and high cost items such as cut flowers should be considered along with more convenient service to Europe and East Asia. Indian airports, despite some recent upgrading, are still far behind other Asian countries in terms of physical facilities and smoothness of entry and exit. Again, the initial emphasis should be on the major cities with secondary focus on smaller urban centers. Better energy policy and greater investment in the power sector are a key policy area for the future.

Reliable and affordable power is critical if India is to make strides in developing its industrial sector further. The existing power grid is augmented by an array of small and expensive generators. These high costs will take a while to overcome and this makes it all the more necessary for the government to allocate resources and also to stimulate the private sector to build new, efficient and environmentally friendly power facilities. This presents a challenge, given the experience of China and the global demands for power facilities that are low in carbon emissions. India is planning on more nuclear capacity but that will not be enough to power an economy growing at up to 8% (see Chapter 6 for further details). Clean coal technology along with development of ethanol and other biofuels is critical. Telecommunications and water supply seem to be less of a bottleneck for further development of the industrial sector and should be secondary priorities for government investment.

Trade policy and development of SEZs

India has to continue to lower tariffs and increase its international competitiveness. There should be increasing emphasis on attracting foreign direct investment into SEZs to concentrate on exports of manufactured goods and also, more generally, to serve the domestic market. There have already been many joint ventures with foreign companies in the automobile, pharmaceuticals and food processing industries and the scope for joint ventures can be expanded. Joint ventures with state governments are also part of the plan for developing SEZs. Private sector involvement also could intensify with further involvement of both mergers and acquisitions and greenfield investments by foreign companies.

Generally, more emphasis should be put on expanding capacity and increasing efficiency in the traded goods sector, particularly for a wider variety of manufactured goods. It is true that India is a latecomer to the international market for many product lines. Nevertheless, given the right mix of foreign and domestic investment, product focus and appropriate government polices designed to increase efficiency and increase flexibility, it is certainly possible for India to become internationally competitive in the electronics sector and also other manufacturing sectors. It is already producing a wide variety of these products for the domestic market. It is our belief that India will not be able to sustain a rapid growth rate of 8% without a vibrant industrial sector.

Accelerated growth in manufacturing will also result in more rapid growth in employment, particularly in those industries focusing on labor-intensive products. This should help to increase the low employment elasticity of output growth which has characterized the Indian economy for many years. In India, growth has not been job-intensive. Purfield (2006) notes that the GDP growth — employment elasticity is only 0.5 for the organized sector. For every one percent increase in GDP, employment increases by only about half a percent. It is also important that the poorer states benefit from some of this proposed new industrial expansion. High-growth and richer states such as Maharashtra and West Bengal have been able to generate job-intensive growth while the poorer states have been less successful (Table 7.14).

Restrictive labor and industrial policies

Despite some relaxation, firms with more than 100 employees must still generally apply to the government to retrench or fire workers. This policy restricts the flexibility of firms and greatly inhibits the entry of new firms while ensuring that existing firms stay in business, and it must be relaxed if the Indian industrial sector is to become more competitive in international markets. Another restrictive policy that inhibits the growth of large-scale labor-intensive industries for export is the reservation of certain labor-intensive products for small firms. This policy essentially has excluded Indian labor-intensive industries from competing in international markets with China and the countries of Southeast Asia that can take advantage of economies of scale to undertake large production runs to be sold in external markets. Small firms in India just generally do not have that cost advantage even though labor costs in India are still extremely low. This policy has distorted the natural comparative advantage of India's

Table 7.15 Comparison of Share of World's Exports — India and China

Commodity	1978/1981 China	1978/1981 India	1998/2000 China	1998/2000 India
Textiles and apparel	9.1	7.0	29.8	7.7
Leather and jewelry	1.5	8.7	14.3	8.0
Manufactured goods and others	1.1	0.3	18.1	0.3

Source: Adopted from Srinivasan (2004).

labor-abundant economy to the extent that most of India's exports are capital- or skill-intensive:

> India is virtually absent from the world market for toys and the vast majority of other light manufacturers. Its share in the US market in apparel is fully matched by a much smaller Bangladesh and is one-fourth that of China (Panagariya, 2006, p. 44).

The disparity in export performance is highlighted in Table 7.15. China's share of world exports has increased dramatically since 1978 while India's has remained stagnant or even fallen.

To be fair, this policy is slowly being revised to open many labor-intensive products to large-scale firms, but many products are still on the list including electrical machines, appliances, paper and plastic products, glass and ceramics, and auto parts and components (see Panagariya, 2006 for further details). It needs to be completely scrapped if India is ever to have a chance to compete against other countries in markets for labor-intensive products. Given recent increases in relative wages in China, India's potential comparative advantage in these products could be even higher than it was 20 years ago.

Financial sector policies

Financial repression continues to be reduced as CCR and SLR ratios have fallen from about 55% combined before liberalization in the early 1990s to just over 30% in 2003. The ratio should fall further if liberalization is to continue. Interest rate controls are still in place in many markets and directed credits are still required for all commercial banks. Furthermore, the government still controls most of the commercial banks. All of this needs to change if India is to obtain the full benefits of financial liberalization and if the banking system is to be able to fully support the real sectors of the economy with financial operations based on comparative advantage, efficient allocation of financial resources to their best use and decisions made on commercial grounds rather than by bureaucratic edict. If appropriate reforms are made in industrial policy, financial institutions and business firms can work together to raise production and productivity. These reforms will be instrumental in bringing the industrial sector back as a key contributor to economic growth and development.

Environment

We have touched on many of the environmental problems facing India in Chapter 6. This section summarizes some of the major policy issues that have to be addressed in the next decade or so. As with China, energy is probably the most pressing issue for India. The main source of energy for electricity for industrial use and commercial consumption is coal. Coal supplies about 50% of India's energy needs. Most of the coal mined in India has high sulfur content and is also high in fly ash. To make matters worse, India's coal-fired plants are generally old and energy-inefficient and they have not generally been retrofitted to remove sulfur and fly ash from smoke stack emissions. There has been an 88% increase in carbon emissions between 1990 and 2004 (see Chapter 6, Table 6.3) and this has made a significant contribution to global warming. India is not far behind China and is expected to make a significant contribution to the world's carbon dioxide emissions in the next decade (US Department of Energy, 2003a).

While carbon dioxide emissions are expected to continue to rise, India has adopted some policies to conserve energy and the volume of carbon emissions is rising only slowly. Furthermore, India's ecological footprint is only half of China's, primarily because its industrial sector is small relative to the economy and the rural sector still relies on non-commercial sources for many of its energy needs (see Chapter 4). Technologies to reduce emissions (particularly sulfur and mercury) or to pump emissions underground have not been explored in India and are unlikely to come into widespread use in the next decade. While the increase in emissions from the Indian economy will be lower than those of China, they will still be substantial; particularly if the rapid growth experienced in the last decade is sustained and industrial sector growth accelerates.

Doubling of carbon emission in the next decade is a distinct possibility unless measures are taken to increase the efficiency of power generation and reduce the reliance on fossil fuels. Per capita emissions will remain small compared with industrial countries, although likely to increase as living standards increase and families become more reliant on commercial energy. It is unlikely that there will be any substantial reduction in the emission of carbon pollutants. A more likely scenario is that India will follow China as the next biggest contributor to greenhouse gases over the next few decades.

India does not have nearly as great a sulfur emission problem as China. The sulfur content in Indian coal is low compared to United States coal, and acid rain due to sulfur dioxide emissions is presently not of great concern. However, increasing coal use or blending Indian coal with imported coal of higher calorific value and greater sulfur content could lead to problems in the future and needs to be monitored.

Currently, half of India's energy needs are met by coal and biomass supplies nearly 40% of the balance of overall energy needs, primarily in the rural sector (see, for example, Chapter 4 where energy consumption of households in Bihar was reviewed). This is more than twice the contribution of biomass to China's energy balance, reflecting greater reliance on commercial energy even in the countryside.

Alternative fuels, conveniently combined together — hydroelectricity, solar, wind, nuclear and geothermal — now supply the balance of around 10% of India's energy needs, primarily electricity generation. In order for these to make a dent in the coal emission story, they would have to increase faster than overall energy use and also make up the added requirement as reliance on biomass declines with further economic development.

Several nuclear power plants are being constructed and India's nuclear power capacity is projected to increase fivefold by 2025. Other alternatives may supply the needed additional capacity although the prospects for additional hydroelectric power are limited aside from purchases from Nepal, and solar power is also limited. Unless additional resources are devoted to alternatives to coal, it is likely that coal will rise as a proportion of total energy use and carbon pollution will rise along with it.

As we suggested in the section dealing with China's energy policy, a number of initiatives are required to deal with the environmental impact of rapid growth in energy usage. Pricing that reflects the externalities of energy use is needed. This has to be combined with heightened awareness on the part of consumers, industry, and the government. Different approaches to offset the use of polluting coal will all help to reduce the growth of carbon emissions.

We turn now to the contribution of the internal combustion engine to air pollution and global warming. Currently the contribution of automobiles and other transportation equipment is even smaller than China. There are only three cars in India for 1000 people as compared to around eight cars per 1000 people capita in China. However, car sales in India are growing at 10%–15% annually as there is ample financing. Around 85% of the cars sold in India are financed. This compares with only 15%–20% in China. Some observers are predicting that India will surpass China by 2050. Emission policies are being implemented and are now focusing on reducing the emissions of two-wheelers and other "belchers" rather than focusing on new car standards such as those recently adopted in China.

The challenge to planners is to slow the increase in automobile vehicular traffic by developing a comprehensive plan that includes more mass transit and expansion of the rail system, which is already focusing on passengers rather than freight. Improving the quality of the rolling stock and improving service to major cities should be a major priority to help forestall the growth in automobile usage. Tolls for traveling in the central business district during rush hour and increased taxes on petroleum and cars while increasing fuel economy and developing alternative fuels such as cellulosic ethanol can also be stressed.

As in China, prospects for bringing about a dramatic slowdown in the increase in automobile production in India are not good. Joint ventures producing a wide array of cars are now operating in India and they are more interested in selling cars than in the environment. If India is really interested in slowing the rate of increase in automobile traffic, there are several other Asian countries that have been able to keep car ownerships down despite having high standards of living. This has been possible by providing an array of alternative transportation modalities including buses, light rail and taxis while making parking expensive and levying high taxes on cars, gasoline and annual car registration. As a result, car ownership per capita is much lower than in other cities of comparable size and income. As noted before, Hong Kong has only 43 private cars per 1000 residents and Singapore 110. In Bangkok, where policies have not been as successful, there are over 200 per 1000 residents in Bangkok, a city with a much lower per capita income than either Singapore or Hong Kong.

There are a host of other environmental problems facing India including air and water pollution, deforestation and soil erosion. These issues were taken up separately in Chapter 6 and analyzed within the context of the challenges facing Asia.

Population and social policy

The poor and populous states of India noted above in the previous section have about 40% of the population, yet they are estimated to contribute to about 55% of the growth in population to 2015. Poor families in rural areas tend to have more children than richer families in cities. There are many reasons for this as noted before in the historical section on population and social policies. To reduce the rate of population growth of these states and more generally in the country as a whole will require concentrated efforts to accelerate the demographic transition. This can be done by raising living standards and human resource endowments and by making temporary birth control methods available and by broadening the reach of sex education and

Table 7.16 Women's Labor Force Participation by Caste Group, 1999 (in percent of Labor Force)

Mode of remuneration	Scheduled tribes	Scheduled castes	Other "backward castes"	Others
Inactive	17	20	24	32
Employed	25	24	14	8
Self-employed	31	16	22	16
Extra-domestic work	26	39	39	43
Unemployed	1	1	1	1

Source: Indian National Statistical Survey (2005).

the use of contraceptives. More resources should be directed toward raising women's access to education and health resources, by encouraging women to participate in the labor force and by directing more resources to women's education and health in these states. Labor force participation of women is only about 35% and, by some accounts, has fallen since the late 1980s. This could be the result of falling labor force participation among higher classes and as family income increases (Table 7.16).

Unlike the experience of industrial countries, women's leisure in India is a superior good, being positively related to income. Therefore, the benefit of increasing women's education for labor force participation may be smaller in India than in other Asian or industrial countries. Still, the benefits of women's education on children's health and education are positive and there is also evidence that more educated women living in cities have smaller families in India for various other reasons. Furthermore, at some point, it is expected that labor force participation rates for educated women will begin to rise as they have in other countries as incomes increase further.

Another interesting phenomenon is that contrary to expectations from human capital theory, girls have relatively lower literacy compared to boys in areas where women's labor force participation is high. This could be because areas with higher women's labor force participation are also areas with higher childhood girls' labor force participation; and that these higher rates of girls' labor depress their literacy and education. One solution is to achieve better enforcement of child labor laws and educational requirements for both sexes (Sundaram and Vanneman, 2003).

Comparison between India and China

Gross output and sector performance

Several things stand out in a review of the pattern of economic growth and sectoral changes in China and India. First is the overwhelming importance of industrial growth in China. Industry in 2005 accounts for two-thirds of GDP, up from a negligible amount only 30 years ago. This growth has not only been spectacular, but it has been concentrated for the most part in a few sectors, primarily SITC categories 5–8, office machinery, electrical machinery and appliances, telecommunications equipment, and miscellaneous manufactured goods. Textiles and apparel were also important but have declined in relative importance in recent years. Much of the focus in manufactured goods began with steel production, which served as the basic input for most of the other products. Some of this production, but by no means all, was financed by foreign direct investment in SEZs and the output was exported. As exports rose, China's role as an export processing center increased and round tripping investment from greater China and other offshore locations increased. Economic growth in the coastal and southern provinces must have exceeded 20% in some years as the region led the industrial advance. The magnitude of the shifts in output is shown in Table 7.17. Notice the shift away from metals to process goods, particularly machinery and electrical machinery. The concentration also increased from 38% of industrial production to 42%.

The focus of exports is even more dramatic. Comparing 1984–1990 with 2001–2004, the focus on a few two-digit categories shows a dramatic increase. SITC codes 75, 76 and 77 (office and data processing machines, telecommunications and sound recording, and reproducing equipment and electrical machinery) went from 5.5% of total exports in the earlier period to nearly 35% of total exports in the

Table 7.17 Manufactured Output Composition — China, 1970–1997 (Percent of Total Manufactured Goods Output)

Category	1970	1980	1990	1997
Metals	16.8	15.6	12.8	7.6
Machinery and vehicles	14.2	12.8	13.7	17.5
Electrical machinery	4.7	5.3	8.3	13.2
Other manufactured goods	3	5.2	6.3	4.5

Source: Nararaj (2005).

Table 7.18 Export Share of Selected Two-Digit Manufacturing Industries — China

Category	1984–1990	2001–2004
SITC 75 — Office machinery and data processing machines	0.4	12.9
SITC 76 — Telecommunications and sound equipment	2.9	10.4
SITC 77 — Electrical machinery	1.2	10.1
SITC 65 and 84 — Apparel and textile yarn, etc.	28.1	17.9
SITC 89 — Misc. manufactured goods	4.9	7.3

Source: Panagariya (2006).

latter period, a remarkable increase in 15 years even as total exports increased rapidly at double-digit rates. These three exports categories, which are closely related product categories, overtook and surpassed the three traditional export leaders of textiles, apparel and miscellaneous exports by the turn of the century (Table 7.18). The shift in export focus in such a short time demonstrates the commitment of China to develop a strong export performance in semi-skilled and labor-intensive manufactured products using steel as the basic input. Much of this output was funded by FDI. By 2003 the total stock of FDI was 35% of GDP, with foreign funded enterprises contributing about half of total exports and having ownership in more than a third of industrial enterprises. The foreign sector has been the main growth driver of the economy in the past few years, contributing about two-thirds of the total annual growth of around 9%.

Table 7.19 Export Share of Selected Two-Digit Manufacturing Industries — India

Category	1984–1990	2001–2004
SITC 66 — Non-metallic minerals manufactures	9.4	14.9
SITC 65 and 84 — Textiles and apparel	20.2	21.0
SITC 33 — Petroleum products	8.5	6.3
SITC 67 — Iron and steel	0.8	5.0
SITC 89 — Misc. manufactured goods	3.0	5.2

Source: Panagariya (2006).

The second observation is the rather anemic performance of the Indian industrial sector. The sector has hardly grown as fast as the rest of the economy and has provided almost no impetus for growth. The sector share is only 26% of GDP, significantly lower than 40% of China's share. Even accounting for some estimation errors, there is still a very big difference. The acceleration in growth since the mid-1990s has been led by the services sector with a small contribution from agriculture. FDI is only one-tenth the level of China and export growth has been slow. Merchandise exports are only about 10% of GDP and are not nearly as well focused on labor-intensive industries. Table 7.19 shows the major Indian exports as a share of total exports. They range from some labor-intensive products such as apparel to capital-intensive products like steel and extractive industries such as petroleum and petroleum products. Furthermore, there have been small shifts in the composition of the major exports, textiles and apparel still accounts for over 20% and the others are much smaller. Furthermore, there is no apparent underlying strategy for industrialization.

What about the future? It looks like China can easily continue to grow at rapid rates and increase its export domination in manufactured goods categories 75–77 and perhaps expanding into 78 — road vehicles. By 2001–2004 this sector's share of output was already 2.6%. Whether this continued growth in industry is in the best interest of the country is another matter. We have argued before that a more appropriate strategy would be to put less stress on industry and more emphasis on services and rural development of the backward interior provinces.

India has the potential to develop a vibrant, efficient and internationally competitive industrial sector. It has the manpower. It has the expertise. It has the organizational and management abilities. So far it has been constrained by a policy environment that has throttled initiative, opposed opening up important sectors of the economy to large-scale industries and refused to allow competitive forces to determine wage rates and employment. And it is only now beginning to address some of the infrastructure bottlenecks. As a result, industry has not been able to mount a successful drive to produce a wide array of labor-intensive products for export nor to generate income and employment to qualify as a takeoff to sustained industrial growth.

There are signs that such a takeoff may be near — however it is still around the corner. FDI growth in manufacturing is still slow, labor markets are still constrained and business investment for new plant and equipment is weak. For the outlook for the next decade to change, the government will have to remove all of these bottlenecks. Then it is definitely possible for the industrial sector to exhibit the dynamism that the IT and related service sector have been showing in the last few years.

International trade

Are India and China likely to come into greater competition with each other in the future? If India is able to liberalize its regulatory environment, it could begin to attract more FDI into labor- and skill-intensive manufacturing industries. Along with growing competitiveness of domestic firms this would present a challenge for Chinese producers. However, China has already developed a wide array of manufactured exports and will continue to develop into new areas. It has such a big jump on India it can easily relinquish some of its markets in low-skilled labor-intensive products in which it may already be beginning to lose comparative advantage as wages rise in the southern provinces and SEZs. India, on the other hand, has a big lead in IT and other skill-intensive software. China is only beginning to enter the services sector. It will be hampered by lack of fluent English speakers and the market penetration already established by India, the Philippines, Sri Lanka and other English speaking countries.

Poverty and income distribution

Comparing the relative success in reducing poverty since 1990, India and China have had different patterns of success (see Table 7.20). Between 1990 and 2000, poverty in China fell by 51% using $1 a day cut off, 47% using a $2 per day cut off and 63% using the national poverty line cut off. In India the comparable percentages are 25%, 9% and 33%, less than half the progress made in China over the decade in reducing poverty.

China has managed to reduce poverty dramatically, primarily because of the rapid growth of income in urban areas in recent years, the phenomenal success of town and village enterprises and the household responsibility system at the beginning of the reforms in the late 1970s and early to mid-1980s. In recent years the thrust to reduce poverty has sputtered somewhat as the coastal provinces have continued to outpace the interior and resources have failed to flow to redress the imbalances created by these dramatic growth differentials. For poverty levels to be reduced further, a strong commitment to the further development of the interior is required, including rural industry, better education and health, infrastructure and removal of migration restrictions that prohibit more out-migration from rural areas to the cities. China has sufficient resources to address these problems if it can muster the political will to do it. Restlessness has been increasing as more demonstrations have been taking place in rural provinces and the distribution of income continues to deteriorate.

As noted above, India has had more modest success in reducing poverty and it is doubtful whether it will be able to achieve the millennium development goal of reducing poverty to half the 2000 level by 2015. Without a dynamic and rapidly growing industrial sector to absorb new entrants into the labor force and a lack of resources to boost education and health expenditures in the poorer states, India has been unable to generate employment and raise incomes enough to make a significant acceleration in poverty reduction. If policies are adjusted to bring about more rapid growth in income by raising industrial sector growth, poverty reduction could accelerate over the next 10 to 15 years. Otherwise, progress will continue to be slow.

Table 7.20 Poverty Levels in India and China — Percent of People in Poverty

Year and poverty line	India	China
$1 per day cut off		
1990	46.6	31.5
1993	51.1	29.0
1996	46.2	16.4
1998	44.2	16.1
2000	34.7	15.4
2002		12.7
$2 per day cut off		
1990	88.2	69.9
1995	82.9	51.6
2000	79.9	44.8
2002		37.3
National poverty line		
1990	38.9	9.4
1995	36.0	7.1
2000	26.1	3.4
2002		3.0

Source: Tables in Chapter 4.

Poverty reduction has also been hampered by the low elasticity of poverty reduction to growth. This is consistent with the low elasticity of employment generation with respect to income growth. Much of the growth in India has been the result of capital- or skill-intensive growth rather than unskilled and semi-skilled labor growth. As a result, employment generation and poverty reduction have been limited. India also has to put an end to widespread discrimination by class and gender.

This contrasts with the Chinese experience where much of the growth in manufacturing has been in labor-intensive manufacturing. As manufacturing value-added accelerates, it should bring with it a more rapid reduction in poverty and further growth in income. Tax revenues will also enable the government to devote more resources to infrastructure and human resource development, creating a virtuous cycle of growth.

Both India and China have been experiencing deterioration in income distribution. From a regional perspective, certain states and provinces have experienced acceleration in growth while others have languished. In China these regional developments are consistent with a pattern of urban and industrial growth that has favored the coastal provinces. In India certain states with limited resource endowments and large minority populations have fallen behind and need to be restored to more rapid growth by a consistent set of policies that increases education and health indicators and creates more employment.

Financial sector performance

The review of financial sectors in India and China show many similarities including dominant control by the government through extensive ownership, domination of the financial system by commercial banks, directed credit to key sectors, historically weak prudential supervision and weak loan supervision, interest rate controls, high reserve requirements, and limited entry of foreign banks. All of these institutional controls are characteristics of financial systems that are "repressed". In the case of China, these institutional weaknesses have resulted in a significant growth in nonperforming loans that were extended to state-owned enterprises and which have resulted in a drag on the banking system, fragile financial indicators, and fear of financial collapse. In the case of India, systemic problems are less severe, partly because commercial banks have to hold large shares of their assets in reserves or government bonds and also because the industrial sector has been growing slowly and has been reluctant to borrow for new investment.

To escape the financial repression trap in the future, both India and China will have to undertake further measures to liberalize their financial systems through appropriate reform measures. We will not go into them in detail here. These measures are standard procedure for financial systems that have been under the control of the government and where competitive forces have not been allowed to develop. The liberalization process has been undertaken by many countries in Asia and other parts of the developing world and there is a wealth of insight to be gained from careful study of these experiences.

There can be many pitfalls in this process and both China and India would do well to study these experiences, particularly those of countries in Southeast Asia and East Asia. Prudential controls are important and institutional safeguards have to be put in place as the liberalization process unfolds. Greater competition, easier entry of foreign banks, relaxation of restrictive regulations on lending, interest rates, and mix of bank asset along with better supervision are all necessary. They should be adopted with caution and after due reflection. The timing and sequencing of reforms also has to be carefully considered and the experience of other countries studied. The process may be drawn out. However, it is necessary and, indeed, inevitable if China and India are to integrate financial and real sectors together in the process of economic development and growth.

Savings, investment and economic growth

China's investment and savings rates are about 40% higher than the levels achieved in India. Investment rate averaged 26.4% of GDP and 36.7% of GDP in India and China respectively, from 2001 to 2004 according to the ADB (2006). To some extent, more investment is responsible for higher rates of growth. Recall in a simple Harrod–Domar growth model — output growth is directly related to the savings rate and the capital to output ratio. In the Solow and related neoclassical growth models, the savings rate is also critical. From growth accounting we can see that the growth in the capital labor ration has accounted for the bulk of growth in China along with total factor productivity (Table 7.21). In fact, increases in the capital to labor ratio have accounted for even more growth in the decade since 1993.

Because the savings rate in China is so high, there are concerns that consumption has been unduly depressed and that is now time to stimulate domestic demand and slowly reduce the rate of investment. Clearly there has been overinvestment in SOEs and probably other parts of the Chinese economy. The central planning imperatives to produce more and more to meet quotes set by the government has not been completely expunged from the minds of managers and businessmen.

Table 7.21 Components of Economic Growth in China (in %)

Component	1978–1993	1993–2005
GDP growth	9.7	9.5
From employment	2.5	1.1
From labor productivity	7.0	8.4
From total factor productiity	3.8	3.0
From increases in capital to labor ratio	3.1	5.3

Source: http://siteresources.worldbank.org/INDIAEXTN/Resources/events/359987-1149066764594/Louis_Kuijs_ppt.pdf.

Saving and investment behavior is likely to change slowly, perhaps driven by a build-up in consumer debt rather than changes in business and government saving. Hire purchase terms are becoming more liberal and Chinese consumers are beginning to go into debt to buy consumer durables.

As indicated above, saving and investment in India are somewhat lower than in China, and growth until recently has been significantly lower, suggesting that capital has been utilized more efficiently in China. However in the last decade India has been catching up. Using a slightly different database which has a lower growth rate for China, Eslake (2005) shows GDP growth in India of 6.2% compared with 8.1% for China from 1991 to 2003 with productivity accounting for 4.2% in the case of India and 6.7% in the case of China.

Future projections by the IMF and Consensus Economics for the next decade 2005–2015 show India growing at 6.9% and China at 8% (Eslake, 2005; growth projections to 2020 in Chapter 1 which forecast growth rates of per capita income in India and China from Table 1.29 in Chapter 1 to be nearly identical at 6.5% and 6.4% respectively between 2011 and 2020). If India continues to liberalize its industrial sector, growth could be pushed even higher.

Environment

India and China face many of the same environmental challenges. Both countries rely on coal as a primary energy source, although India has also extensively relied on biomass in the past (see section on India's environment). However, this will change as incomes grow and commercial energy becomes more widely available. Because it has a much smaller industrial sector relative to the size of the economy, India's energy intensity is lower than China's. This will also change as industrial sector growth accelerates. India has more arable land per capita than China and this should provide some room for changes in land use policy to accommodate further population growth, which is projected to be higher than China's for the next few decades. Nevertheless, both countries will have to face environmental challenges posed by expanding population and pollution caused by the process of growth and further industrialization. We discussed these challenges in this chapter and also in Chapter 6. What is striking is that both countries are faced with very similar issues — carbon emissions from factories and from a growing fleet of automobiles and trucks, shrinking water resources, air pollution in major cities, destruction of wildlife habitats, and threats to endangered species. The most pessimistic estimates of global warming suggest further widespread destruction and disruption in both countries.

Because they face so many similar problems and also because economic growth in both countries is projected to remain at high levels and perhaps accelerate further in India, it is incumbent upon both India and China to address these environmental challenges immediately. They are the two most populous countries in the world and, over the next decades, the biggest polluters as well. Joint approaches and interaction between policy makers and the private sectors in both countries would be useful to raise awareness of the variety of environmental challenges and issues and also to introduce policy changes to address these issues. Cooperation between India and China to reduce pollution and heal the environment is critical for the welfare of the entire global ecosystem.

Bibliography

Anderson, K. (2003). Agriculture and Agricultural Incentives in China and India Post-Uruguay Round. Paper presented at the International Conference on Agricultural Policy Reforms and the WTO: Where are We Heading? Capri, Italy.

Asia Monitor Resource Center (2006). Asian Labor Update. http://www.amrc.org.hk/5304.htm [Retrieved 12 January 2006].

Asian Development Bank (ADB) (2006). *Asian Development Outlook 2006*. Manila: Asian Development Bank.

Banga, R. and B. N. Goldar (2004). Contribution of Services to Output Growth Productivity in Indian Manufacturing: Pre and Post Reform. ICRIER Working Paper No. 139, August.

Barro, R., G. Mankiw and X. Sala-i-Martin (1995). Capital Mobility in Neo-Classical Models of Economic Growth. *American Economic Review*, 85(1), 103–105.

Bergsten, C. F., B. Gill, N. R. Lardy and D. Mitchell (2006). *China: The Balance Sheet*. New York: Public Affairs.

Bhattarai, M. and A. Narayanamoorthy (2002). Impact of Irrigation on Agricultural Growth and Poverty Alleviation: Macro Level Analyses. India Water Policy Research Highlight, IWMI-Tata Water Policy Program. http://www.iwmi.org/iwmi-tata.

Bradsjer, K. and D. Barboza (2006). Pollution from Chinese Coal Casts a Global Shadow. *New York Times*, 11 January.

Chinese Central Bureau of Statistics (2006).

Dhesi, A. S. (1998). Caste, Class Synergies and Discrimination in India. *International Journal of Social Economics*, 25(678), 1030–1048.

Dowling, J. M. and R. Valenzuela (2004). *Economic Development in Asia*. Thomson Publishing.

Energy Information Administration (EIA) (2006). Country Analysis Briefs: China Environment. http://www.eia.doe.gov/emeu/cabs/China/Environment.html [Retrieved 26 January 2006].

Erskine, A. (2004). The Rise in China's FDI: Myths and Realities. Paper presented at Australia-China Free Trade Agreement Conference, Sydney, Australia.

Eslake, S. (2005). China and India in the World Economy and Implications for Australia. ABERU Discussion Paper No. 20. Melbourne, Australia: Monash University.

Evanson, R., C. Pray and M. Rosegrant (2002). Agricultural Research and Productivity Growth in India. Research Report No. 109. Washington, DC: International Food Policy Research Institute.

Fei, J. C. H. and G. Ranis (1961). A Theory of Economic Development. *American Economic Review*, 51(4), 533–565.

Foster, A. D. and M. R. Rosenzweig (2003). Agricultural Productivity Growth, Rural Economic Reforms: India 1970–2000. Paper presented at D. Gale Johnson Memorial Conference, 24 October.

Gajwani, K. and R. Kanbur (2006). Patterns of Spatial Convergence and Divergence in India and China. Paper presented at Annual Bank Conference on Development Economics, St. Petersburg, Russia.

Groves, T., Y. Hong, J. McMillan and B. Naughton (1995). Productivity in Chinese State-Run Industry. In F. Dong, C. Lin and B. Naughton (eds.), *Reform of China's State-Owned Enterprises*. London: MacMillan.

Gunter, F. R. (2004). Capital Flight from China: 1984–2001. *China Economic Review*, 15(1), 63–85.

Gupta, S., B. Clements and E. Tiongson (1998). Public Spending on Human Development. *Finance and Development*, 35(3).

Hesketh, T., L. Lu and W. X. Zhu (2005). The Effect of China's One–Child Family Policy After 25 Years. *The New England Journal of Medicine*, 252, 1171–1176.

Heytens, P. and H. Zebregs (2003). How Fast Can China Grow? In W. Tseng and M. Rodlauer (eds.), *China Competing in the World Economy*. IMF.

Huang, C. C. (2006). A Brief Chronology of China's Intellectual Property Protection. http://www.american.edu/TED/hpages/ipr/cheng.htm [Retrieved 23 January 2006].

Huang, Y., T. Saich and E. Steinfeld (eds.) (2005). *Financial Sector Reforms in China*. Cambridge, MA: Harvard University Asian Center.

Indian SEZ website (2006). http://www.sezindia.nic.in [Retrieved 18 January 2006].

International Labor Organization (ILO) (2003). Key Indicators of the Labour Market. Geneva.

International Monetary Fund (IMF) (2005). Country Report 05/85.

International Water Management Institute (IWMI) (2006). http://www.iwmi.org/iwmi-tata [Retrieved 2 February 2006].

Jackson, R. and N. Howe (2004). *Graying of the Middle Kingdom*. Center for Strategic and International Studies. Washington, DC: Prudential Foundation.

Kang, C. and Y. Wang (2003). Sex Ratios at Birth. In *Theses Collection of 2001 National Family Planning and Reproductive Health Survey*. Beijing: China Population Publishing House.

Kim, J. U. (2001). Empirics for Economic Growth and Convergence in Asian Economies: A Panel Data Approach. *Journal of Economic Development*, 26(2), 49–59.

Kletzer, K. M. (2003). Liberalizing Capital Flows in India: Financial Repression, Macroeconomic Policy and Gradual Reform. Brookings NCAER Conference on the Indian Economy, New Delhi.

Kuijs, L. and T. Wang (2005). China's Pattern of Growth: Moving to Sustainability and Reducing Inequality. World Bank China Office Research Working Paper No. 2.

Lewis, A. (1954). Economic Development with Unlimited Supplies of Labor. *Manchester School of Economics and Social Studies*, 22, 131–191.

Macroscan website (2003). Per Capita Income Growth in the States of India. 10 August 2003. http://www.macroscan.com/fet/aug03/fet100803SDP_1.htm [Retrieved 5 December 2005].

Mako, W. P. and C. Zhang (2003). Management of China's State-Owned Enterprises Portfolio: Lessons from International Experience. Mimeo, World Bank Resident Office, Beijing.

McGuckin, R. H. and M. Spiegelman (2004). China's experience with Productivity and Jobs: Benefits and Costs of Change. Report No. 1352-04-RR, The Conference Board.

McKinsey and Company (2000). *India: The Growth Imperative.* http://www.mckinsey.com/mgi/publications/india.asp.

Mellor, J. and M. Mudahar (1992). Agriculture in Economic Development: Theories, Findings and Challenges in the Asian Context. In R. Martin Lee (ed.), A Survey of Agricultural Economics Literature, Vol. 4, *Agriculture in Economic Development.* University of Minnesota Press.

Morgan Stanley (2006). http://www.morganstanley.com/GEFdata/digests/20060814-mon.html#anchor3.

Nararaj, R. (2005). Industrial Growth in China and India: A Preliminary Comparison. *Economic and Political Weekly*, 40(21), 21–27.

Naughton, B. (1995). *Growing Out of the Plan: Chinese Economic Reform 1978–1993.* Cambridge: Cambridge University Press.

Naughton, B., T. Buoye, K. Denton and B. Dickson (2002). *China: Adapting the Past, Confronting the Future.* Center for Chinese Studies, University of Michigan.

Panagariya, A. (2006). India and China: Trade and Foreign Investment. Paper presented at the Pan Asia Conference on Challenges of Economic Policy Reform in Asia.

Pandita, S. (2006). Caste Based Discrimination in India 2006: Hidden Apartheid for Dalits. Asian Labor Update. http://www.amrc.org.hk/5304.htm.

Pathak, K. B., G. Feeney and N. Y. Luther (1998). Alternative Contraceptive Methods and Fertility Decline in India, National Family Health survey Subject Reports No. 7, International Institute for Population Sciences and East-West Center Program on Population, Honolulu, Hawaii.

Purfield, C. (2006). Mind the Gap — Is Economic Growth in India Leaving Some States Behind? IMF Working Paper No. 103.

Putterman, L. (1997). On the Past and Future of China's Township and Village-Owned Enterprises. *World Development*, 25(10), 1639–1655.

Raiser, M. (1998). Subsidizing Inequality: Economic Reforms, Fiscal Transfer and Convergence across Chinese Provinces. *Journal of Development Studies*, 34(3), 1–26.

Sachar, R. (2006). Justice Rajinder Sachar Committee Report on Status of Muslims; chaired by Justice R. Sachar. Government of India.

Sehrt, K. (1998). Banks versus Budgets: Credit Allocation in the People's Republic of China 1978–1996. PhD Dissertation, University of Michigan.

Solow, R. M. (1956) A Contribution to the Theory of Economic Growth. *Quarterly Journal of Economics*, 70(1), 65–94.

Srinivasan, T. N. (2004). China and India: Economic Performance, Competition and Cooperation: An Update. Mimeo.

Sundaram, A. and R. Vanneman (2003). Gender Differentials in Literacy in India: The Intriguing Relationship with Women's Labor Force Participation. Mimeo, University of Maryland.

United Nations (2004). World Population Prospects: The 2004 Revision Population Database. http://esa.un.org/unpp [Retrieved 24 January 2006].

United Nations Development Programme (UNDP) (2005). *Human Development Report.* New York: United Nations.

US Department of Energy (2003a). China: Environmental Issues. http://www.eia.de.gov.emeu/cabs/chinaenv.html.

US Department of Energy (2003b). Research News. http://www.eurekalert.org/doe/features.php?topic=energy&start=50 [Retrieved 21 February 2006].

US Department of State (2005). Investment Climate Statement — China: US-China Business Council.

US International Trade Administration website (2006). Financial Services: US Commercial Service China. http://www.buyusa.gov/china/en/bank.html [Retrieved 1 February 2006].

Waldman, A. (2005). All Roads Lead to Cities, Transforming India. *New York Times*, 7 December.

Whalley, J. and X. Yui (2006). Rural Income Volatility and Inequality in China. Mimeo.

World Bank (2002). Improving the Investment Climate in India. Private Sector Development Team, South Asia Region and Investment Climate Unit, World Bank and Confederation of Indian Industries.

World Bank (2006a). *Quarterly Update.* Beijing: World Bank office.

World Bank (2006b). Saving and Investment in China. Workshop on the International Financial Integration of China and India, 26 May 2006, New Delhi. worldbank.org/INDIAEXTN/Resources/events/359987-1149066764594/Louis_Kuijs_ppt.pdf [Retrieved 16 July 2006].

World Business Environment Survey (2000). http://info.worldbank.org/governance/wbes/index2.html.

Xiao, G. (2004). People's Republic of China: Round Tripping FDI: Scale, Causes and Implications. http://www.hiebs.hku.hk/working_paper_updates/pdf/wp1137.pdf [Retrieved 9 April 2006].

Zhao, Y. (1997). Rural to Urban Labor Migration in China. Mimeo, China Center for Economic Research, Beijing University.

Author Index

Subject Index